T0390832

Thin Film Nanophotonics

Nanophotonics Series

Thin Film Nanophotonics

Conclusions from the Third International Workshop on Thin Films for Electronics, Electro-Optics, Energy and Sensors (TFE3S)

Edited by

GURU SUBRAMANYAM

Center of Excellence for Thin-Film Research and Surface Engineering (CETRASE), University of Dayton, Dayton, OH, United States

PARTHA P. BANERJEE

Department of Electro-Optics and Photonics, University of Dayton, Dayton, OH, United States

KARL S. GUDMUNDSSON

Department of Electrical Engineering, University of Iceland, Reykjavík, Iceland

AKHLESH LAKHTAKIA

Department of Engineering Science and Mechanics, Pennsylvania State University, University Park, PA, United States

Series Editor

AKHLESH LAKHTAKIA

ELSEVIER

Elsevier
Radarweg 29, PO Box 211, 1000 AE Amsterdam, Netherlands
The Boulevard, Langford Lane, Kidlington, Oxford OX5 1GB, United Kingdom
50 Hampshire Street, 5th Floor, Cambridge, MA 02139, United States

Notices
Knowledge and best practice in this field are constantly changing. As new research and experience broaden our understanding, changes in research methods, professional practices, or medical treatment may become necessary.

Practitioners and researchers must always rely on their own experience and knowledge in evaluating and using any information, methods, compounds, or experiments described herein. In using such information or methods they should be mindful of their own safety and the safety of others, including parties for whom they have a professional responsibility.

To the fullest extent of the law, neither the Publisher nor the authors, contributors, or editors, assume any liability for any injury and/or damage to persons or property as a matter of products liability, negligence or otherwise, or from any use or operation of any methods, products, instructions, or ideas contained in the material herein.

British Library Cataloguing-in-Publication Data
A catalogue record for this book is available from the British Library

Library of Congress Cataloging-in-Publication Data
A catalog record for this book is available from the Library of Congress

ISBN: 978-0-12-822085-6

For Information on all Elsevier publications
visit our website at https://www.elsevier.com/books-and-journals

Publisher: Matthew Deans
Acquisitions Editor: Simon Holt
Editorial Project Manager: Rafael G. Trombaco
Production Project Manager: Prasanna Kalyanaraman
Cover Designer: Greg Harris

Typeset by MPS Limited, Chennai, India

Contents

3. Structural, electrical, and electromagnetic properties of nanostructured vanadium dioxide thin films

65

Guru Subramanyam, Eunsung Shin, Prudhvi Ram Peri, Ram Katiyar, Golali Naziripour and Sandwip Dey

4. Photoactive ZnO nanostructured thin films modified with TiO₂, and reduced graphene oxide

91

Pierre G. Ramos, Luis A. Sánchez and Juan M. Rodriguez

5. Orthotropic friction at the edges and interior of graphene and graphene fluoride and frictional anisotropy of graphene at the nanoscale   123

Sergei F. Lyuksyutov, Liudmyla V. Barabanova, Alper Buldum and Jeffrey A. McCausland

Part II Applications   137

6. Optical manipulation of nanoparticles with structured light   139

Guanghao Rui, Ying Li, Bing Gu, Yiping Cui and Qiwen Zhan

9. Thin film solar cells with graded-bandgap photon-absorbing layer — **239**

Faiz Ahmad, Akhlesh Lakhtakia and Peter B. Monk

List of contributors

Faiz Ahmad
Department of Engineering Science and Mechanics, Pennsylvania State University, University Park, PA, United States

Hammid Al-Ghezi
Department of Electro-Optics and Photonics, University of Dayton, Dayton, OH, United States

Pandurang Ashrit
Thin Films and Photonic Research Group (GCMP), Department of Physics and Astronomy, University of Moncton, Moncton, NB, Canada

Rajab Y. Ataai
Department of Electrical and Computer Engineering, University of Dayton, Dayton, OH, United States

Partha P. Banerjee
Department of Electro-Optics and Photonics, University of Dayton, Dayton, OH, United States

Liudmyla V. Barabanova
Physics Department, University of Akron, Akron, OH, United States; Department of Chemistry, University of Akron, Akron, OH, United States

Alper Buldum
Department of Mechanical Engineering, University of Akron, Akron, OH, United States

Monish R. Chatterjee
Department of Electrical and Computer Engineering, University of Dayton, Dayton, OH, United States

Yiping Cui
Advanced Photonics Center, Southeast University, Nanjing, P.R. China

Sandwip Dey
School of Engineering of Matter, Transport and Energy, Arizona State University, Tempe, AZ, United States

Dean R. Evans
Air Force Research Laboratory, Materials and Manufacturing, Wright-Patterson Air Force Base, OH, United States

Rudra Gnawali
Applied Optimization, Inc., Fairborn, OH, United States

Bing Gu
Advanced Photonics Center, Southeast University, Nanjing, P.R. China

Ram Katiyar
Department of Physics, University of Puerto Rico Rio Piedras, San Juan, Puerto Rico

Akhlesh Lakhtakia
Department of Engineering Science and Mechanics, Pennsylvania State University, University Park, PA, United States

Ying Li
Advanced Photonics Center, Southeast University, Nanjing, P.R. China

Sergei F. Lyuksyutov
Physics Department, University of Akron, Akron, OH, United States

Jeffrey A. McCausland
Physics Department, University of Akron, Akron, OH, United States; Molecular Biology and Microbiology, School of Medicine, Case Western Reserve University, Cleveland, OH, United States

Peter B. Monk
Department of Mathematical Sciences, University of Delaware, Newark, DE, United States

Golali Naziripour
Indiana University Purdue University Indianapolis, Indianapolis, IN, United States

Prudhvi Ram Peri
School of Engineering of Matter, Transport and Energy, Arizona State University, Tempe, AZ, United States

Pierre G. Ramos
Center for the Development of Advanced Materials and Nanotechnology, National University of Engineering, Lima, Perú

Victor Reshetnyak
Physics Faculty, Taras Shevchenko National University of Kyiv, Kyiv, Ukraine

Juan M. Rodriguez
Center for the Development of Advanced Materials and Nanotechnology, National University of Engineering, Lima, Perú

Guanghao Rui
Advanced Photonics Center, Southeast University, Nanjing, P.R. China

Luis A. Sánchez
Center for the Development of Advanced Materials and Nanotechnology, National University of Engineering, Lima, Perú

Andrew M. Sarangan
Department of Electro-Optics and Photonics, University of Dayton, Dayton, OH, United States

Eunsung Shin
Center of Excellence for Thin-Film Research and Surface Engineering (CETRASE), University of Dayton, Dayton, OH, United States

Jonathan Slagle
Air Force Research Laboratory, Materials and Manufacturing, Wright-Patterson Air Force Base, Dayton, OH, United States

Tran-Vinh Son
Thin Films and Photonic Research Group (GCMP), Department of Physics and Astronomy, University of Moncton, Moncton, NB, Canada

Guru Subramanyam
Center of Excellence for Thin-Film Research and Surface Engineering (CETRASE), University of Dayton, Dayton, OH, United States

Qiwen Zhan
School of Optical-Electrical and Computer Engineering, University of Shanghai for Science and Technology, Shanghai, P.R. China

Introductory remarks

The international workshop series on Thin Films for Electronics, Electro-Optics, Energy and Sensors (TFE3S) was initiated in 2015 by the faculty and staff of the Center of Excellence for Thin Film Research and Surface Engineering (CETRASE) at the University of Dayton. The first workshop was held in the University of Dayton's China Institute in Suzhou, PR China in July 2015. The workshop was started with a new approach for technical exchange and to boost technical and educational collaboration within the thin film research community around the world. The TFE3S is a unique workshop that is primarily based on invited speakers from around the world, addressing various topics including, but not limited to

- thin film microelectronics;
- multifunctional oxide thin films;
- flexible and printable thin films;
- thin film metamaterials;
- optical thin films;
- organic, biological thin films and their applications;
- thin films of phase-change materials;
- thin films for energy storage and harvesting;
- thin film sensors; and
- novel processing and characterization techniques of thin films.

The workshop is organized biennially and rotated between Asia, North America, and Europe to cover all geographic regions, and the attendance is limited to less than 100. After the first one in Suzhou, PR China in 2015, the second workshop was held in 2017 in Dayton, OH. The third TFE3S workshop, on which this book is based, was organized by CETRASE in collaboration with the Pennsylvania State University, and the University of Iceland in July 2019.

Why Iceland and how? After the 2nd TFE3S workshop at Dayton, the idea was casually suggested by Akhlesh Lakhtakia who also works on thin films and was brought on board by Guru Subramanyam as one of the future meeting coorganizers. This would have the European rotation for the workshop and also be an attractive and convenient place for participants to travel, both from North America and mainland Europe. Partha P. Banerjee knew *one* Icelander, Robert Magnusson, Professor at the University of Texas at Arlington, who he bounced the idea off at a SPIE Photonics West meeting. Robert Magnussen is from the former group of Tom Gaylord at Georgia Tech and specializes in rigorous coupled wave theory, which Partha had worked on. Robert promised to attend the conference if it was held in

Iceland and offered to give some references of professors at some of the universities. When Partha contacted them, one of those references was Karl S. Gudmundsson who promptly responded. To his surprise, Partha discovered that Karl obtained his PhD (and MS and BS) from Wright State University, another university in Dayton, in the area of optics, and his advisor was a graduate of Electro-Optics from the University of Dayton! The rest is history—Karl gladly offered his assistance in organizing the workshop. Together, we worked to make sure that the TFE3S workshop became a reality in panoramic Iceland, in the main campus of the University of Iceland in Reykjavik.

After the successful TFE3S workshop, which included 34 invited speakers from around the world, Akhlesh proposed another idea to consider: publishing a special edited book on *Thin Films for Nano-Photonics*, based on the many photonics related presentations at the workshop. With the photonics area gaining importance in the society such as for high-speed networks, autonomy, and sensing, we aimed to bring together articles that addressed fundamental research to application oriented. The idea was informed to all participants, and sure enough, we got about a dozen who indicated that they could contribute to the edited book. This is how the special edited volume on thin film nanophotonics was conceived. Unfortunately, with the COVID-19 pandemic, a few of the authors could not complete their chapters in time to be included in this volume; thus we went from a dozen to the final nine.

The book is organized with chapters in the basic science of nanophotonics (Chapter 1: Optical Propagation Through Metamaterial Structures With Multilayered Metallo-Dielectrics: Hyperbolic Dispersion and Transmission Filters, and Chapter 2: Spectral Characteristics of a Thin Lithographic Tri-Layer Chiral Slab Resonator), thin film fabrication, and determining optical properties of nanostructured thin films (Chapters 3—5), advanced characterization techniques (Chapter 6: Optical Manipulation of Nanoparticles With Structured Light) and applications of nanophotonics (Chapters 7—9). Our sincere thanks to all the authors who contributed to this volume, despite the many difficulties during the pandemic year. Special thanks to Ms. Julie Luanco from Elsevier for putting this book together and to Karl's friend and photographer for providing the stunning one-of-a-kind cover graphics. Also, our special thanks to Mr. Simon Holt for giving us an opportunity to consider publishing this book.

Guru Subramanyam,
Partha P. Banerjee,
Akhlesh Lakhtakia and
Karl S. Gudmundsson

PART I

Basics

CHAPTER 1

Optical propagation through metamaterial structures with multilayered metallo-dielectrics: Hyperbolic dispersion and transmission filters

Partha P. Banerjee[1], Rudra Gnawali[2], Hammid Al-Ghezi[1], Dean R. Evans[3], Jonathan Slagle[3] and Victor Reshetnyak[4]

[1]Department of Electro-Optics and Photonics, University of Dayton, Dayton, OH, United States
[2]Applied Optimization, Inc., Fairborn, OH, United States
[3]Air Force Research Laboratory, Materials and Manufacturing, Wright-Patterson Air Force Base, OH, United States
[4]Physics Faculty, Taras Shevchenko National University of Kyiv, Kyiv, Ukraine

1.1 Introduction to metamaterials

The propagation of electromagnetic (EM) waves in a medium is determined by its electric permittivity ϵ and magnetic permeability μ. These appear in the constitutive relations expressed as $D = \epsilon E, B = \mu H$ for simplicity, where E, H, D, and B denote the electric field, magnetic field, electric displacement, and magnetic flux density, respectively. Rigorously speaking, the constitutive relations as written above are valid in the frequency domain; also, E, H, D, and B are complex vectors, and ϵ and μ are complex tensors. Materials can be classified on the basis of the sign of (the real part of) ϵ and μ. As shown in Fig. 1.1, most dielectric materials have $\epsilon > 0, \mu > 0$. Metals and doped semiconductors often exhibit $\epsilon < 0, \mu > 0$. Some ferrites have $\epsilon > 0, \mu < 0$. A negative index material (NIM) is the one that has simultaneously negative values of permittivity ϵ and permeability μ [1−5]. Chiral materials can, for instance, impart negative index through the coupling of the constitutive relations, $D = \epsilon E - j\kappa H, B = j\kappa E + \mu H$, via the chirality parameter κ [6].

A simplified treatment of wave and beam propagation in these NIMs (sometimes called double NIMs) can be developed on the basis of the negative refractive index resulting from negative values of the permittivity and permeability ($\epsilon < 0, \mu < 0$). In simple terms, the refractive index

$$n \propto \sqrt{\epsilon}\sqrt{\mu} = \sqrt{|\epsilon|e^{j\pi}}\sqrt{|\mu|e^{j\pi}} = \sqrt{|\epsilon|e^{\frac{j\pi}{2}}}\sqrt{|\mu|e^{\frac{j\pi}{2}}} = \sqrt{|\epsilon||\mu|}e^{j\pi} = -\sqrt{|\epsilon||\mu|}, \quad (1.1)$$

Thin Film Nanophotonics
DOI: https://doi.org/10.1016/B978-0-12-822085-6.00009-1

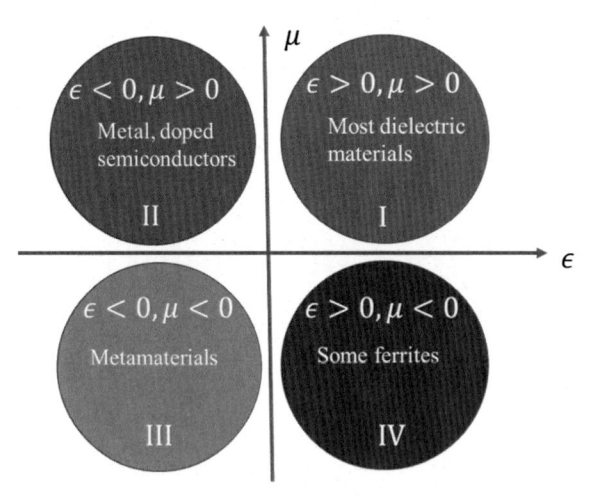

Figure 1.1 Characterization of material parameters on the basis of electric permittivity and permeability. Quadrants: (I) most dielectric materials ($\epsilon > 0$, $\mu > 0$), (II) metals and doped semiconductor ($\epsilon < 0$, $\mu > 0$), (III) no natural materials so-called metamaterials ($\epsilon < 0$, $\mu < 0$), and (IV) some ferrites ($\epsilon > 0$, $\mu < 0$) [5].

where $j = \sqrt{-1}$ is negative, implying negative phase velocity. More exactly, as discussed by Pendry [1,2], it can be shown that causality forces us to choose a negative sign for the refractive index when both ϵ and μ are negative, since both ϵ and μ have positive imaginary parts due to losses, however small, in physical systems. Imagine that as the frequency is varied, the real part of ϵ passes through zero, where there is a branch point for $\sqrt{\epsilon}$. By virtue of the small positive imaginary part of ϵ, the causal solution forces a trajectory above the branch point, giving a positive imaginary value to $\sqrt{\epsilon}$ when $\epsilon < 0$. A similar argument holds for μ and, therefore, $n \propto -\sqrt{|\epsilon||\mu|}$ when $\epsilon < 0$ and $\mu < 0$. The group velocity can, however, be positive.

A simple dispersion relation describing the behavior of the angular frequency ω to the propagation constant k, namely,

$$\omega \propto -1/k \tag{1.2}$$

accommodates negative phase velocity and positive group velocity, characteristic of NIMs [3,4]. Using this dispersion relation, properties of spatiotemporal pulses have been investigated, including solitary waves in nonlinear, negative index media [7,8]. Gaussian beam propagation has been analyzed and reconciled with the ray transfer matrix approach as applied to propagation in NIMs [7]. For instance, linear focusing of a Gaussian beam can occur in NIMs [7]. A more generalized complex dispersion relation based on causality has been developed for NIMs [9]. However, all of these analyses have not accounted for the polarization of the waves and beams.

In reality, the situation is a bit more complex. In general, ϵ and μ are tensor quantities [10–14]. The anisotropy in these materials can be expressed as a diagonal matrix of ϵ and μ with their principal components having different values [10−13]. Hyperbolic metamaterials (HMMs) are a form of an anisotropic material where the dielectric tensor elements have opposite signs [10−13]. There are two types of HMMs that can be distinguished by the signs of the principal elements of the diagonal permittivity matrix. One type has $\epsilon_{zz} < 0; \epsilon_{xx}, \epsilon_{yy} > 0$, while the other type has $\epsilon_{xx}, \epsilon_{yy} < 0; \epsilon_{zz} > 0$ [10−13]. Dispersion in these metamaterials occurs when the principal components of diagonalized matrices change signs [10−13]. As will be shown, effects like negative refraction can be achieved through the hyperbolic dispersion of these materials [10]. Assuming that the effective permittivity can be expressed as a matrix of the form

$$[\epsilon_{\text{eff}}] = \begin{bmatrix} \epsilon_{xx} & 0 & 0 \\ 0 & \epsilon_{yy} & 0 \\ 0 & 0 & \epsilon_{zz} \end{bmatrix}, \quad \epsilon_{xx} = \epsilon_{yy}, \tag{1.3}$$

dispersion in HMMs can be explained from the topology of the isofrequency surface. The dispersion relation for this anisotropic metamaterial is [10−13]

$$\frac{k_x^2}{\epsilon_{zz}} + \frac{k_y^2}{\epsilon_{zz}} + \frac{k_z^2}{\epsilon_{xx}} = \frac{k_0^2}{\epsilon_0}, \tag{1.4}$$

where k_x, k_y, and k_z are the transverse and longitudinal components of the wave vector, respectively. Note that for a simple isotropic material, $\epsilon_{xx} = \epsilon_{zz} > 0$, and the plot of Eq. (1.4) is a sphere, as shown in Fig. 1.2A. For a uniaxial anisotropic material

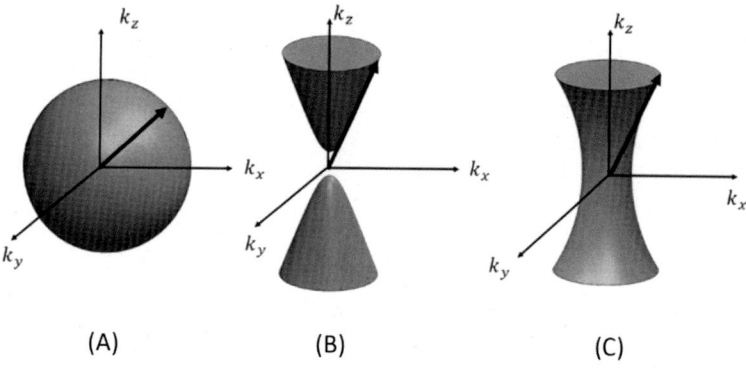

(A)   (B)   (C)

Figure 1.2 Sketches of isofrequency surfaces, (A) Spherical isofrequency surface for an isotropic dielectric. (B) Hyperbolic isofrequency surfaces for (B) Type I (C) Type II HMM. The tip of the arrow denoting the wavevector on the sphere in (A) is indicative of the operating frequency for the isofrequency surface. Similar arrows are drawn in (B) and (C). HMM, Hyperbolic metamaterials. *Adapted after P. Shekhar, J. Atkinson, Z. Jacob, Hyperbolic metamaterials: fundamentals and applications, Nano Converg. 1 (2014) 1−17.*

where $\epsilon_{xx} \neq \epsilon_{zz}; \epsilon_{xx}, \epsilon_{zz} > 0$, the sphere becomes an ellipsoid. If, however, the material is anisotropic with $\epsilon_{zz} < 0$ and $\epsilon_{xx} > 0$ (Type I HMM) or $\epsilon_{zz} > 0$ and $\epsilon_{xx} < 0$ (Type II HMM), the dispersion relation is hyperbolic, as shown in Fig. 1.2B and C, respectively.

HMMs have many potential applications in EM and optics. HMMs are used in several optical applications, such as waveguiding, imaging, sensing, quantum, and thermal engineering [5,15,16]. More on this is discussed later. HMMs can be constructed, for instance, as a multilayer structure consisting of alternating layers of dielectric and metal [17], and modeled as an anisotropic bulk medium (BM), based on effective medium theory (EMT) [18], as shown in Section 1.2.

1.2 Anisotropic metamaterials with hyperbolic dispersion: Effective medium theory

Consider a multilayered metallo–dielectric (MD) stack, such as that shown in Fig. 1.3, where the MD patterning has a spatial scale that is much smaller than the radiation wavelength. One can treat the system as an anisotropic BM with an effective dielectric permittivity

$$[\epsilon_{\text{eff}}] = \begin{bmatrix} \epsilon_{xx} & 0 & 0 \\ 0 & \epsilon_{yy} & 0 \\ 0 & 0 & \epsilon_{zz} \end{bmatrix}, \tag{1.5}$$

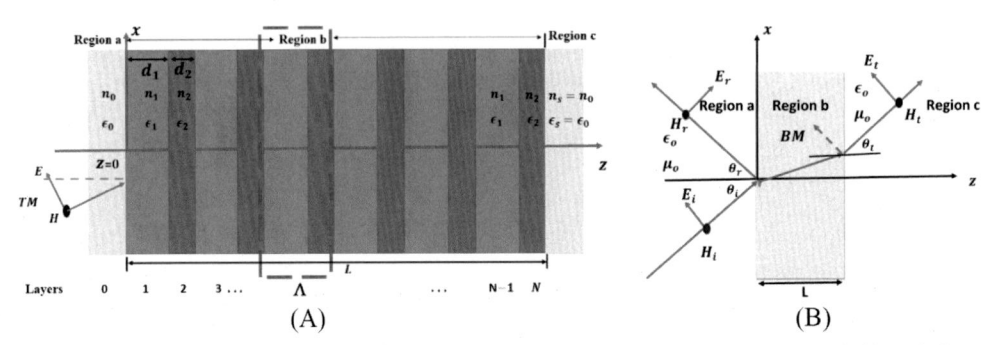

(A)   (B)

Figure 1.3 (A) Geometry of multilayered dielectric (red (dark in print version))—metal (blue (light in print version)) metamaterial structure; (B) same as (A) but with the multilayered structure modeled as a BM using EMT. (B) also shows oblique plane wave incidence onto the structure. The angle of incidence is θ_i, the angle of reflection is $\theta_r = \theta_i$, and the angle of transmission is $\theta_t = \theta_i$. The incident and transmitted media (regions a and c, respectively) can be assumed to be free space for simplicity [17]. In later discussions, region c is taken to be the substrate. *Adapted from R. Gnawali, P.P. Banerjee, J.W. Haus, V. Reshetnyak, D.R. Evans, Berreman approach to optical propagation through anisotropic metamaterials: application to metallo-dielectric stacks, Opt. Commun. 425 (2018) 71—79.*

$$\epsilon_{xx} = \epsilon_{yy} = \epsilon_0 \frac{\epsilon_1 d_1 + \epsilon_2 d_2}{d_1 + d_2}, \tag{1.6a}$$

$$\epsilon_{zz} = \epsilon_0 \frac{(\epsilon_1 \epsilon_2)(d_1 + d_2)}{(d_1 \epsilon_2 + d_2 \epsilon_1)}, \tag{1.6b}$$

where d_1 and d_2 are the thicknesses of the dielectric and metal layers, respectively, ϵ_0 is the permittivity of free space and $\epsilon_1 = n_1^2$, $\epsilon_2 = n_2^2$ are relative permittivities of the dielectric and the metal, respectively [10,18−20]. The derivation of this is based on EM boundary conditions for the tangential and normal components of the electric field, starting from two layers, say, 1 and 2, as shown in Born and Wolf [21]. Extending this to the N layers as shown in Fig. 1.3, it can be shown that the effective permittivities written in Eqs. (1.6a) and (1.6b) generalize to

$$\epsilon_{xxeff} = \frac{\sum_{i=1}^{i=N} \epsilon_i d_i}{\sum_{i=1}^{i=N} d_i} = \frac{\sum_{i=1}^{i=N} \epsilon_i d_i}{L} = \epsilon_{xx}, \tag{1.7a}$$

and

$$\epsilon_{zzeff} = \frac{\sum_{i=1}^{i=N} d_i}{\sum_{i=1}^{i=N} \frac{d_i}{\epsilon_i}} = \frac{L}{\sum_{i=1}^{i=N} \frac{d_i}{\epsilon_i}} = \epsilon_{zz}. \tag{1.7b}$$

For future reference, it is useful to note that as long as the ratio d_1/d_2 is constant, the values of ϵ_{xx} and ϵ_{zz} remain unchanged. The dispersion relation for this anisotropic metamaterial is

$$\frac{k_x^2}{\epsilon_{zz}} + \frac{k_z^2}{\epsilon_{xx}} = \frac{k_0^2}{\epsilon_0}, \tag{1.8}$$

where k_x and k_z are the transverse and longitudinal components of the wave vector, assuming one transverse dimension (1TD). If the condition $\epsilon_{zz} < 0$ and $\epsilon_{xx} > 0$, the dispersion relation is hyperbolic [10]. It is remarked that assuming real values for ϵ_1 and ϵ_2 for simplicity, hyperbolic dispersion can be obtained when the condition $\frac{\epsilon_1}{\epsilon_2} \geq -\frac{d_1}{d_2}$ is satisfied [10,22,23].

The concept of combining dissimilar isotropic or anisotropic materials to form a BM using EMT can be further extended for anisotropic multilayer structures. In fact, it can be shown that the effective permittivity tensor $\overline{\overline{\epsilon}}_{eff}$ for the overall BM comprising N uniaxial anisotropic layers as in Fig. 1.4 can be expressed as [24],

$$\bar{\bar{\epsilon}}_{\text{eff}} = \begin{bmatrix} \epsilon_{xx\text{eff}} & 0 & 0 \\ 0 & \epsilon_{xx\text{eff}} & 0 \\ 0 & 0 & \epsilon_{zz\text{eff}} \end{bmatrix}, \tag{1.9}$$

where

$$\epsilon_{xx\text{eff}} = \frac{\sum_{i=1}^{i=N} \epsilon_{xxi} d_i}{\sum_{i=1}^{i=N} d_i} = \frac{\sum_{i=1}^{i=N} \epsilon_{xxi} d_i}{L}, \tag{1.10a}$$

and

$$\epsilon_{zz\text{eff}} = \frac{\sum_{i=1}^{i=N} d_i}{\sum_{i=1}^{i=N} \frac{d_i}{\epsilon_{zzi}}} = \frac{L}{\sum_{i=1}^{i=N} \frac{d_i}{\epsilon_{zzi}}}, \tag{10b}$$

Indeed, an anisotropic uniaxial "layer" can be made from two isotropic layers, for example, metal (M) and dielectric (D) [24].

In the limit as $d_i \to 0$, Eqs. (1.10a) and (1.10b) reduce to

$$\epsilon_{xx\text{eff}} = \frac{\int_0^L \epsilon_{xx}(z) dz}{L}, \tag{1.11a}$$

$$\epsilon_{zz\text{eff}} = \frac{L}{\int_0^L \frac{dz}{\epsilon_{zz}(z)}}, \tag{1.11b}$$

where L is the total thickness of the effective BM. Eqs. (1.11a) and (1.11b) should prove useful in the analysis of a spatially inhomogeneous anisotropic medium with continuously varying permittivities in the longitudinal dimension [24].

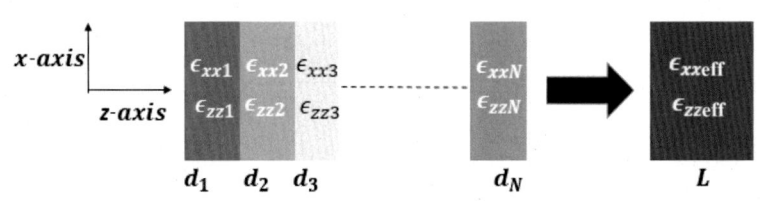

Figure 1.4 Multilayer anisotropic metamaterial structure. *Adapted from H. AL-Ghezi, R. Gnawali, P.P. Banerjee, L. Sun, J. Slagle, D. Evans, A 2×2 anisotropic transfer matrix approach for optical propagation in uniaxial transmission filter, Opt. Exp. 28 (2020) 35761–35783.*

1.3 Methods of analysis

Three different methods have been use to analyze optical propagation in this work, namely, the Berreman matrix method (BMM), transfer matrix method (TMM), and finite element methods (FEMs). All three of these methods are explained next.

1.3.1 Berreman matrix method

Consider a plane wave obliquely incident from an isotropic ambient medium (assumed to be free space) onto the anisotropic medium, finally exiting into free space once again, as shown in Fig. 1.3. It is remarked that this technique can be readily extended to the case of arbitrary incident and transmitted media. The plane of incidence is the x-z plane, and we assume there is no variation in y-direction and wave is propagating in the x-z-direction. The effective medium varies from $z = 0$ to $z = L$. The x-variation of all fields in all regions (a,b,c) is in the form $\exp(-jk_x x)$ where $k_x = k_0 \sin \theta_i$, where k_0 is the propagation constant in free space, and θ_i is the angle of incidence from free space onto the medium, as shown in Fig. 1.3B. This is because the momentum of the waves along the x-direction is unchanged since there is no interface normal to the x-direction. The incident magnetic field for *TM* polarization can be written as

$$\boldsymbol{H_i} = \hat{a}_y H_a^+ e^{-jk_0(x\sin \theta_i + z\cos \theta_i)}, \tag{1.12}$$

where H_a^+, k_0, and θ_i are the amplitude of the incident magnetic field, the free space wavenumber, and the angle of incidence, respectively [17,25]. Also, $\hat{a}_i, i = x,y,z$ denotes the unit vectors along the *x,y,z*-directions, respectively. Throughout the text, the engineering convention for traveling waves has been used. The reflected and transmitted magnetic fields, $\boldsymbol{H_r}$ and $\boldsymbol{H_t}$, respectively, can be represented in a similar way. The corresponding incident, reflected and transmitted electric fields, $\boldsymbol{E_i}$, $\boldsymbol{E_r}$, and $\boldsymbol{E_t}$ can be similarly written as [17,25,26]

$$\boldsymbol{E_i} = (\hat{a}_x \cos \theta_i - \hat{a}_z \sin \theta_i) E_a^+ e^{-jk_0(x\sin \theta_i + z\cos \theta_i)}, \tag{1.13a}$$

$$\boldsymbol{E_r} = (\hat{a}_x \cos \theta_i + \hat{a}_z \sin \theta_i) E_a^- e^{-jk_0(x\sin \theta_r - z\cos \theta_r)}, \tag{1.13b}$$

$$\boldsymbol{E_t} = (\hat{a}_x \cos \theta_t - \hat{a}_z \sin \theta_t) E_c^+ e^{-jk_0(x\sin \theta_t + (z-L)\cos \theta_t)}. \tag{1.13c}$$

Now the electric and magnetic fields inside the BM can be found explicitly writing Maxwell's curl equations and using the constitutive relations. This leads, after straightforward algebra, to a set of coupled differential equations relating E_x, E_y, H_x and H_y, which can be conveniently written in matrix form as [14,27]

$$\frac{\partial}{\partial z}\begin{bmatrix} E_x \\ H_y \\ E_y \\ -H_x \end{bmatrix} = M_B \begin{bmatrix} E_x \\ H_y \\ E_y \\ -H_x \end{bmatrix},$$ (1.14)

where the Berreman matrix for the anisotropic BM, assuming magnetic isotropy and $\mu = \mu_0$ for simplicity, along with $\epsilon_{xx} = \epsilon_{yy} \neq \epsilon_{zz}$, can be written as [17,25,26]

$$M_B = j \begin{bmatrix} 0 & -\omega_0\mu_0 + \dfrac{k_x^2}{\omega_0\epsilon_{zz}} & 0 & 0 \\ -\omega_0\epsilon_{xx} & 0 & 0 & 0 \\ 0 & 0 & 0 & -\omega_0\mu_0 \\ 0 & 0 & -\omega_0\epsilon_{xx} + \dfrac{k_x^2}{\omega_0\mu_0} & 0 \end{bmatrix}.$$ (1.15)

In the general case, the solution of Eq. (1.14), symbolically expressed as

$$\begin{bmatrix} E_x(z) \\ H_y(z) \\ E_y(z) \\ -H_x(z) \end{bmatrix} = [\exp(M_B z)] \begin{bmatrix} E_x(0) \\ H_y(0) \\ E_y(0) \\ -H_x(0) \end{bmatrix},$$ (1.16)

determines the z dependence of the EM fields inside the anisotropic material. The matrix operation $\exp(M_B z)$ can be achieved, for instance, by first determining the eigenvalues of the Berreman matrix and then the eigenvectors, or by other techniques for matrix exponentiation [27]. More on this follows below. The eigenvalues can be found as

$$jk_{zTM} = j\sqrt{\epsilon_{xx}\omega_0^2\mu_0 - \frac{k_x^2\epsilon_{xx}}{\epsilon_{zz}}},$$ (1.17a)

$$jk_{zTE} = j\sqrt{\epsilon_{xx}\omega_0^2\mu_0 - k_x^2}.$$ (1.17b)

The subscripts TM (transverse magnetic) and TE (transverse electric) have been added for reasons explained in Gnawali et al. [17,27]. It can be argued that Eq. (1.17a) gives the pertinent set of eigenvalues for TM incidence, while Eq. (1.17b) respond to TE incidence.

Alternatively, to solve Eq. (1.14) with Eq. (1.15), one can explicitly write the coupled equations as

$$\frac{\partial E_x}{\partial z} = j\left(\frac{k_x^2}{\omega_0 \epsilon_{zz}} - \omega_0 \mu\right) H_y, \tag{1.18a}$$

$$\frac{\partial H_y}{\partial z} = -j\omega_0 \epsilon_{xx} E_x, \tag{1.18b}$$

$$\frac{\partial E_y}{\partial z} = -j\omega_0 \mu H_x, \tag{1.18c}$$

$$\frac{\partial H_x}{\partial z} = j\left(\omega_0 \epsilon_{xx} - \frac{k_x^2}{\omega_0 \epsilon_{zz}}\right) E_y. \tag{1.18d}$$

For the *TM* case, only the E_x and H_y components are nonzero [26]. Upon solving this set of coupled equations directly, the EM fields inside the BM can be expressed as

$$E_x = \left(E_{bx}^- e^{jk_{zTM}z} + E_{bx}^+ e^{-jk_{zTM}z}\right) e^{-jk_x x}, \tag{1.19a}$$

$$H_y = \left(H_{by}^- e^{jk_{zTM}z} + H_{by}^+ e^{-jk_{zTM}z}\right) e^{-jk_x x}, \tag{1.19b}$$

where E_{bx}^+ and E_{bx}^- are the forward and backward propagating electric fields, respectively; and H_{by}^+ and H_{by}^- are the forward and backward propagating magnetic fields, respectively, inside the BM [17,26].

The longitudinal (z) components of $\boldsymbol{E}$ and $\boldsymbol{H}$ in the Berreman approach can be written in terms of their transverse components using Maxwell's equations. In the *TM* case, these simplified to

$$E_{bz} = -\frac{k_x}{\omega_0} H_{by}/\epsilon_{zz}, \quad H_{bz} = 0. \tag{1.20}$$

Using EM boundary conditions, for example, continuity of the tangential components of the electric and magnetic fields, and the normal components of the electric displacement and magnetic flux density (assuming no sources) at the boundaries, all EM field amplitudes inside the material can be calculated. After extensive but straightforward algebra, the transmission coefficient (t) and the reflection coefficient (r) for *TM* incidence can be written as [17,26]

$$t = \frac{E_c^+}{E_a^+} = \frac{4\eta_0 k_{zTM}\omega_0 \epsilon_{xx}\cos\theta_i e^{-jk_{zTM}L}}{\left(k_{zTM} + \omega_0 \epsilon_{xx}\eta_0 \cos\theta_i\right)^2 - \left(k_{zTM} - \omega_0 \epsilon_{xx}\eta_0 \cos\theta_i\right)^2 e^{-2jk_{zTM}L}}, \tag{1.21}$$

$$r = \frac{E_a^-}{E_a^+} = \frac{\left(k_{zTM} - \omega_0\epsilon_{xx}\eta_0\cos\theta_i\right)\left(k_{zTM} + \omega_0\epsilon_{xx}\eta_0\cos\theta_i\right)\left(1 - e^{-2jk_{zTM}L}\right)}{\left(k_{zTM} + \omega_0\epsilon_{xx}\eta_0\cos\theta_i\right)^2 - \left(k_{zTM} - \omega_0\epsilon_{xx}\eta_0\cos\theta_i\right)^2 e^{-2jk_{zTM}L}}, \quad (1.22)$$

where

$$\cos\theta_i = \sqrt{1 - \sin^2\theta_i} = \sqrt{1 - \left(k_x/k_0\right)^2}. \quad (1.23)$$

This is the essence of the BMM that will be used in this work.

1.3.2 Simplified transfer matrix method: Anisotropic and isotropic cases

The BMM discussed above can be used to analyze propagation through layered structures. We will summarize now an alternative technique, the TMM, which can be used to analyze propagation through multilayer isotropic and anisotropic structures. For simplicity, we will restrict ourselves to the 2×2 TMMs, which can be used for isotropic and anisotropic uniaxial multilayer structures. While BMM will be used for effective media which is valid if the constituent layer thicknesses are much smaller compared to the wavelength, TMM can be useful if any or multiple constituent layers have arbitrary thicknesses. For brevity, we will derive the anisotropic 2×2 BMM, which reduces to the isotropic 2×2 BMM that is commonly encountered in the literature.

Consider, therefore, a schematic example of three uniaxial layers, UN_1, UN_2, and UN_3 as shown in Fig. 1.5, where D_i^+ and B_i^+ are the electric displacement and magnetic flux density, respectively, of the forward propagating EM wave, respectively, and D_i^- and B_i^- are the electric displacement and magnetic flux density, respectively, for the backward propagating wave with $i = 0, 1, 2, 3, t$, (0: incident medium, t: transmitted medium). Also, $D_{0,1}'$, $D_{1,2}'$, $D_{2,3}'$, and $D_{3,0}'$ are dynamical matrices at interfaces $1, 2, 3$, and

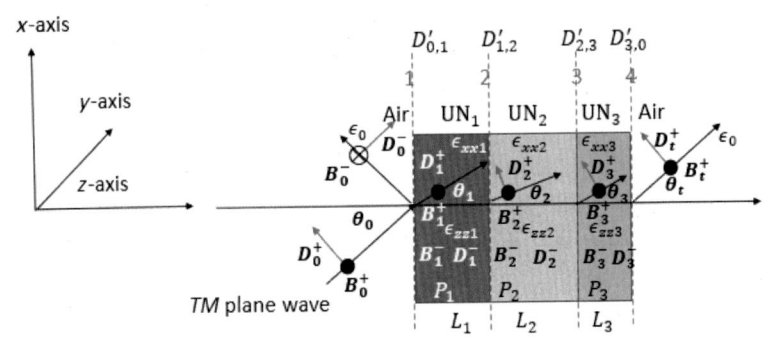

Figure 1.5 Schematic of three uniaxial layers with air as the incident medium and the exit medium for oblique *TM* incidence with the incident angle θ_0. *Adapted from H. AL-Ghezi, R. Gnawali, P.P. Banerjee, L. Sun, J. Slagle, D. Evans, A 2 × 2 anisotropic transfer matrix approach for optical propagation in uniaxial transmission filter, Opt. Exp. 28 (2020) 35761–35783.*

4, respectively. The pertinent propagation constants in the z-direction are denoted by k_{zi}, $i = 0, 1, 2, 3, t$ [24]. Each uniaxial layer is characterized by its permittivity matrix of the form given in Eq. (1.4), where we have omitted the subscript eff for simplicity. We consider the case of *TM* polarization here; *TE* can be analyzed in a similar way [24].

The dynamical matrix is usually derived from the continuity of the tangential components of the electric and magnetic fields across an interface. *Noting that Maxwell's equations are not independent of each other, the continuity of the tangential component of the electric field and the normal component of the electric displacement can instead be used.* This has been verified for the case of the simple *TM* incidence case for isotropic layers. In the anisotropic case, assume, for instance, the *TM* plane wave incidence at the interface 2, shown in Fig. 1.5 [24]. From the continuity of the tangential components of the electric field (E_x) across the two uniaxial layers UN$_1$ and UN$_2$,

$$e^{-jk_{z1}L_1}E_{1x}^+ + e^{+jk_{z1}L_1}E_{1x}^- = E_{2x}^+ + E_{2x}^-, \tag{1.24}$$

which can be rewritten as

$$e^{-jk_{z1}L_1}\frac{D_1^+}{\epsilon_{xx1}}\cos\theta_1 + e^{jk_{z1}L_1}\frac{D_1^-}{\epsilon_{xx1}}\cos\theta_1 = \frac{D_2^+}{\epsilon_{xx2}}\cos\theta_2 + \frac{D_2^-}{\epsilon_{xx2}}\cos\theta_2. \tag{1.25}$$

The subscript *TM* has been omitted from the z-components of the propagation constants k_{zi} for brevity; the expression for these propagation constants can be found in Eq. (1.17a). Also, at interface 2, continuity of the normal components of the electric displacement (D_z) gives

$$e^{-jk_{z1}L_1}D_1^+\sin\theta_1 - e^{jk_{z1}L_1}D_1^-\sin\theta_1 = D_2^+\sin\theta_2 - D_2^-\sin\theta_2. \tag{1.26}$$

Eqs (1.25) and (1.26) can be rewritten in matrix form as

$$\begin{bmatrix} \dfrac{\cos\theta_1}{\epsilon_{xx1}} & \dfrac{\cos\theta_1}{\epsilon_{xx1}} \\ \sin\theta_1 & -\sin\theta_1 \end{bmatrix} \begin{bmatrix} e^{-jk_{z1}L_1} & 0 \\ 0 & e^{+jk_{z1}L_1} \end{bmatrix} \begin{bmatrix} D_1^+ \\ D_1^- \end{bmatrix} = \begin{bmatrix} \dfrac{\cos\theta_2}{\epsilon_{xx2}} & \dfrac{\cos\theta_2}{\epsilon_{xx2}} \\ \sin\theta_2 & -\sin\theta_2 \end{bmatrix} \begin{bmatrix} D_2^+ \\ D_2^- \end{bmatrix}. \tag{1.27}$$

With straightforward algebra, Eq. (1.27) reduces to

$$\begin{bmatrix} D_1^+ \\ D_1^- \end{bmatrix} = \frac{\cos\theta_2}{2\cos\theta_1} \begin{bmatrix} e^{+jk_{z1}L_1} & 0 \\ 0 & e^{-jk_{z1}L_1} \end{bmatrix} \begin{bmatrix} \left(\dfrac{\epsilon_{xx1}}{\epsilon_{xx2}} + \dfrac{\tan\theta_2}{\tan\theta_1}\right) & \left(\dfrac{\epsilon_{xx1}}{\epsilon_{xx2}} - \dfrac{\tan\theta_2}{\tan\theta_1}\right) \\ \left(\dfrac{\epsilon_{xx1}}{\epsilon_{xx2}} - \dfrac{\tan\theta_2}{\tan\theta_1}\right) & \left(\dfrac{\epsilon_{xx1}}{\epsilon_{xx2}} + \dfrac{\tan\theta_2}{\tan\theta_1}\right) \end{bmatrix} \begin{bmatrix} D_2^+ \\ D_2^- \end{bmatrix}. \tag{1.28}$$

In Eq. (1.28), $\tan \theta_{1,2}$ and $\cos \theta_{1,2}$ can be expressed in terms of the components of the wavevectors inside the respective anisotropic media as [24]

$$\cos \theta_{1,2} = \frac{k_{z1,2}}{\sqrt[2]{k_{x1,2}^2 + k_{z1,2}^2}}, \quad \tan \theta_{1,2} = \frac{k_{x1,2}}{k_{z1,2}}. \qquad (1.29)$$

Now, using Eq. (1.29), Eq. (1.28) becomes

$$\begin{bmatrix} D_1^+ \\ D_1^- \end{bmatrix} = \frac{1}{2} \frac{\epsilon_{xx1} k_{z2}}{\epsilon_{xx2} k_{z1}} \left[\frac{k_{z1}^2 + k_x^2}{k_{z2}^2 + k_x^2} \right]^{\frac{1}{2}} \begin{bmatrix} e^{+jk_{z1}L_1} & 0 \\ 0 & e^{-jk_{z1}L_1} \end{bmatrix} \begin{bmatrix} \left(1 + \dfrac{\epsilon_{xx2} k_{z1}}{\epsilon_{xx1} k_{z2}}\right) & \left(1 - \dfrac{\epsilon_{xx2} k_{z1}}{\epsilon_{xx1} k_{z2}}\right) \\ \left(1 - \dfrac{\epsilon_{xx2} k_{z1}}{\epsilon_{xx1} k_{z2}}\right) & \left(1 + \dfrac{\epsilon_{xx2} k_{z1}}{\epsilon_{xx1} k_{z2}}\right) \end{bmatrix} \begin{bmatrix} D_2^+ \\ D_2^- \end{bmatrix}. $$

$$(1.30)$$

Eq. (1.30) includes the dynamical matrix and the propagation matrix. The new *anisotropic* dynamical matrix $D'_{1,2}$ at interface 2 can be written as [24]

$$D'_{1,2} = \frac{1}{2} \frac{\epsilon_{xx1} k_{z2}}{\epsilon_{xx2} k_{z1}} \left[\frac{k_{z1}^2 + k_x^2}{k_{z2}^2 + k_x^2} \right]^{\frac{1}{2}} \begin{bmatrix} \left(1 + \dfrac{\epsilon_{xx2} k_{z1}}{\epsilon_{xx1} k_{z2}}\right) & \left(1 - \dfrac{\epsilon_{xx2} k_{z1}}{\epsilon_{xx1} k_{z2}}\right) \\ \left(1 - \dfrac{\epsilon_{xx2} k_{z1}}{\epsilon_{xx1} k_{z2}}\right) & \left(1 + \dfrac{\epsilon_{xx2} k_{z1}}{\epsilon_{xx1} k_{z2}}\right) \end{bmatrix}, \qquad (1.31a)$$

with the remaining matrix

$$P_1 = \begin{bmatrix} e^{+jk_{z1}L_1} & 0 \\ 0 & e^{-jk_{z1}L_1} \end{bmatrix}$$

being the propagation matrix through layer 1. Similarly, other dynamical matrices can be written as [24]

$$D'_{0,1} = \frac{1}{2} \frac{\epsilon_0 k_{z1}}{\epsilon_{xx1} k_{z0}} \left[\frac{k_{z0}^2 + k_x^2}{k_{z1}^2 + k_x^2} \right]^{\frac{1}{2}} \begin{bmatrix} \left(1 + \dfrac{\epsilon_{xx1} k_{z0}}{\epsilon_0 k_{z1}}\right) & \left(1 - \dfrac{\epsilon_{xx1} k_{z0}}{\epsilon_0 k_{z1}}\right) \\ \left(1 - \dfrac{\epsilon_{xx1} k_{z0}}{\epsilon_0 k_{z1}}\right) & \left(1 + \dfrac{\epsilon_{xx1} k_{z0}}{\epsilon_0 k_{z1}}\right) \end{bmatrix}, \qquad (1.31b)$$

$$D'_{2,3} = \frac{1}{2} \frac{\epsilon_{xx2} k_{z3}}{\epsilon_{xx3} k_{z2}} \left[\frac{k_{z2}^2 + k_x^2}{k_{z3}^2 + k_x^2} \right]^{\frac{1}{2}} \begin{bmatrix} \left(1 + \dfrac{\epsilon_{xx3} k_{z2}}{\epsilon_{xx2} k_{z3}}\right) & \left(1 - \dfrac{\epsilon_{xx3} k_{z2}}{\epsilon_{xx2} k_{z3}}\right) \\ \left(1 - \dfrac{\epsilon_{xx3} k_{z2}}{\epsilon_{xx2} k_{z3}}\right) & \left(1 + \dfrac{\epsilon_{xx3} k_{z2}}{\epsilon_{xx2} k_{z3}}\right) \end{bmatrix}, \qquad (1.31c)$$

$$D'_{3,0} = \frac{1}{2}\frac{\epsilon_{xx3}k_{z0}}{\epsilon_0 k_{z3}}\left[\frac{k_{z3}^2+k_x^2}{k_{z0}^2+k_x^2}\right]^{\frac{1}{2}}\left[\begin{array}{cc}\left(1+\dfrac{\epsilon_0 k_{z3}}{\epsilon_{xx3}k_{z0}}\right) & \left(1-\dfrac{\epsilon_0 k_{z3}}{\epsilon_{xx3}k_{z0}}\right)\\[4mm]\left(1-\dfrac{\epsilon_0 k_{z3}}{\epsilon_{xx3}k_{z0}}\right) & \left(1+\dfrac{\epsilon_0 k_{z3}}{\epsilon_{xx3}k_{z0}}\right)\end{array}\right]. \tag{1.31d}$$

The propagation matrices P_l, $l = 1, 2, 3$ can be expressed as

$$P_1 = \begin{bmatrix} e^{jk_{z1}L_1} & 0 \\ 0 & e^{-jk_{z1}L_1} \end{bmatrix}, \tag{1.32a}$$

$$P_2 = \begin{bmatrix} e^{jk_{z2}L_2} & 0 \\ 0 & e^{-jk_{z2}L_2} \end{bmatrix}, \tag{1.32b}$$

$$P_3 = \begin{bmatrix} e^{jk_{z3}L_3} & 0 \\ 0 & e^{-jk_{z3}L_3} \end{bmatrix}. \tag{1.32c}$$

The overall modified 2×2 *anisotropic* transfer matrix for anisotropic uniaxial layers is then given as

$$M'_T = D'_{0,1}P_1 D'_{1,2}P_2 D'_{2,3}P_3 D'_{3,0} = \begin{bmatrix} M'_{11} & M'_{12} \\ M'_{21} & M'_{22} \end{bmatrix}. \tag{1.33}$$

The relation between the electric displacements, D_0^+ and D_0^-, for the forward and backward traveling waves, respectively, in the incident medium and the transmitted electric displacement D_t^+ is given by [24]

$$\begin{bmatrix} D_0^+ \\ D_0^- \end{bmatrix} = \begin{bmatrix} M'_{11} & M'_{12} \\ M'_{21} & M'_{22} \end{bmatrix}\begin{bmatrix} D_t^+ \\ 0 \end{bmatrix}. \tag{1.34}$$

The transmission (t) and the reflection (r) coefficients for the *TM* incidence equal the ratio between the reflected-to-incident *electric* fields $\frac{E_i^-}{E_i^+}$ and the transmitted-to-incident electric fields $\frac{E_t^+}{E_i^+}$ in air, respectively. In our case, the electric fields are related to the electric displacements simply by

$$D_i^{\mp} = \epsilon_0 E_i^{\mp}, \tag{1.35a}$$

$$D_t^+ = \epsilon_0 E_t^+. \tag{1.35b}$$

Thus t and r can be given by

$$t = \frac{1}{M'_{11}}, \tag{1.36}$$

$$r = \frac{M'_{21}}{M'_{11}}. \tag{1.37}$$

The results from the anisotropic TMM (aTMM) can be reduced to those of the usually encountered isotropic TMM (iTMM) if the anisotropic layers are replaced by isotropic layers. This is discussed in detail in AL-Ghezi et al. [24,28]. As an example, for *TM* polarization, the dynamical matrix $D_{n,n+1}$, which accounts for propagation across the interface between isotropic layers n and $n+1$, is [29,30]:

$$D_{n,n+1} = \frac{k_{z(n+1)}\sqrt{\epsilon_n}}{2k_{zn}\sqrt{\epsilon_{n+1}}} \begin{pmatrix} 1 + \dfrac{\epsilon_{n+1}k_{zn}}{\epsilon_n k_{z(n+1)}} & 1 - \dfrac{\epsilon_{n+1}k_{zn}}{\epsilon_n k_{z(n+1)}} \\ 1 - \dfrac{\epsilon_{n+1}k_{zn}}{\epsilon_n k_{z(n+1)}} & 1 + \dfrac{\epsilon_{n+1}k_{zn}}{\epsilon_n k_{z(n+1)}} \end{pmatrix}, \tag{1.38}$$

while the propagation matrix P_n, which accounts for propagation through layer n, is the same as discussed earlier, viz.,

$$P_n = \begin{pmatrix} e^{jk_{zn}d_n} & 0 \\ 0 & e^{-jk_{zn}d_n} \end{pmatrix}. \tag{1.39}$$

In Eq. (1.39), d_n is thickness of layer n. Also, in this discussion,

$$k_{zn} = \left[\omega_0{}^2\mu_0\epsilon_n - k_{xn}^2\right]^{\frac{1}{2}} = \left[\omega_0{}^2\mu_0\epsilon_n - k_x^2\right]^{\frac{1}{2}} \tag{1.40}$$

for layer n, since $k_{xn} = k_x = k_0\sin\theta_i$, as the tangential component of the $\boldsymbol{k}$ vector is unchanged. All layers are assumed to be isotropic and nonmagnetic with permittivity ϵ_n for layer n. The product is an isotropic transfer matrix (iTM) that takes into account propagation from the beginning of layer n to the beginning of layer $n+1$ through the interface between layers n and $n+1$. Multiplication of all these iTMs accounts for propagation from one end of the stack to the other, yielding the overall iTM [24]:

$$M_T = \begin{bmatrix} M_{11} & M_{12} \\ M_{21} & M_{22} \end{bmatrix}. \tag{1.41}$$

1.3.3 Finite element methods

The FEM is a method to solve the partial differential equations in 2 or 3D. When using the FEM one has to discretize a system under the investigation by meshing it into small relatively simple parts or subregions like triangles, quadrilaterals (2D and 3D boundaries), tetrahedra, pyramids, prisms, and hexahedra. These relatively simple parts contain a finite number of points and are called finite elements. The initial boundary value problem finally becomes a problem of solving a system of algebraic equations.

FEM is widely used to solve problems in complex geometries, inhomogeneous and anisotropic media, particularly in EMs [31]. There are currently many software, both free and commercial, available to implement the FEM, for instance, ANSYS, COMSOL Multiphysics, Code-Aster, FreeFem.

A comparison of computation times using TMM, BMM, and COMSOL is presented in the following section.

1.4 Beam propagation in hyperbolic metamaterials

Thus far, we have examined plane wave propagation in anisotropic bulk and layered media using BMM and TMM. Using the BMM approach, it has been shown that the electric and magnetic fields can be expressed in terms of a matrix solution as in Eq. (1.16). However, as Eq. (1.16) shows, the matrix $\exp(M_B z)$ is expressed in terms of the transverse spatial frequency k_x that is related to the angle of propagation of the plane wave. In general, for a collection of plane waves, the solution can be visualized as the superposition of the response of the system to each plane wave component. This leads to the concept of the propagation of the angular plane wave spectrum, which is discussed below.

1.4.1 Transfer function for propagation

As explained above, the matrix $\exp(M_B z)$ can be regarded as the *spatial transfer function matrix for propagation* $\tilde{h}(k_x; z)$ in the anisotropic material relating the angular plane wave spectra of the initial and final EM field components during propagation. Eq. (1.16) should, therefore, be rewritten as

$$
\begin{bmatrix} \tilde{E}_x(k_x;z) \\ \tilde{H}_y(k_x;z) \\ \tilde{E}_y(k_x;z) \\ -\tilde{H}_x(k_x;z) \end{bmatrix} = [\exp(M_B(k_x)z)] \begin{bmatrix} \tilde{E}_x(k_x;0) \\ \tilde{H}_y(k_x;0) \\ \tilde{E}_y(k_x;0) \\ -\tilde{H}_x(k_x;0) \end{bmatrix}, \tag{1.42}
$$

where the tildes refer to the *spectra* of the respective field components and $\tilde{h}(k_x; z) = \exp(M_B z)$ is the transfer function matrix for propagation. Exponentiation has been performed using analytical techniques such as the well-known Cayley—Hamilton theorem for the case of 1TD and cross-checked with results from symbolic math packages like Mathematica [26,32]:

$$
\tilde{h}(k_x;z) = \exp(M_B z) = \begin{bmatrix} a_{11} & a_{12} & 0 & 0 \\ a_{21} & a_{22} & 0 & 0 \\ 0 & 0 & a_{33} & a_{34} \\ 0 & 0 & a_{43} & a_{44} \end{bmatrix}, \tag{1.43}
$$

where

$$a_{11} = \cos(k_{zTM}z) = a_{22}, a_{12} = -j\frac{k_{zTM}}{\omega_0\epsilon_{xx}}\sin(k_{zTM}z), \quad a_{21} = -j\frac{\omega_0\epsilon_{xx}}{k_{zTM}}\sin(k_{zTM}z), \quad (1.44a)$$

$$a_{33} = \cos(k_{zTE}z) = a_{44}, a_{34} = j\frac{\omega_0\epsilon_{xx}}{k_{zTE}}\sin(k_{zTE}z), a_{43} = j\frac{k_{zTE}}{\omega_0\epsilon_{xx}}\sin(k_{zTE}z), \quad (1.44b)$$

and where k_{zTM} and k_{zTE} are functions of the spatial frequency and given by Eqs. (1.17a) and (1.17b).

For unidirectional propagation as in a semiinfinite anisotropic material, the spatial evolution of the angular plane wave spectrum can be described by simplified transfer functions which have the form $\exp(jk_z z)$. For *TM* polarization, eigenvalues jk_{zTM} determine the requisite transfer function for propagation, while for TE polarization, jk_{zTE} are the pertinent eigenvalues. In the isotropic case, $\exp(jk_{zi}z), i = 1, 2, 3, 4$ reduces to the transfer function for propagation derived in Banerjee and Poon [33]. For unidirectional propagation (along $+z$) and 1TD, and for the *TM* case, the simplified paraxial transfer function relating $\tilde{E}_x(k_x;0)$ to $\tilde{E}_x(k_x;z)$ can be written, using Eq. (1.43), as

$$\tilde{h}_{TM}(k_x;z) \propto \exp\left(-j\frac{k_x^2}{2k_1}\left|\frac{\epsilon_{xx}}{\epsilon_{zz}}\right|z\right), \quad (1.45)$$

where k_1 is a propagation constant given by $k_1 = \sqrt{\epsilon_{xx}\omega_0^2\mu_0}$. The situation is similar to the expression for a propagating (scalar) Gaussian in a NIM modeled by a simple dispersion relation by Banerjee and Nehmetallah [8]. As shown in Ref. [8], if indeed a Gaussian with diverging phase fronts can be introduced in a NIM, it should focus after some distance of propagation. This is shown rigorously for a *TM* polarized Gaussian in the next subsection. A similar paraxial transfer function can be calculated for TE polarization [27]:

$$\tilde{h}_{TE}(k_x;z) \propto \exp\left(j\frac{k_x^2}{2k_1}z\right). \quad (1.46)$$

1.4.2 Linear self-focusing

Focusing is an important property of lenses. The focusing of a Gaussian beam can be readily understood by monitoring the variation of the q parameter of a Gaussian beam which is incident on a lens, after it propagates a distance z behind the lens. The q parameter of a Gaussian beam determines the width and radius of curvature of the wavefronts, and is expressed as

$$\frac{1}{q} = \frac{1}{R} + \frac{1}{jk_1w^2/2} \quad (1.47)$$

where R denotes the radius of curvature and w is the width, and where k_1 denotes the appropriate propagation constant.

Consider a Gaussian beam in 1TD; its spectrum at $z = 0$ can be written in terms of the q parameter as

$$\tilde{g}(k_x;0) \propto \exp\left[j\frac{k_x^2}{2k_1}q(0)\right], \tag{1.48}$$

and has an inverse transform

$$g(x;0) \propto \exp\left[-j\frac{k_1 x^2}{2q(0)}\right]. \tag{1.49}$$

The paraxial transfer function of propagation has the form

$$\tilde{h}(k_x;z) = \exp\left(j\frac{k_x^2}{2k_1}\alpha z\right), \tag{1.50}$$

where α can be a scaling parameter, to be made more precise later, and where the subscript TM has been dropped for notational convenience. From Eqs. (1.48) and (1.50), it is clear that during propagation, the q parameter simply transforms as [27]

$$q(z) = q(0) + \alpha z. \tag{1.51}$$

The q parameter may also change due to lensing and transmission across an interface. In general, the change in the q parameter can be described through the bilinear transformation $q \rightarrow \frac{Aq+B}{Cq+D}$. For instance, due to lensing by a (thin) lens of focal length f located at $z = 0$, the transformed q is $q(0^+) = \frac{q(0^-)}{1 - q(0^-)/f}$. While traveling across an interface at $z = 0$ from a medium of refractive index n_1 to a medium of refractive index n_2, the transformed q is $q(0^+) = (n_2/n_1)q(0^-)$. Defining

$$z_R(z) = \frac{k_1 w^2(z)}{2} \tag{1.52}$$

for notational simplicity, Eq. (1.47) can be rewritten as [27]

$$q = \frac{z_R^2 R}{R^2 + z_R^2} + j\frac{z_R R^2}{R^2 + z_R^2}. \tag{1.53}$$

When $R = \infty$, z_R has the connotation of Rayleigh range corresponding to a waist w, and the Gaussian beam is said to be in focus. During propagation, for example, through air, the radius of curvature and the width of the Gaussian beam are coupled, namely, through the relation $\frac{1}{R} = \frac{1}{w}\frac{dw}{dz}$. Otherwise, R can be changed independent of w, for instance, when the Gaussian beam passes through a lens or across an interface.

Assume now that a *TM* polarized Gaussian beam with $q(0)$ exists in a material with effective permittivities $\epsilon_{xx} > 0$ and $\epsilon_{zz} < 0$ at $z = 0^+$. The spectrum of the beam is

$$\tilde{g}(k_x;0) = \exp\left[j\frac{k_x^2}{2k_1}q(0)\right] = \exp\left[j\frac{k_x^2}{2k_1}\frac{z_R^2(0)R(0)}{R^2(0) + z_R^2(0)}\right]\exp\left[-\frac{k_x^2}{2k_1}\frac{z_R(0)R^2(0)}{R^2(0) + z_R^2(0)}\right].$$

$$(1.54)$$

The paraxial transfer function is $\tilde{h}(k_x;z) = \exp\left(-j\frac{k_x^2}{2k_1}\left|\frac{\epsilon_{xx}}{\epsilon_{zz}}\right|z\right)$ [27]. After propagating through a distance z in the medium, the spectrum of the Gaussian beam is

$$\tilde{g}(k_x;z) = \tilde{g}(k_x;0)\tilde{h}(k_x;z) = \exp\left[j\frac{k_x^2}{2k_1}q(0)\right]\exp\left[-j\frac{k_x^2}{2k_1}\left|\frac{\epsilon_{xx}}{\epsilon_{zz}}\right|z\right] = \exp\left[j\frac{k_x^2}{2k_1}q(z)\right],$$

$$(1.55)$$

where

$$q(z) = q(0) - \left|\frac{\epsilon_{xx}}{\epsilon_{zz}}\right|z,$$

$$(1.56)$$

as in Eq. (1.51). Then,

$$\frac{1}{q(z)} \equiv \frac{1}{R(z)} + \frac{1}{jz_R(z)} = \frac{1}{q(0)} \cdot \frac{1}{1 - \left|\frac{\epsilon_{xx}}{\epsilon_{zz}}\right|\frac{z}{q(0)}} = \left(\frac{1}{R(0)} + \frac{1}{jz_R(0)}\right) \cdot \frac{1}{1 - \left|\frac{\epsilon_{xx}}{\epsilon_{zz}}\right|z\left(\frac{1}{R(0)} + \frac{1}{jz_R(0)}\right)}.$$

$$(1.57)$$

After some algebra,

$$\frac{1}{R(z)} = \frac{1}{R(0)} \cdot \frac{1 - \left|\frac{\epsilon_{xx}}{\epsilon_{zz}}\right|\frac{z}{R(0)}\left(1 + R^2(0)/z_R^2(0)\right)}{\left(1 - \left|\frac{\epsilon_{xx}}{\epsilon_{zz}}\right|\left(\frac{z}{R(0)}\right)\right)^2 + \left(\left|\frac{\epsilon_{xx}}{\epsilon_{zz}}\right|\left(\frac{z}{z_R(0)}\right)\right)^2},$$

$$(1.58)$$

$$\frac{1}{z_R(z)} = \frac{1}{z_R(0)} \frac{1}{\left(1 - \left|\frac{\epsilon_{xx}}{\epsilon_{zz}}\right|\left(\frac{z}{R(0)}\right)\right)^2 + \left(\left|\frac{\epsilon_{xx}}{\epsilon_{zz}}\right|\left(\frac{z}{z_R(0)}\right)\right)^2}.$$

$$(1.59)$$

From (1.59) the location where $z_R(z)$, and hence, $w^2(z)$, will be an extremum can be determined. This is done by setting $\frac{dz_R(z)}{dz} = 0$. After straightforward algebra, this gives

$$z = \left|\frac{\epsilon_{zz}}{\epsilon_{xx}}\right|\frac{z_R^2(0)R(0)}{z_R^2(0) + R^2(0)}.$$

$$(1.60)$$

The second derivative, $\frac{d^2 z_R(z)}{dz^2} \propto \left|\frac{\epsilon_{xx}}{\epsilon_{zz}}\right|^2 \left(\frac{1}{z_R{}^2(0)} + \frac{1}{R^2(0)}\right) > 0$, indicating that the function $z_R(z)$, and hence, $w^2(z)$, is a minimum, implying focusing. Therefore,

$$z = z_f = \left|\frac{\epsilon_{zz}}{\epsilon_{xx}}\right| \frac{z_R{}^2(0) R(0)}{z_R{}^2(0) + R^2(0)}, \tag{1.61}$$

where z_f is the focusing distance [27]. The focal spot size can be found by substituting (1.59) into (1.57). This gives

$$z_R(z_f) \equiv k_1 w_f{}^2/2 = \frac{z_R(0) R^2(0)}{z_R{}^2(0) + R^2(0)}. \tag{1.62}$$

Also, it can be readily checked that at $z = z_f$, $\frac{1}{R(z_f)} = 0$, indicating infinite radius of curvature. Furthermore, for $z < z_f$, $R(z) > 0$, that is, the wavefronts are *diverging*, and for $z > z_f$, $R(z) < 0$, that is, the wavefronts are *converging*. Indeed, it is easy to check that for this case when $\epsilon_{33} < 0$, $R(z)$ is related to $z_R(z)$ according to

$$\frac{1}{R(z)} = -\frac{1}{2} \left|\frac{\epsilon_{zz}}{\epsilon_{xx}}\right| \cdot \frac{1}{z_R(z)} \frac{dz_R(z)}{dz}. \tag{1.63}$$

It is important to note that this is contrary to propagation and focusing in a conventional medium with positive permittivities, where there is no negative sign on the right hand side (RHS) for the expression of the radius of curvature, as in Eq. (1.63). Numerical examples of *TM* polarized 1TD Gaussians in a HMM are shown below, using both the nonparaxial and paraxial transfer functions, for purposes of comparison [27].

As discussed earlier, along with the x-component of the polarization, there exists a z-component as well that evolves during propagation. From the Berreman approach, the z-component is given by

$$\tilde{E}_z = -\frac{k_x \tilde{H}_y}{\omega_0 \epsilon_{zz}}. \tag{1.64}$$

Propagation of $\tilde{E}_z$ can therefore be modeled by using the same transfer function of propagation $\tilde{h}(k_x; z)$. This is in agreement with the results from the q-parameter approach to focusing of radially polarized beams [34], where it is shown that $\tilde{E}_z \propto k_x \tilde{E}_x$. The spectrum $\tilde{E}_z$ after propagation by a distance z is

$$\tilde{E}_z(k_x; z) \propto k_x \exp\left[j \frac{k_x{}^2}{2k_1} q(z)\right], \tag{1.65}$$

where $q(z)$ is defined in Eq. (1.56). The z-component of the optical field has the form of the spatial derivative (in x) of the x-polarized Gaussian, and is a first-order

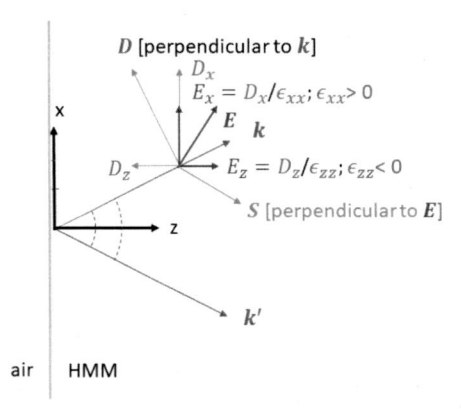

Figure 1.6 Simplified picture explaining the focusing of beams in HMMs. *HMM*, Hyperbolic metamaterials.

Hermite-Gaussian. Strong enhancement of the z-polarized field may be achievable using a low value of ϵ_{zz}; this will be investigated in the future.

A similar calculation can be performed for the *TE* case. The results are similar to propagation in a conventional positive index medium and are not shown here for the sake of conciseness.

A heuristic argument for linear self-focusing in HMMs can be advanced as follows. Assume *TM* polarization, with $\boldsymbol{H}$ along y, and nominal propagation along z, with x denoting the other transverse dimension, as shown in Fig. 1.6. Assume two "rays" or propagation vectors denoted by $\boldsymbol{k},\boldsymbol{k}'$, showing diverging wavefronts incident in the HMM. These rays correspond to two representative plane wave spectral components of a beam. The displacement $\boldsymbol{D}$ is always perpendicular to $\boldsymbol{k}$. The components D_x and D_z are shown. The corresponding E_x and E_z can now be found. It is assumed that $\epsilon_{xx} > 0$; so E_x is in the same direction as D_x. However, if $\epsilon_{zz} < 0$, the direction of E_z is opposite to the direction of D_z. The resultant $\boldsymbol{E}$ can be now drawn. The Poynting vector $\boldsymbol{S}$ is always perpendicular to $\boldsymbol{E}$, and hence directed toward the z-axis, as shown. A similar argument can be applied for the other propagation vector $\boldsymbol{k}'$. This means that the beam with diverging wavefronts will tend to "focus" in a HMM.

As can be easily checked, this type of focusing is neither possible in media that do not exhibit hyperbolic dispersion, nor for the case of *TE* polarization in HMMs.

1.4.3 Negative refraction

Negative refraction is a characteristic of a medium that has a negative refractive index. In a HMM, it is observed that negative refraction occurs only for *TM* polarization, and not for *TE* polarization [10]. The effect of negative refraction can be determined by monitoring the spatial shift of a beam when it emerges from a HMM. As shown in Fig. 1.7, a ray incident at the interface of the anisotropic HMM should be shifted

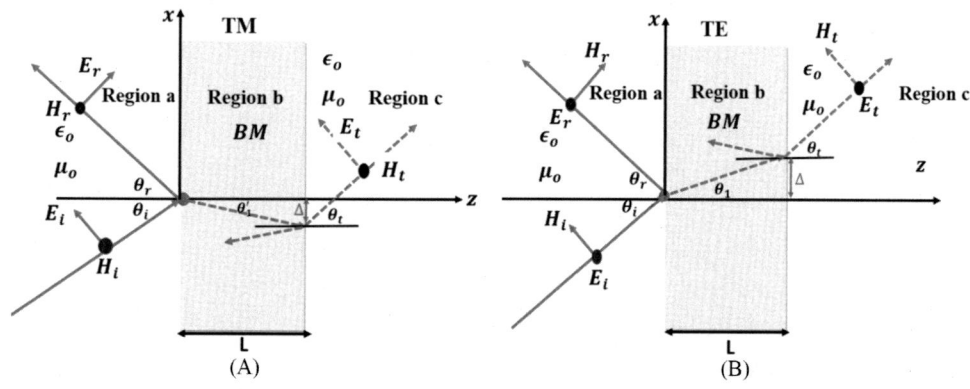

Figure 1.7 Gaussian beam propagation through a BM slab (A) for *TM* polarization (anisotropic HMM); (B) for *TE* polarization (represents as normal material). *BM*, Bulk medium; *HMM*, hyperbolic metamaterials. *Adapted from R. Gnawali, A. Kota, P.P. Banerjee, J.W. Haus, V. Reshetnyak, D. Evans, A simplified transfer function approach to beam propagation in anisotropic metamaterials, Opt. Commun. 461 (2020) 1−10.*

toward the negative direction (see Fig. 1.7A) inside the material for *TM* polarization [27]. For the *TE* case, no negative refraction is expected, as shown in Fig. 1.7B. A simple heuristic estimation of the spatial shift Δ can be found by using Snell's law. In Fig. 1.7A for HMM,

$$n_1 \sin \theta_i = n_2 \sin \theta'_1, \tag{1.66}$$

where θ'_1 is the angle of refraction. For $n_2 < 0$, $\theta'_1 < 0$; hence the spatial shift, given by $\Delta = L \tan \theta'_1$, is negative. The opposite is true for propagation through a BM with $n_2 > 0$, as shown in Fig. 1.7B [27].

Our estimation of the spatial shift during beam propagation in a HMM BM is based on rewriting the transmission coefficient as a function of the spatial frequency [17,34−36]:

$$\tilde{t}_{TM}(k_x, L) = \frac{4\eta_0 k_{zTM}\omega_0\epsilon_{xx}\cos\theta_i e^{-jk_{zTM}L}}{\left(k_{zTM}+\omega_0\epsilon_{xx}\eta_0\cos\theta_i\right)^2 - \left(k_{zTM}-\omega_0\epsilon_{xx}\eta_0\cos\theta_i\right)^2 e^{-2jk_{zTM}L}}, \tag{1.67}$$

where k_{zTM} is given in Eq. (1.17a) with $k_y = 0$ for 1TD and where θ_i now denotes the nominal angle of incidence, with the angular spectrum centered about this angle. This can be regarded as the *transmission transfer function* of the HMM.

Now, a Gaussian beam with initially plane wavefronts can be expressed as

$$g(x, 0) = E_0 \exp\left(-\frac{x^2}{w_0^2}\right). \tag{1.68}$$

Its spectrum is

$$\tilde{g}(k_x, 0) = \pi w_0^2 \exp(-k_x^2 w_0^2/4). \qquad (1.69)$$

The spectrum of a Gaussian traveling nominally at an angle θ_{i0} w.r.t. the z-axis can be expressed as

$$\tilde{g}'(k_x; 0) = \pi w_0^2 \exp\left(-\frac{(k_x - K)^2 w_0^2}{4}\right), \qquad (1.70)$$

where $\frac{K}{k_0} = \sin\theta_i$. The output spectrum after the HMM is the product of the incident spectrum and the transmission transfer function [27]:

$$\tilde{g}'(k_x; L) = \tilde{g}'(k_x; 0) t\tilde{}_{TM}(k_x, L). \qquad (1.71)$$

The TE case can be examined in a similar way [34–36], using the transmission transfer function

$$\tilde{t}_{TE}(k_x, L) = \frac{4\eta_0 k_{zTE}\omega_0 \mu \cos\theta_i e^{-jk_{zTE}L}}{\left(\eta_0 k_{zTE} + \omega_0\mu\cos\theta_i\right)^2 - \left(\omega_0\mu\cos\theta_i - \eta_0 k_{zTE}\right)^2 e^{-2jk_{zTE}L}}, \qquad (1.72)$$

where k_{zTE} is given in Eq. (1.17b) with $k_y = 0$ for 1TD, and where, as before, θ_i denotes the nominal angle of incidence [17].

Eqs. (1.67), (1.71), and (1.72) are used to study the beam shift as it traverses a HMM. Simulation results are presented below. The results show agreement with the simple heuristic picture described above.

1.5 Illustrative examples

The real and imaginary parts of the complex refractive index $\tilde{n} = n - jk$ for the constituent metal silver (Ag) and dielectrics titanium dioxide (TiO$_2$) and zinc oxide (ZnO) are shown in Fig. 1.8. In one of the examples to follow, the MD structure is assumed to be deposited on a quartz substrate. The dispersion of noncrystalline quartz is not plotted since the change of refractive index is minimal [37] and is taken to be $n_s \sim 1.476$. From the complex refractive index, the real and imaginary parts of the permittivity of the material can be readily calculated [24].

1.5.1 Example 1: Validation—comparison of BMM, TMM, and COMSOL

In this example, we show the validation of using EMT with BMM to the problem of MD structures by comparing with results from iTMM and from finite element techniques such as COMSOL. The ratio of the thicknesses of the dielectric (TiO$_2$) and the metal (Ag) is assumed to be $d_1/d_2 = 2$ (see Fig. 1.3) [17]. The variations of the real

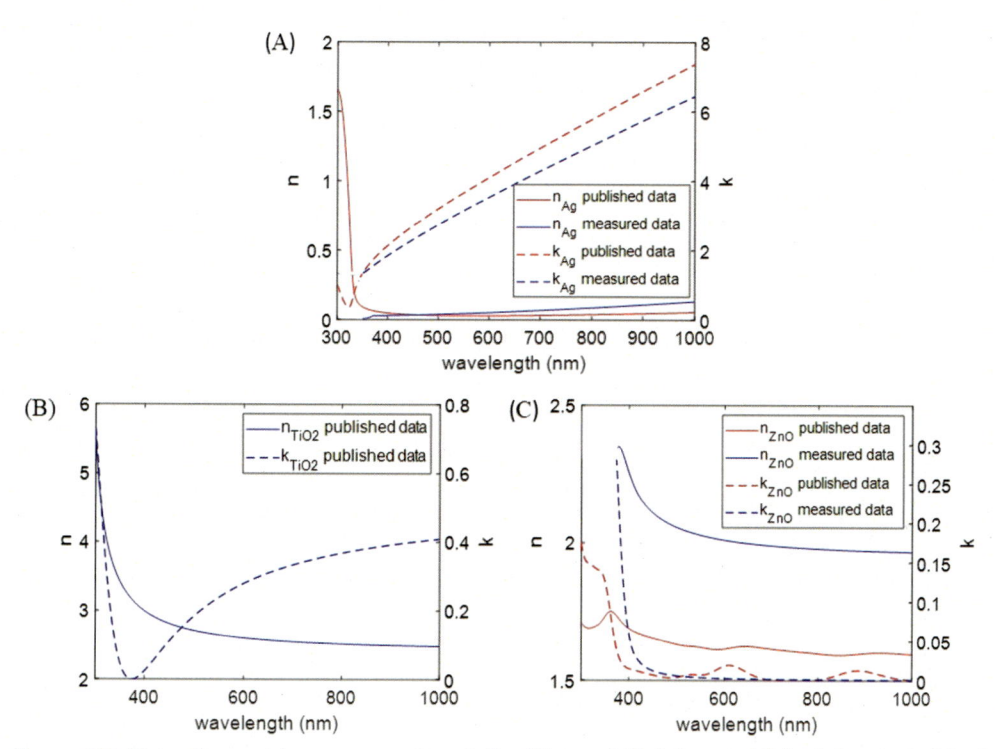

Figure 1.8 Plots of *n* and *k* versus wavelength for (A) metal (Ag) from published data in Ref. [38] and measured data reported in Ref. [24]; (B) dielectric (TiO$_2$) from published data in Ref. [39], (C) dielectric (ZnO) from published data in Ref. [40] and measured data reported in Ref. [24]. Measured data in Ref. [24] are between 375 and 1000 nm. The measured data for Ag pertain to layer 2 of a ZnO$-$[Ag$-$ZnO]^4 type multilayer MD structure deposited on quartz, as shown in Figure 7 of Ref. [24].

and imaginary parts of ϵ_{xx} and ϵ_{zz} with wavelength, needed for BMM computations, are shown in Fig. 1.9A and B, respectively over the range of wavelengths from 300 to 500 nm. Note that the imaginary part of ϵ_{xx} is approximately zero for wavelengths larger than 350 nm. The real and imaginary parts of ϵ_{zz} are typical of the Lorentz model, with the real and imaginary parts related through the Kramers$-$Kronig relations. The effective medium exhibits hyperbolic dispersion over the wavelength range approximately between 325 and 395 nm.

Fig. 1.10 shows our validation for using BMM with EMT, compares BMM with *i*TMM and COMSOL and finally uses BMM to find the dependence of transmittance on the incidence angle. All computations are performed for *TM* incidence. Fig. 1.10A shows that BMM results get closer to *i*TMM results as the number of layers are increased. Fig. 1.10B shows that the results of BMM with sufficient number of layers as indicated from Fig. 1.10A are also almost identical to the results obtained from COMSOL. This shows that EMT does indeed yield an accurate model of the MD structure as the layer

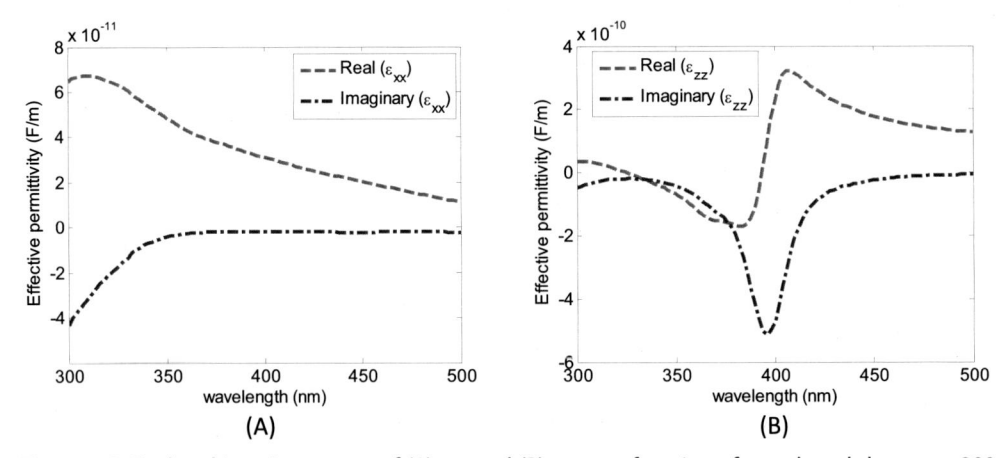

Figure 1.9 Real and imaginary parts of (A) ϵ_{xx} and (B) ϵ_{zz} as a function of wavelength between 300 and 500 nm, and for ratio of the thicknesses of the dielectric TiO$_2$ (d_1) and the metal Ag (d_2) taken to be $d_1/d_2 = 2$, derived from Fig. 1.8. *Adapted from R. Gnawali, P.P. Banerjee, J.W. Haus, V. Reshetnyak, D.R. Evans, Berreman approach to optical propagation through anisotropic metamaterials: application to metallo-dielectric stacks, Opt. Commun. 425 (2018) 71–79.*

thicknesses become increasingly smaller than the wavelength. As reported in Gnawali et al. [17], typical computation times are on the scale of a few seconds for BMM, a few minutes for TMM, and over an hour for COMSOL, using, for instance, a PC with Intel (R) core(TM) i5−2430M CPU @ 2.40 GHz with 4 GB memory, 64 bit operating system, which brings out the computational advantage of BMM with EMT. With the advantages of BMM with EMT now established, Fig. 1.10C shows the predicted transmittance results for different angles of incidence [17]. The results show that the performance of this transmission filter (TF) is virtually independent on the angle of incidence, at least for the case when there is no substrate assumed [17]. Results of transmittance with a substrate are discussed in subsequent examples.

It should be noted that despite the fact that COMSOL is less effective in computing time for the system discussed above, utilizing COMSOL gives a possibility to study multilayered systems when the layers are optically anisotropic with arbitrary orientation of the optical axes, for example, not in the plane of the layers. In the latter case, one cannot use TMM or BMM as developed in this chapter.

1.5.2 Example 2: Focusing and negative refraction in hyperbolic metamaterials

In this example, we first show simulation results for a *TM* polarized Gaussian beam in 1 transverse dimension propagating through a MD stack modeled as a BM using EMT. Simulations are performed assuming Gaussian beams of different widths and radii of curvatures in a HMM. The initial width and radius of curvature of the

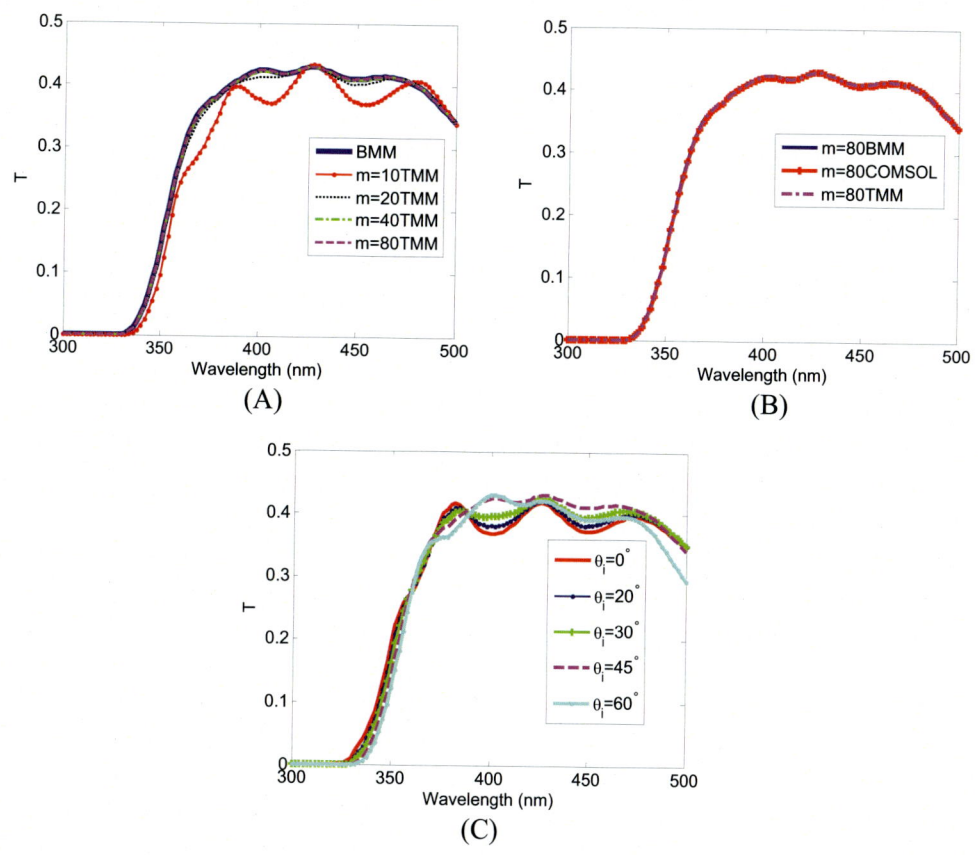

Figure 1.10 Numerical results showing transmittance as a function of wavelength for *TM* incidence and for a metamaterial with total thickness 375 nm composed of TiO$_2$ (thickness d_1) and Ag (thickness d_2) stacks with different number of layers (m) and thicknesses, keeping the total thickness constant at 375 nm and with $d_1/d_2 = 2$ (see Figs. 1.1 and 1.9). (A) BMM for (1) $m = 10$; $d_1 = 25$ nm, $d_2 = 12.50$ nm, (2) $m = 20$; $d_1 = 12.50$ nm, $d_2 = 6.25$ nm, (3) $m = 40$; $d_1 = 6.25$ nm, $d_2 = 3.125$ nm, and (4) $m = 80$; $d_1 = 3.125$ nm, $d_2 = 1.5625$ nm. Results from *i*TMM are also superimposed (in blue) for comparison. (B) Comparison of COMSOL with BMM and TMM for $m = 80$; $d_1 = 3.125$ nm, $d_2 = 1.5625$ nm. Incidence at 45 degrees is assumed for (A) and (B). (C) Transmittance as a function of wavelength for different incident angles, using BMM. *BMM*, Berreman matrix method; *TMM*, transfer matrix method. *Adapted from R. Gnawali, P.P. Banerjee, J.W. Haus, V. Reshetnyak, D.R. Evans, Berreman approach to optical propagation through anisotropic metamaterials: application to metallo-dielectric stacks, Opt. Commun. 425 (2018) 71−79.*

Gaussian beam in the metamaterial can be chosen independently. It has been analytically shown using the q-parameter of Gaussian beams that a Gaussian beam can self-focus in a HMM. We show simulations first by using the paraxial transfer function (on which the q-parameter approach is based), followed by the nonparaxial transfer function. The nonparaxial transfer function can be readily derived starting from the exact expression for the complex propagation constant, given by Eq. (1.14a).

Fig. 1.11A shows the propagation of a Gaussian beam with initial width $w(0^+) = 0.8$ μm, and with initially diverging wavefront with radius of curvature $R(0^+) = 1$ μm in a HMM with $\epsilon_{xx}, \epsilon_{zz}$ equal to 4.6×10^{-11} F/m and -8.33×10^{-11} F/m, respectively. These correspond to the real parts of the complex permittivities of an effective medium comprising MD structures made from TiO_2 and Ag with $d_1/d_2 = 2$ at an operating wavelength 354 nm (see Fig. 1.9 for the dispersion curves). The imaginary part is neglected to highlight the occurrence

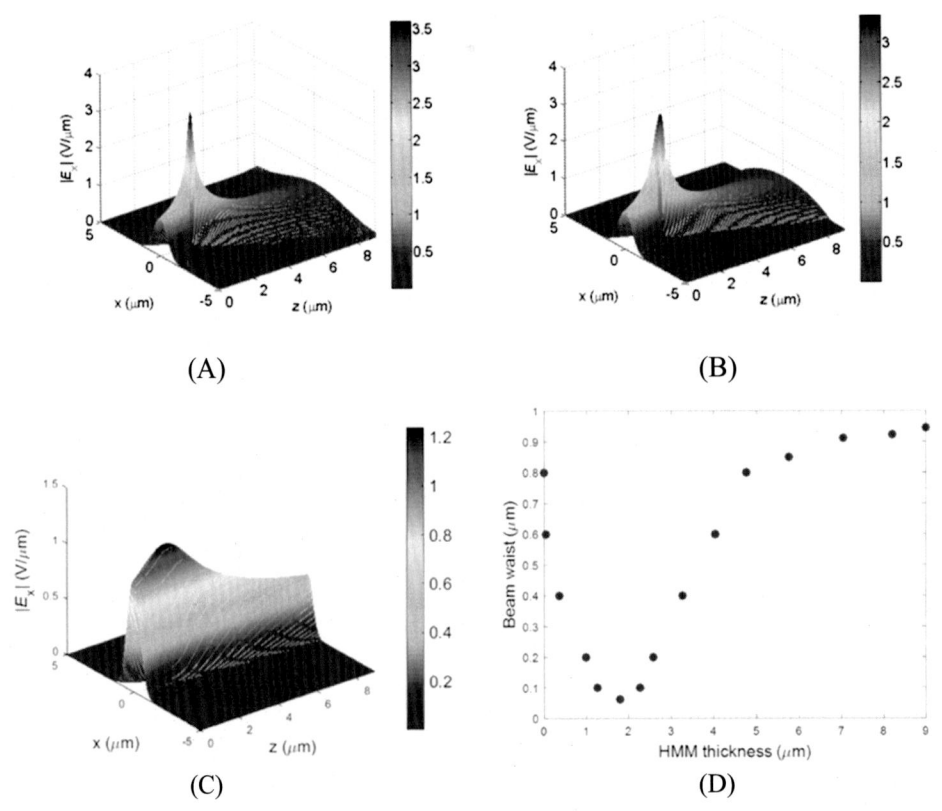

Figure 1.11 Propagation of *TM* polarized Gaussian beams with initial width $w(0) = 0.8$ μm and with initially diverging wavefronts, $R(0) = 1$ μm. (A) paraxial approximation in lossless HMM with $\epsilon_{rxx} = 5.1977$ and $\epsilon_{rzz} = -9.4124$, showing focusing; (B) same as in (A), but using nonparaxial propagation; (C) using complex permittivities $\epsilon_{xx} = (4.6 - j0.3468)10^{-11}$ F/m and $\epsilon_{zz} = (-8.33 - j5.1961)10^{-11}$ F/m, showing focusing with reduced maximum on-axis amplitude; (D) variation of beam width with respect to propagation distance inside the HMM, with a minimum value of 0.062μm at around 1.79 μm, in agreement with the lossless case. *HMM*, Hyperbolic metamaterials. *Adapted from R. Gnawali, A. Kota, P.P. Banerjee, J.W. Haus, V. Reshetnyak, D. Evans, A simplified transfer function approach to beam propagation in anisotropic metamaterials, Opt. Commun. 461 (2020) 1−10.*

of the on-axis maximum (which also implies minimum width), as predicted by paraxial theory (which neglected absorption). In Fig. 1.11B, we reproduce the propagation but now using the nonparaxial transfer function for propagation. Fig. 1.11C shows that the addition of absorption, implying the exact complex permittivities $\epsilon_{xx} = (4.6 - j0.3468)10^{-11}$ F/m and $\epsilon_{zz} = (-8.33 - j5.1961)10^{-11}$ F/m still maintains the minimum width but shows a lower maximum on- axis amplitude, as expected. Finally, Fig. 1.11D shows the variation of the beam width as a function of propagation distance in the HMM for the case in Fig. 1.11C. Comparison with results from Eq. (1.62) shows that the paraxial calculations using the q-parameter approach for focusing that assumes lossless HMMs gives a good estimate of the focusing distance and focal spot size, even for the more realistic case of complex permittivities. The percentage errors between the paraxial and nonparaxial calculations is of the order of 1%, a detailed analysis of the errors for different initial widths and radii of curvature can be found in Gnawali et al. [27].

For the case of *TE* propagation, no such self-focusing is observed, as is to be expected from the transfer function for propagation, as in Eq. (1.46).

Negative refraction in a HMM is now demonstrated using Gaussian beams in 1TD. An anisotropic slab of thickness $L = 6.35$ μm is considered. The wavelength is 0.354 μm, initial beam waist size incident on the slab is $w_0 = 0.6$ μm, and the values of the permittivity, as before, are $\epsilon_{xx} = (4.6 - j0.3468)10^{-11}$ F/m and $\epsilon_{zz} = (-8.33 - j5.1961)10^{-11}$ F/m [42,43] corresponding to $\epsilon_{rxx} = 5.1977 - j0.3918$, $\epsilon_{rzz} = -9.4124 - j5.8712$, consistent with the requirements of a HMM. Since the initial radius of curvature is infinite, there is no self-focusing of the beam inside the HMM. The slab thickness is chosen to be slightly larger than the Rayleigh range of the initial beam. The Gaussian beam profiles showing the beam shift (Δ), are plotted for different angles of incidence and are shown in Fig. 1.12 for *TM* and *TE* polarizations. As expected, the shift is negative for *TM* polarization and positive for *TE* polarization (see Fig. 1.7). The magnitudes of the shifts increase with increasing angle of incidence, which can be inferred from the heuristic picture in Fig. 1.7 [27].

1.5.3 Example 3: Transmission filters using metallo-dielectric structures

Consider a TF comprising alternating thin films of dielectric zinc oxide (ZnO) and metal (Ag), deposited on a quartz substrate. Such a structure has previously been fabricated using pulsed laser deposition and modulated pulse power magnetron sputtering techniques according to $ZnO(\frac{d_1}{2})[Ag(d_2) | ZnO(d_1)]^3 Ag(d_2) | ZnO(\frac{d_1}{2})$ with $d_1 = 82$ nm, $d_2 = 20$ nm, as shown in Fig. 1.13, deposited on a 1mm quartz substrate and experimentally tested by Sun et al. [41].

Fig. 1.14 compares the computed transmittance as a function of wavelength with experimental results. Since the thickness of the substrate is many orders of magnitude compared with the wavelength, BMM using EMT cannot be used for the entire

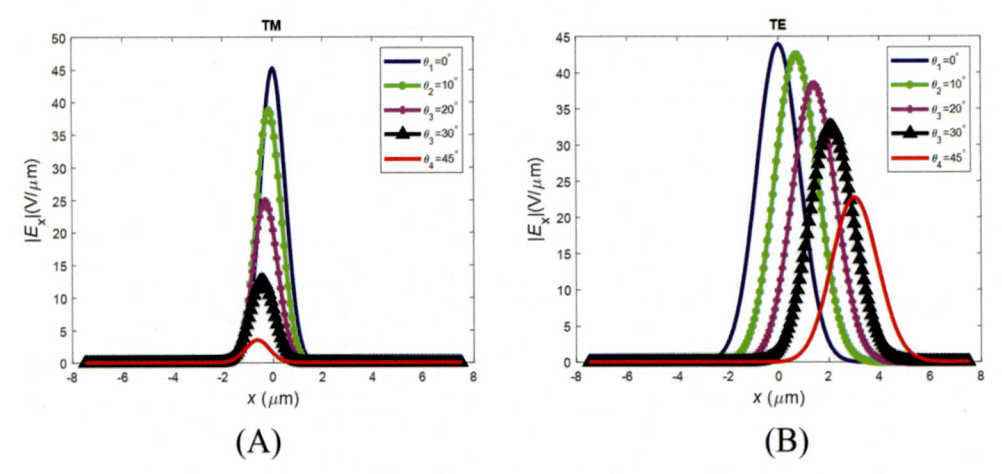

(A)   (B)

Figure 1.12 Typical exit beam profiles for a Gaussian beam with $w_0 = 0.6$ μm for various angles of incidence after propagation through an anisotropic slab of thickness 6.35 μm with $\epsilon_{xx} = (4.6 - j0.3468)10^{-11}$ F/m and $\epsilon_{zz} = (-8.33 - j5.1961)10^{-11}$ F/m (A) *TM*, (B) *TE* polarizations. *Adapted from R. Gnawali, A. Kota, P.P. Banerjee, J.W. Haus, V. Reshetnyak, D. Evans, A simplified transfer function approach to beam propagation in anisotropic metamaterials, Opt. Commun. 461 (2020) 1−10.*

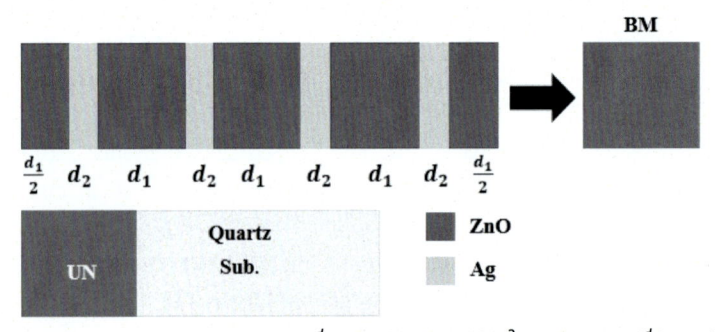

Figure 1.13 MD structure comprising $\text{ZnO}(\frac{d_1}{2})[\text{Ag}(d_2)\,|\,\text{ZnO}(d_1)]^3\text{Ag}(d_2)\,|\,\text{ZnO}(\frac{d_1}{2})$ with $d_1 = 82$ nm, $d_2 = 20$ nm for application as a transmission filter. *MD, Metallo-dielectric. Adapted from H. AL-Ghezi, R. Gnawali, P.P. Banerjee, L. Sun, J. Slagle, D. Evans, A 2×2 anisotropic transfer matrix approach for optical propagation in uniaxial transmission filter, Opt. Exp. 28 (2020) 35761−35783.*

structure. Instead, EMT is used to find the anisotropic BM to replace the MD structure, and *a*TMM is used for the anisotropic BM—(isotropic) substrate combination. As shown in Fig. 1.14A, the transmittance of a $\text{ZnO}(\frac{d_1}{2})[\text{Ag}(d_2)\,|\,\text{ZnO}(d_1)]^3\text{Ag}(d_2)\,|\,\text{ZnO}(\frac{d_1}{2})$ MD structure with $d_1 = 82$ nm, $d_2 = 20$ nm is computed using *a*TMM. Both experimentally measured n and k values (explained in the caption of Figs. 1.8 and 1.14) and also published n, k data in Refs. [38,40]. The computations are done using no substrate, and also a $L_s = 1 - $ mm-thick quartz substrate of refractive index $n_s \sim 1.476$. As is seen in

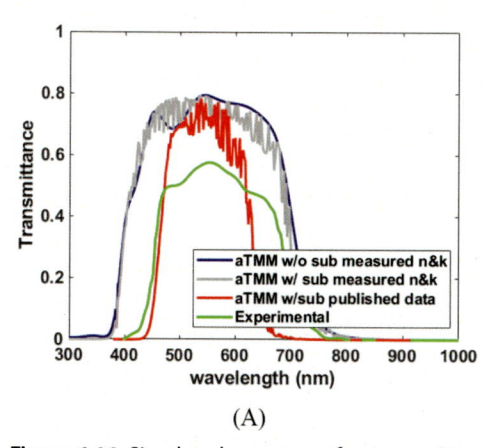
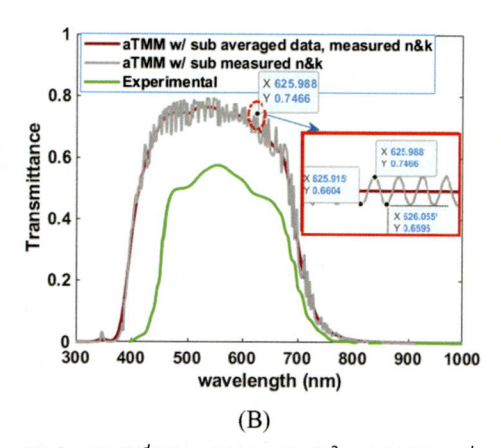

(A) (B)

Figure 1.14 Simulated spectrum for transmittance (T) for $ZnO(\frac{d_1}{2})[Ag(d_2)|ZnO(d_1)]^3Ag(d_2)|ZnO(\frac{d_1}{2})$ TF compared with experimentally observed spectrum for a similar $ZnO - [Ag - ZnO]^4$ structure, taken from [41]. Details of ZnO and Ag layer thicknesses deposited on quartz for experimentally observed spectrum is given in AL-Ghezi et al. [24]. (A) Transmittances are calculated for the $ZnO(\frac{d_1}{2})[Ag(d_2)|ZnO(d_1)]^3Ag(d_2)|ZnO(\frac{d_1}{2})$ structure with air as the transmitted medium (referred to as w/o sub), along with the case when the MD structure is assumed to be deposited on a $1 - $ mm-thick quartz substrate (referred to as w/sub). Published data refer to n and k values from the literature [38,40], as shown in Fig. 1.8. Measured n and k, as explained in the caption for Fig. 1.8, refer to measured n and k data for ZnO, and measured n and k data for Ag for layer 2 of a $ZnO - [Ag - ZnO]^4$ type multilayer MD structure deposited on quartz, as shown in Fig. 1.7 of Ref. [24]. (B) Transmittance spectrum in (A) for the case computed with the measured n and k and with substrate to show the period of the oscillations, as well as the result of averaging over 20 periods of the oscillations [24]. The averaging is done to show the expected transmittance if a spectrometer with a resolution of $2 - 3$ nm is used. *Adapted from H. AL-Ghezi, R. Gnawali, P.P. Banerjee, L. Sun, J. Slagle, D. Evans, A 2×2 anisotropic transfer matrix approach for optical propagation in uniaxial transmission filter, Opt. Exp. 28 (2020) 35761-35783.*

Fig. 1.14A, the experimentally observed transmittance spectrum reported in Ref. [41] lies between these two computed transmittance spectra. As noted, the computed transmission spectrum without a substrate does not show the rapid oscillations as in the cases when the quartz substrate is assumed. It is also noted that the rapid oscillations are not observed in the experimentally measured transmission spectrum.

The possible reason the oscillations in our simulations for the MD-quartz structure can be attributed to the Fabry–Perot effect. To quantify this, note that the optical path length (OPL) for one round trip in quartz and MD layers is

$$OPL = 2(n_{xxeff}L_{UN} + n_sL_s), \tag{1.73}$$

where n_{xxeff} is an estimate of the real part of the refractive index of the uniaxial MD structure (labeled UN in Fig. 1.13) and L_{MD} is the total thickness of the MD structure.

An estimate of the wavelength separation between successive Fabry−Perot maxima can be found using the relation

$$\Delta\lambda = \frac{\lambda^2}{\text{OPL}} \sim \frac{\lambda^2}{2(n_{xxeff} L_{UN} + n_s L_s)}, \tag{1.74}$$

which can be readily derived from Eq. (1.73). For $\lambda = 625.988$ nm, $n_{xxeff} = 1.004$, and using $L_{UN} = 408$ nm, $L_s = 1$ mm, along with $n_s = 1.476$, it turns out that $\Delta\lambda = 0.135$ nm. This compares favorably with the computed separation of $\Delta\lambda = 0.14$ nm around the same wavelength, as shown through the inset in Fig. 1.14B. The inset is a plot of the variation of the transmittance over a very narrow wavelength range with step size of 0.02 nm.

The occurrence of weak Fabry−Perot modes in a geometry such as in Fig. 1.11 has been substantiated by varying the substrate thickness. The results (not plotted here) show that the periods of these oscillations are in accordance with Eq. (1.74); the period increases substantially for substrate thicknesses hypothetically in the order of microns [24].

It is interesting to speculate on the reason for the absence of the oscillations in measured data. It is noted that the measurements have been performed using a spectrometer, which has a typical bandwidth of around 1 nm. In fact, averaging the computed spectral data over several oscillation periods removes the oscillations, as shown superposed in Fig. 1.14B when the data are averaged over 20 oscillation periods, and the resulting spectrum is smoother, similar to experimental data, and also similar to the case when there is no substrate. It is to be noted that one can expect larger oscillations and with smaller periods as the refractive index of the substrate increases, due to the fact that the MD-substrate "cavity" has a larger quality factor or finesse [24].

Once again, a note on computation times: typical run times on a i7-5600U core 2.60 GHz processor with 8 GB RAM and with 64 bit operating system and x64-based processor is in the order of a few seconds.

1.6 Selected applications

As stated in the Introduction, multilayer, MD structures have long been employed as EM/optical band filters, either in transmission or in reflection modes. Experimental results have verified their application as TFs [41]. Multilayer structures made from phase change materials like vanadium dioxide (VO_2) instead of metal, such as TiO_2−VO_2 thin films, have been projected to perform as optical transmission switches, around the transition temperature (68°C) of VO_2 [42]. This temperature-activated switching behavior may serve as a basis for fabrication of a new class of tunable filters, thermal regulation, camouflage and other applications in the IR spectral region. Depositing MD layers on an electro-optic (EO) substrate such as barium tita-nate ($BaTiO_3$) can lead to further electrical tuning by varying the bias across the EO material, as shown through simulation results in AL-Ghezi et al. [24].

In Gnawali et al. [17], it has been shown that for MD filters made from TiO_2 and Ag, for wavelengths around 330 nm and for *TM* polarization, the reflectance and transmittance are both small at angles of incidence of around 60 degrees, so that the light is almost entirely absorbed. In general, by minimizing transmission and reflection, one can enhance absorption or trapped radiation. The trapped radiation leads to heat-ing, a form of energy harvesting. It has been proposed that can be achieved through bi-layer screens, where the top (front) screen is made of metallic hole-array and the bottom (back) screen is made of metallic disk-array, with the gap between them filled with an array of dielectric spheres [43]. The spheres are embedded in a dielectric host material, which is made of either a heat-insulating (air, polyimide) or heat-conducting (MgO) layer. EM trapping of 97% is projected when a $0.15 - \mu m$ gap is filled with MgO and Si spheres, which are treated as pure dielectrics (namely, with no added absorption loss). Envisioned applications are antifogging surfaces, EM shields, and energy harvesting structures.

MD structures simulating a HMM can be used for subdiffraction-limit imaging method. The "hyperlens imager" is constructed in the form of a cylindrical metama-terial comprising periodic metal and dielectric layers. This anisotropic metamaterial has electric permittivities in opposite signs in two orthogonal directions, which not only enables waves with large tangential wave vectors to propagate in the medium but also achieves an image magnification. Resolution in the order of 150 nm, which is beyond the diffraction limit of conventional microscopes has been reported [16,44].

The continued miniaturization of electronic and optoelectronic devices, and the drive to optically probe ever-smaller materials and devices, requires the ability to guide and/or focus light down to the nanoscale range for sensing, security, and imaging applications. However, the diffraction limit poses a limitation to focusing light down to less than a wavelength. It has been suggested that highly anisotropic waveguiding structures, called *photonic funnels*, have the potential to serve as optical "bridges" between the subdiffraction-limited nanoscale and macroscopic length scales associated with free space or dielectrics [45]. Photonic funnels in the long-wavelength infrared have been achieved using all-semiconductor HMMs. In the all-semiconductor HMM used, highly doped semiconductor layers serve as the plasmonic (metallic) components in an alternating metal/dielectric stack [46]. The HMM, which is grown lattice-matched to a semiinsulating InP substrate by molecular beam epitaxy, is composed of alternating highly n–doped ($n^{++} \sim 10^{25}$ m^{-3}) InGaAs and undoped AlInAs layers with equal thicknesses of 80 nm. The total thickness of the HMM material is $4 \mu m$. Enhanced transmission through single funnels with aperture diameter $\lambda/20$ is demon-strated, in excellent agreement with simulations. Details can be found in Li et al. [47].

With the metamaterial market worldwide projected to grow by United States $5.6 billion, driven by a compounded growth of 56.4% by 2027 [48], these and other innovations should pave the way for potential benefits in the commercial market for real-life applications.

1.7 Conclusion

This chapter has aimed to present an overview of selected methods to analyze optical propagation in uniaxial anisotropic metamaterials, including those exhibiting hyperbolic dispersion. Such structures can be made from alternating layers of metal and dielectric. We have shown that BMM using EMT provides a reliable and accurate description of the optical characteristics of the composite material. Both *TM* and *TE* polarizations have been considered. Exact calculations of multilayer MD stacks using TMM and finite element techniques are also provided for comparison to show the convergence of the BMM results as the number of layers are increased. For MD multilayer structures, it is observed that the transmittance spectrum shows only small shifts as a function of incident angle, while the reflectance spectrum is much more sensitive to the angle of incidence.

In general, BMM applies to a wide range of systems with effective anisotropic dielectric constants. It is therefore an efficient and fast technique that can be first used for exploring parameters in inhomogeneous systems with either a gradient in the effective dielectric constants or multiple stacks with different material compositions and physical parameters. More complex geometries are amenable to rapid exploration of the parameter space using BMM before advanced, and more accurate, simulations are adopted.

BMM with EMT can readily lead to the concept of the transfer function matrix for propagation in anisotropic materials through which the angular plane wave spectra of all components of the electric and magnetic fields can be computed. The transfer function for propagation for *TM* and *TE* polarized beams and their corresponding z-polarizations have been derived starting from the Berreman matrix for a uniaxial HMM, along with the transmission transfer function. The transfer function concept has been used to analyze self-focusing and negative refraction in HMMs. Our analytical derivation of transfer functions for anisotropic metamaterial using BMM is unique, and in agreement with a simple physical model. The paraxial transfer function has been derived from the nonparaxial transfer function and employed to analyze Gaussian beam propagation using the q-parameter approach. We have shown that the paraxial approach can yield results close to the more exact nonparaxial approach. Furthermore, we have shown that the paraxial theory and the associated q-parameter approach derived for lossless HMMs can yield results for focusing which are in good agreement with more rigorous results based on the exact complex permittivities of the HMMs.

The analysis of two-dimensional Gaussian beams using these transfer functions will be reported in the future. The concept of the transmission transfer function has been used to determine beam shifts from negative and positive refraction in a HMM for *TM* and *TE* polarizations, respectively.

In structures containing multilayer anisotropic (uniaxial) materials of arbitrary thickness, BMM with EMT cannot be used. We have developed a simplified 2×2 *a*TMM to analyze EM/optical propagation through multilayer uniaxial stacks of arbitrary thicknesses. The 2×2 *a*TMM reduces to the (traditional) *i*TMM for multilayer isotropic stacks. The 2×2 *a*TMM can be substantiated by analyzing TM propagation through anisotropic stacks comprised of dissimilar MD structures [24]. The 2×2 *a*TMM is applied to uniaxial structures comprising multilayer MDs deposited on a substrate, and computational results are reconciled with experimental observations. As shown in AL-Ghezi et al. [24], the 2×2 *a*TMM is applied to similar multilayer uniaxial structures deposited on an EO substrate. Conditions under which electrical tuning of the transmission spectrum can be achieved have been determined. Our 2×2 *a*TMM can be conveniently applied to any uniaxial stacks, offers computational rigor over traditional 2×2 *i*TMM techniques and offers computational simplicity over 4×4 TMM techniques.

Our simplified 2×2 *a*TMM technique, along with some prior information about the cutoff frequencies, should prove to be computationally simpler to implement in deep learning methods which will allow for optimal design of the MD layers for a given spectral response for TFs. Furthermore, as discussed in AL-Ghezi et al. [24], use of the 2×2 *a*TMM technique along with measurement of the transmittance spectrum can give valuable information about the complex refractive indices of the constituent metal and dielectric within the fabricated MD structure.

Acknowledgment

PPB would like to acknowledge partial support through a contract from AFRL (FA8650−16−D−5404).

References

[1] J.B. Pendry, Negative refraction makes a perfect lens, Phys. Rev. Lett. 85 (2000) 3966−3969.
[2] J.B. Pendry, D.R. Smith, Reversing light with negative refraction, Phys. Today 57 (2004) 37−43.
[3] S.A. Ramakrishna, Physics of negative refractive index materials, Rep. Prog. Phys. 68 (2005) 449−521.
[4] S.A. Ramakrishna, T.M. Grzegorczyk, Physics and Applications of Negative Refractive Index Materials, SPIE Press, 2008.
[5] Y. Liua, X. Zhang, Metamaterials: a new frontier of science and technology, Chem. Soc. Rev. 40 (2011) 2494−2507.
[6] T.G. Mackay, A. Lakhtakia, Negatively refracting chiral metamaterials: a review, SPIE Rev. 1 (2010) 018003.

[7] P.P. Banerjee, G. Nehmetallah, Linear and nonlinear propagation in negative index materials, J. Opt. Soc. Am. B 23 (2006) 2348−2355.

[8] P.P. Banerjee, G. Nehmetallah, Spatial and spatiotemporal solitary waves and their stabilization in nonlinear negative index materials, J. Opt. Soc. Am. B 24 (2007) A69−A76.

[9] P.P. Banerjee, R. Aylo, G. Nehmetallah, Baseband and envelope propagation in media modeled by a class of complex dispersion relations, J. Opt. Soc. Am. B 25 (2008) 990−994.

[10] C. Argyropoulos, N. Estakhri, F. Monticone, A. Alù, Negative refraction, gain and nonlinear effects in hyperbolic metamaterials, Opt. Exp. 21 (2013) 15037−15047.

[11] A. Poddubny, I. Iorsh, P. Belov, Y. Kivshar, Hyperbolic metamaterials, Nat. Photonics 7 (2013) 948−957.

[12] V. Drachev, V. Podolskiy, A. Kildishev, Hyperbolic metamaterials: new physics behind a classical problem, Opt. Exp. 21 (2013) 15048−15064.

[13] P. Shekhar, J. Atkinson, Z. Jacob, Hyperbolic metamaterials: fundamentals and applications, Nano Converg. 1 (2014) 1−17.

[14] V. Veselago, The electrodynamics of substances with simultaneously negative values of ϵ and μ, Sov. Phys. Usp. 10 (1968) 509−514.

[15] X. Hu, C. Chan, Photonic crystals with silver nanowires as a near-infrared superlens, Appl. Phys. Lett. 85 (2004) 1520−1522.

[16] Z. Liu, H. Lee, Y. Xiong, C. Sun, X. Zhang, Far-field optical hyperlens magnifying sub-diffraction-limited objects, Science 315 (2007) 1686.

[17] R. Gnawali, P.P. Banerjee, J.W. Haus, V. Reshetnyak, D.R. Evans, Berreman approach to optical propagation through anisotropic metamaterials: application to metallo-dielectric stacks, Opt. Commun. 425 (2018) 71−79.

[18] B. Wood, J. Pendry, D. Tsai, Directed subwavelength imaging using a layered metal-dielectric system, Phys. Rev. B 74 (115116) (2006) 1−8.

[19] S.M. Rytov, Electromagnetic properties of laminated medium, Zh. Eksp. Teor. Fiz. 29 (1955) 605−616 [Sov. Phys. JETP 2 (1956) 466−475].

[20] V.M. Agranovich, V. Ginzburg, Crystal Optics With Spatial Dispersion and Excitons, second ed., Springer-Verlag, New York, 1984.

[21] M. Born, E. Wolf, Principle of Optics, Cambridge University Press, Cambridge, 1980.

[22] S. Guenneau, S.A. Ramakrishna, Negative refractive index, perfect lenses and checkerboards: trapping and imaging effects in folded optical spaces, C. R. Phys. 10 (2009) 352−378.

[23] X. Zhang, Y. Wu, Effective medium theory for anisotropic metamaterials, Sci. Rep. 7892 (2015) 1−7.

[24] H. AL-Ghezi, R. Gnawali, P.P. Banerjee, L. Sun, J. Slagle, D. Evans, A 2x2 anisotropic transfer matrix approach for optical propagation in uniaxial transmission filter, Opt. Exp. 28 (2020) 35761−35783.

[25] D. Berreman, Optics in stratified and anisotropic media: 4×4-matrix formulation, J. Opt. Soc. Am. 62 (1972) 502−510.

[26] R. Gnawali, Berreman Approach to Optical Propagation Through Anisotropic Metamaterials (Ph.D. dissertation), University of Dayton, 2018.

[27] R. Gnawali, A. Kota, P.P. Banerjee, J.W. Haus, V. Reshetnyak, D. Evans, A simplified transfer function approach to beam propagation in anisotropic metamaterials, Opt. Commun. 461 (2020) 1−10.

[28] H. AL-Ghezi, R. Gnawali, P.P. Banerjee, Optical propagation through multilayered anisotropic media, Proc. SPIE 10743 (2018) 1−7.

[29] P. Yeh, Optical Waves in Layered Media, Wiley, 1998.

[30] M. Troparevsky, A. Sabau, A. Lupini, Z. Zhang, Transfer matrix formalism for the calculation of optical response in multilayer system: from coherent to incoherent interference, Opt. Express 18 (2010) 24175−24721.

[31] M.N.O. Sadiku, Numerical Techniques in Electromagnetics with MATLAB, third ed., CRC Press, 2009.

[32] R. Gnawali, P.P. Banerjee, J.W. Haus, D.R. Evans, Transfer function for electromagnetic propagation through anisotropic metamaterials, Proc. SPIE 10526 (2018) 1−6.

[33] P.P. Banerjee, T.C. Poon, Principles of Applied Optics, CRC Press, 1991.

[34] J.A. Fleck, M.D. Feit, Beam propagation in uniaxial anisotropic media, J. Opt. Soc. Am. 73 (1983) 920—926.

[35] H. Liu, Q. Lv, H. Luo, S. Wen, W. Shu, D. Fan, Focusing of vectorial fields by a slab of indefinite media, J. Opt. A Pure Appl. Opt 11 (2009) 105103.

[36] V. Kivijärvi, M. Nyman, A. Karrila, P. Grahn, A. Shevchenko, M. Kaivola, Interaction of metamaterials with optical beams, N. J. Phys. 17 (2015) 1—14.

[37] L. Gao, F. Lemarchand, M. Lequime, Exploitation of multiple incidences spectrometric measurements for thin film reverse engineering, Opt. Exp. 20 (2012) 15734—15751.

[38] Y. Jiang, S. Pillai, M.A. Green, Realistic silver optical constants for plasmonics, Sci. Rep. 30605 (2016) 1—7.

[39] S. Ratzsch, E.B. Kley, A. Tunnermann, A. Szeghalmi, Influence of the oxygen plasma parameters on the atomic layer deposition of titanium dioxide, Nanotechnology 26 (2015) 1—11.

[40] C. Stelling, R.S. Chetan, M. Karg, T.A.F. Kong, M. Thelakkat, M. Retsch, Plasmonic nanomeshes: their ambivalent role as transparent electrodes in organic solar cells, Sci. Rep. 42530 (2017) 1—13.

[41] L. Sun, N. Murphy, J. Jones, J. Grant, R. Jakubiak, ZnO/Ag multilayer stack for induced transmission filters, Proc. OSA MB. 4 (2013) 1—3.

[42] R. Gnawali, P.P. Banerjee, J.W. Haus, V. Reshetnyak, D.R. Evans, Optical properties of titanium dioxide—vanadium dioxide multilayer thin-film structures, IEEE RAPID (2018) 1—2.

[43] J.P. Walker, V. Swaminathan, A.S. Haynes, H. Grebel, Periodic metallo-dielectric structures: electromagnetic absorption and its related developed temperatures, Materials 12 (2019) 2108.

[44] H. Lee, Z. Liu, Y. Xiong, C. Sun, X. Zhang, Development of optical hyperlens for imaging below the diffraction limit, Opt. Exp. 15 (2007) 15886—15891.

[45] A.A. Govyadinov, V.A. Podolskiy, Metamaterial photonic funnels for sub-diffraction light compression and propagation, Phys. Rev. B 73 (2006) 1—5.

[46] A.J. Hoffman, L. Alekseyev, S.S. Howard, K.J. Franz, D. Wasserman, V.A. Podolskiy, et al., Negative refraction in semiconductor metamaterials, Nat. Mat. 6 (2007) 946—950.

[47] K. Li, E. Simmons, A. Briggs, J. Xu, Y. Cheng, R.T. Chen, et al., Hyperbolic metamaterial photonic funnels, in: Proc. Conference on Lasers and Electro-Optics, OSA Paper FM1B.4, 2020.

[48] Reporter link: Global metamaterials industry, ID: 5899903 (2021) 1—298.

CHAPTER 2

Spectral characteristics of a thin lithographic tri-layer chiral slab resonator

Rajab Y. Ataai and Monish R. Chatterjee
Department of Electrical and Computer Engineering, University of Dayton, Dayton, OH, United States

2.1 Introduction

The problem of plane electromagnetic (EM) wave propagation (assuming electric field perpendicular to the plane of incidence, or s-polarization) through and from a chiral slab has been analyzed by various groups [1—5]. In the process, amplitude reflection and transmission coefficients [i.e., the Fresnel coefficients (FCs)] may be numerically derived for the achiral/chiral (ACC) and the chiral/achiral (CAC) boundaries. In this chapter, greater emphasis is applied to the derivation of FCs relative to the CAC boundary. In this chapter, for completeness, some discussion is presented relative to lossless achiral and chiral materials (including slab resonators), and also cases involving lossy materials (both achiral and chiral) with the understanding that most chiral metamaterials tend to have intrinsic dielectric losses. Then the FCs are applied to obtain the corresponding reflectances and transmittances of a general slab resonator as functions of the chirality parameter, and also other material parameters such as especially the phase refractive index (note that the conventional refractive index, ideally written n_p), represents a ratio of the *phase* velocity relative to the free-space velocity c. To distinguish this notation from the ratio of the *group* velocity to the free-space velocity, we have adopted the terminology of "phase refractive index" or simply "phase index" for n_p of the chiral region (denoted as n_{p2}) and the angle of incidence at the slab input plane. In a related study [2,3,6,7], typical slab thicknesses were assumed to be around 1 μm and higher, thus placing the slab in a thick film regime. The results generally indicated that the transverse mode spectra transition from a purely periodic pattern controlled by the single-pass reflection coefficient (for the case of zero chirality, that is, the so-called dimensionless chirality coefficient $\tilde{\kappa} = 0$ case) to a generally nonperiodic pattern, with the peak amplitudes being nonuniform when chirality was introduced. More importantly, it was found that for the lossless chiral slab problem, the sum of the $\mathcal{R}$ (reflectance) and T (transmittance) for the slab undergoes minor fluctuations around 1; these fluctuations, however, are likely a consequence of computational artifacts arising from resolution in the computational routines. In this context, it must be

Thin Film Nanophotonics
DOI: https://doi.org/10.1016/B978-0-12-822085-6.00003-0

mentioned that a chiral medium realistically tends to have dielectric losses, a feature introduced in this chapter; further developments for the physical characteristics of lossy chiral slabs (both micro and nanometer thickness ranges) will be considered in follow-up studies. Typically, for the thick resonator problem, the sum of $\mathcal{R}$ and T fluctuated between 0 and 1 with varying deviations (which would be considered compatible with expectation in that energy cannot be gained); in some cases (where the FC amplitudes exceeded unity for specific angles of incidence), the $\mathcal{R} + T$ sum tended to exceed unity via minor deviations, attributable to numerical computational artifacts. It is believed that the dielectric constant of a chiral material is generally complex (A. Lakhtakia, personal communication, September 2017), implying an inherent loss as mentioned earlier. In this work, attention is focused on the specific limit where the slab approaches thin film thicknesses (a thickness range approximately 100 nm or less). Moreover, the thin film slab problem is further subdivided into two structural categories where the thin film slab is immersed in (1) the same external medium and (2) in different media above and below (or left and right, depending on the slab orientation). Of the two configurations, it turns out that the purely lithographic structure (where a slab material is layered between two *dissimilar* host materials, such that overall we have a tri-layer device that resembles a typical lithographically constructed integrated circuit) involving a chiral layer has not been studied at length unlike other similar structures. As discussed in related work [2,3,5−11], FCs at an ACC interface exhibit anomalous behavior within different chiral bands and also depend on the phase index of the two dielectrics. The anomalies include total internal reflection (TIR) even for propagation from lower to higher phase index, and Brewster effects for *s*-polarized incident light. For the slab problem, extensions to the case of two interfaces (with ACC and CAC in front and back respectively) resulting in entirely different optical characteristics relative to a standard Fabry−Perot (F−P) type chiral slab resonator. For the ACC interface, it turns out that for certain angles of incidence and material parameters, the FCs may exceed unity in amplitude [5]; this is not unusual for a single achiral interface problem, since amplitude FCs are rather often complex in nature and their magnitudes do not necessarily follow any conservation law. It needs to be mentioned that propagation in a chiral medium inherently involves the generation of two modes, one right circularly polarized (RCP) and the other left circularly polarized (LCP) in polarization. The characteristics across the single interfaces described above may be significantly different for the two modes [5,12]. In the present work, further analysis is carried out for a chiral slab (say thickness d) which conforms to the thin film regime, under the two configurations mentioned. The slab thickness d (placed between an achiral surface and an achiral substrate) is considered to be in the range 10−100 nm. Overall, when the slab thickness approaches the thin film limit, it is found that the (nonuniform) transverse mode spectra for a lossless chiral slab become considerably less dense numerically. Furthermore, it is found that the degree of fluctuations of $\mathcal{R} + T$ around 1 varies based on slab thicknesses and frequency bands; however, since the system is lossless, all these numerical results indicate minor computational artifacts. The more realistic lossy chiral dielectric analysis is carried out first in some detail for the case of an achiral—chiral slab (F—P resonator), where the complex slab permittivity is incorporated into the FC and reflectance and transmittance calculations. These results are analytically derived. Finally, the case of a lossy chiral slab resonator is introduced via a few basic case studies. Since analytical results for chiral interfaces and slabs are not tractable, the numerical results are considerably more complex, and further investigations for such a realistic chiral slab within or outside the thin film limit are currently pending.

The organization of this chapter is as follows. In Section 2.2, we introduce a rectangular slab made of a thin film isotropic chiral material. The FCs for a CAC interface with s-polarized incident plane wave are derived by considering a lithographic configuration. We note that the analysis for an ACC interface is essentially similar, has been presented in detail in several references [5,12—14], and is not covered here. One important procedural difference is that for the ACC problem, the incident wave arrives from an achiral region, and its angle and amplitude are needed for the FC analysis. For the CAC problem, there are two incident waves (one RCP and the other LCP) whose angles and amplitudes are dictated by the transmission conditions at the first (ACC) interface. In Section 2.3, amplitude-based FCs at each of the (ACC and CAC) interfaces are discussed in some detail, in particular taking into account multiple reflections and transmissions within the chiral slab in the thin film limit. One matter to be noted here is that in a standard chiral F—P etalon analysis, following the use of the FCs at each slab, effective interface power coefficients (reflectance and transmittance) for a single traversal are usually derived and later used in the development of the slab reflectance and transmittance as a whole. In the present case, however, since the FCs at either interface are not analytically derivable, we proceed (as described in Section 2.3) to carry out the Poynting vector analysis *en route* to developing the slab reflectance and transmittance, such that individual single-traversal reflectance and transmittance results are not separately derived; these are implicit in the eventual calculations. Another matter to be noted is that the slab problem may consist of two different environments, namely, one where the chiral slab is surrounded on both sides by the same external medium [described symbolically as an $(\varepsilon_1, \varepsilon_2, \varepsilon_1)$-configuration in terms of the phase indices of the materials] [6], and another where the slab is in an environment with two dissimilar materials on either side [described symbolically as an $(\varepsilon_1, \varepsilon_2, \varepsilon_3)$-configuration]—a situation more likely to be manifested for a thin film F—P-type slab with a top medium and a substrate [7,15,16]. As discussed, Section 2.4 along with Section 2.3 present some detail regarding the effective reflected and transmitted Poynting vectors based on the multiple internal transits through the slab in the thin film limit, leading eventually to the slab reflectance and transmittance coefficients. The final reflectance and transmittance behavior for the chiral F—P are quantitatively analyzed relative to (1) monochromatic propagation frequency; (2) chirality $\tilde{\kappa}$, and slab thickness d. Some numerical simulations and results for the reflectance and transmittance of chiral slab in the thin film limit are reported in some detail in Section 2.4; more details of the mathematical derivations may be found in Refs. [6,16]. Finally, concluding remarks and ideas to further expand the research are presented in Section 2.5.

2.2 Isotropic lossless chiral rectangular slab

In this section, a regular isotropic lossless chiral slab and its structural and physical characteristics will be discussed in some detail.

2.2.1 Phasor electromagnetic analysis *en route* to Fresnel coefficients for lithographic (ε_1, ε_2, ε_3)-configuration

The slab is placed for this problem in a lithographic setup with two dissimilar media on either side. A plane EM wave of frequency f is incident at an *oblique* angle θ_i upon the front face of the slab. The incident plane wave traverses the second interface to emerge from the right face, as shown in Fig. 2.1. In the figure, εs, μs, and $\tilde{\kappa}$ represent the permittivity, permeability, and chirality coefficient for the three layers (with only the medium in the middle being chiral). Incidentally, in the details of the mathematical derivations, we make use of the relation between the relative permittivity ε_r and the phase refractive index ($n_p = \sqrt{\varepsilon_r}$) for nonmagnetic materials throughout. Also, ks are the wavenumbers of the individual wavefronts. The distance between the two interfaces in this work is ≤ 100 nm. In what follows, the amplitude reflection and transmission coefficients at the front interface (propagation from n_{p1} toward n_{p2}) will be represented as $\Gamma_\perp$ and $T_\perp$. Likewise, the coefficients at the back interface (propagation from n_{p2} toward n_{p3}) are represented by $\tilde{\Gamma}_\perp$ and $\tilde{T}_\perp$. In finding the FCs for the interfaces of the slab, as indicated, the second interface is

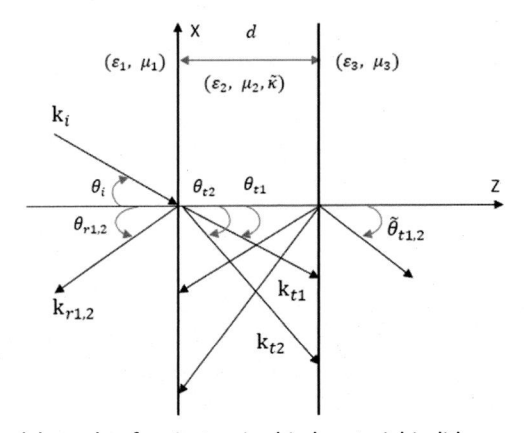

Figure 2.1 Rectangular slab made of an isotropic chiral material in lithographic configuration [2].

analytically more complicated than the first [4,7]. The FCs for EM propagation across the ACC boundary (first interface) remain identical to that presented in the literature [5], and hence will not be discussed here. It must be noted that in all analyses, any EM propagation from a lower-to-higher phase index will be considered R/D (Rarer to Denser), while that from higher-to-lower phase index will be considered D/R (Denser to Rarer). As described in recent work [2,5], FCs are usually derived (one for reflection and one for transmission) and their properties analyzed under various material conditions, leading to standard (in some cases anomalous) phenomena such as the Brewster effect, TIR, and possible appearance of evanescent propagation modes leading to complex angles of transmission.

The chiral problem is studied using the Maxwell and Helmholtz equations deriving therefrom the corresponding wave numbers (four in the chiral medium), together with phase matching applied through the (tangential) boundary conditions (BCs), of which only the two forward modes (RCP and LCP) are considered in the analysis. The specifically novel aspect of this work involves the examination of two different configurations, namely, (1) where the slab is placed in a homogeneous background, and (2) where the configuration is made lithographic to conform to a typical thin film layer between dissimilar materials. Of these, of course, the case (1) is very similar in structure to that reported in Ref. [6], and hence the corresponding FC derivations will not be considered here (except the impact of the slab thickness on the overall power coefficients will be discussed later). Case (2), however, has a different CAC interface (with n_{p2} and n_{p3} being 1.5 and 1.772 in the majority of cases); hence, the FC characteristics for this interface will be discussed briefly in what follows to highlight the main issues.

2.2.2 Phasor electromagnetic reflection and transmission across a lithographic chiral—achiral interface

As usual, bimodality adds to the complexity of the problem, since the incident wave at the boundary is now transformed into an RCP and an LCP mode in the chiral region, propagating in different directions. Fig. 2.2 shows a schematic of the physical configuration with bimodal incidence, reflection and transmission at the CAC interface butted against a substrate (such as sapphire, diamond or silicon dioxide). The physical symbols shown in Fig. 2.2 appear in the equations that follow.

The analysis for FCs relative to a CAC interface has been presented in recent work [4,6,7,16] emphasizing a uniform background surrounding the slab. For the lithographic setup, however, the medium of the transmitted EM wave is different from that at the ACC interface on the left of the slab. Hence, it becomes important to examine the FCs pertinent to the waves *transmitted* to the substrate. In the chiral medium, one may define an effective wave impedance (expressed in terms of the forward wave numbers) as:

$$1/\tilde{\eta}_{2\text{eff}} = \frac{1}{\tilde{\mu}_2 \omega}\left(k_{z1,2} \pm \frac{\omega\tilde{\kappa}}{c}\right). \tag{2.1}$$

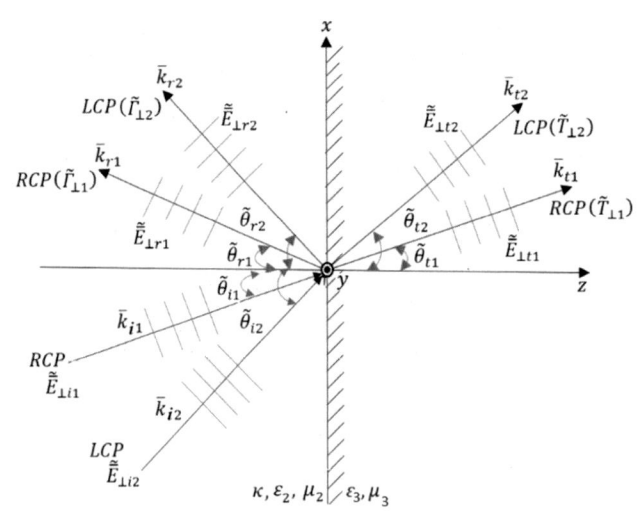

Figure 2.2 Circularly polarized uniform plane waves incident at a CAC interface showing RCP and LCP incident, reflected and transmitted modes.

$\tilde{T}_1$ and $\tilde{T}_2$ are the desired transmission FCs for the CAC interface. Note that the wavenumbers k_{z1} and k_{z2} refer to forward RCP and LCP modes for both the incident and reflected waves [5]; note that in Eq. (2.1) the " $+$ " k_{z1}, and " $-$ " correspond to k_{z1} and k_{z2}, respectively, resulting in two possible wave impedances for the *forward* modes; on the other hand, the wavenumbers k_{t1} and k_{t2} shown in Fig. 2.2 corresponding to the transmitted RCP and LCP modes are actually equal (being in an achiral environment), and expressed by the well-known unbounded quantity $\omega\sqrt{\mu_3\varepsilon_3}$ past the CAC interface, with the medium on the right of the slab being lithographically different to that on the left; hence, the material parameters on the right are denoted by a subscript "3."

The equations used in the analyses include pairs of the incident, angles of reflection or transmission due to the bimodality $(\tilde{\theta}_{i1,2}, \tilde{\theta}_{r1,2}, \tilde{\theta}_{t1,2})$ in the three regions, and an *effective* wave impedance $(\tilde{\eta}_{2\text{eff}})$ in the chiral region defined in terms of the mode wavenumbers (two for incident RCP and LCP, and two for corresponding reflected RCP and LCP) and the chirality parameter $\tilde{\kappa}$, as was mentioned earlier (see Eq. (2.1)).

2.2.2.1 Boundary conditions and phase matching for a lithographic chiral–achiral interface

Upon applying the tangential BCs for the $\tilde{\bar{E}}$ and $\tilde{\bar{H}}$ fields at the CAC interface along with related phase matching, after some algebra one obtains the following relations (which are essentially similar to those obtained for the ACC case, except the order of the coefficients within the chirality-based Snell's laws (note that we use the widely accepted descriptor "Snell's Law" instead of "Snel's Law" used in certain discourses) are interchanged; also the reflected modes in this case while being equal to the corresponding angles of incidence, are actually divergent relative to each other) [16]:

$$\tilde{\theta}_{i1} = \tilde{\theta}_{r1} \text{ and } \tilde{\theta}_{i2} = \tilde{\theta}_{r2}, \text{(law of reflection)}, \tag{2.2a}$$

$$\tilde{\theta}_{t1} = \tilde{\theta}_{t2} = \tilde{\theta}_{t}, \text{(co-propagating transmitted fields)}, \tag{2.2b}$$

$$\left(\sqrt{\varepsilon_{r2}} - \tilde{\kappa}\right)\sin\tilde{\theta}_{i1} = \sqrt{\varepsilon_{r3}}\sin\tilde{\theta}_{t1}, \text{ and} \tag{2.3a}$$

$$\left(\sqrt{\varepsilon_{r2}} + \tilde{\kappa}\right)\sin\tilde{\theta}_{i2} = \sqrt{\varepsilon_{r3}}\sin\tilde{\theta}_{t2} \text{ (Snell's laws under chirality)}. \tag{2.3b}$$

We note also that because here the media to the left and right of the slab are dissimilar, the angle of transmission no longer matches the angle of incidence for the chiral rectangular slab. Hence,

$$\theta_i \neq \tilde{\theta}_t, \tag{2.4}$$

and we note also that the two angles of incidence at the CAC interface match the two angles of transmission from the ACC interface via the geometry. Eqs. (2.2)−(2.4) provide the necessary conditions for numerically deriving the bimodal FCs and field amplitudes. It must be noted that the algebraic symbols in Eqs. (2.2)−(2.4) are represented in Fig. 2.2.

Finally, the FCs (composite symbols) are obtained as MATLAB-derived functions:

$$\tilde{\Gamma}_1 = f_1(E_{o1}, E_{o2}, \eta_3, \tilde{\eta}_{2\text{eff}}, \tilde{\theta}_{i1}, \tilde{\theta}_{i2}, \tilde{\theta}_{t1}, \tilde{\theta}_{t2}), \tag{2.5a}$$

$$\tilde{\Gamma}_2 = f_2(E_{o1}, E_{o2}, \eta_3, \tilde{\eta}_{2\text{eff}}, \tilde{\theta}_{i1}, \tilde{\theta}_{i2}, \tilde{\theta}_{t1}, \tilde{\theta}_{t2}), \tag{2.5b}$$

$$\tilde{T}_1 = f_3(E_{o1}, E_{o2}, \eta_3, \tilde{\eta}_{2\text{eff}}, \tilde{\theta}_{i1}, \tilde{\theta}_{i2}, \tilde{\theta}_{t1}, \tilde{\theta}_{t2} \text{ and } (b_1 - b_5)), \tag{2.5c}$$

$$\tilde{T}_2 = f_4(E_{o1}, E_{o2}, \eta_3, \tilde{\eta}_{2\text{eff}}, \tilde{\theta}_{i1}, \tilde{\theta}_{i2}, \tilde{\theta}_{t1}, \tilde{\theta}_{t2} \text{ and } (b_1 - b_5)). \tag{2.5d}$$

The four FCs are therefore functions (f_1 through f_4) of the variables defined above. Symbols $b_1 - b_5$ are composite parameters consisting of $\eta_3, \tilde{\eta}_{2\text{eff}}, \tilde{\theta}_{i1}, \tilde{\theta}_{i2}, \tilde{\theta}_{t1}, \tilde{\theta}_{t2}, \tilde{\kappa}, \varepsilon_{r1}$, and ε_{r2}; for details, please refer to Appendix A.

2.3 Analysis of lossless chiral Fabry−Perot slab in the thin film limit

The problems involving chiral slabs and interfaces have been studied at length in the late 1980s and early 1990s by several investigators [2,3,17,18]. In our ongoing work, attention is focused upon details of phenomena associated with propagation through

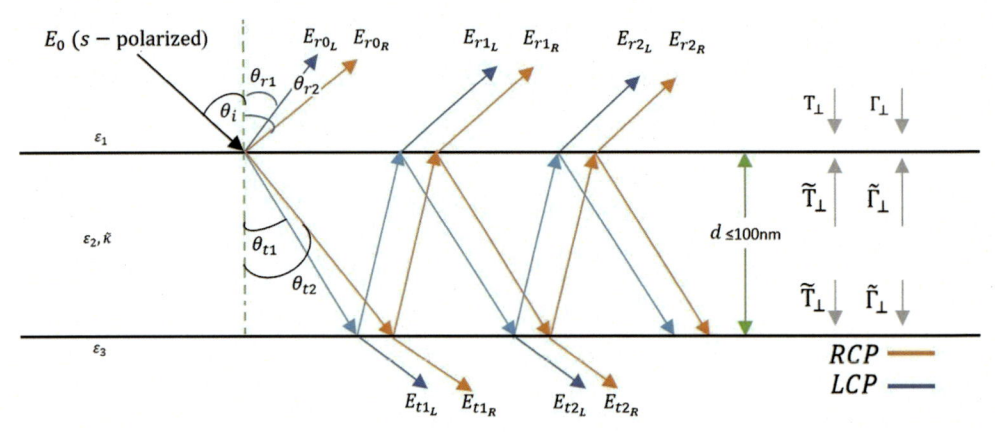

Figure 2.3 Oblique incidence on a slab of chiral material in a lithographic setup. In the chiral slab, the E_R and E_L waves are respectively left- and right-circularly polarized.

chiral regions and structures (such as slab resonators) leading to possible applications evolving from sideband dispersion, negative index and also magnetic effects. The intent in this work is to derive results for the chiral slab in the thin film regime and compare them with the (known) purely achiral results [19,20] especially for thick films. The problem will be extended to the more realistic lossy chiral resonator. Analyses will be carried out for the RCP and LCP modes separately. Thereafter, the results will be added together to obtain the total reflected and transmitted fields, and additionally the total reflected and transmitted intensity coefficients for the slab.

Fig. 2.3 shows a schematic for the various reflected and transmitted (partial) waves in a chiral F−P. Details of the EM propagation across the two interfaces (ACC and CAC) relative to bimodal propagation paths and transmitted and reflected fields angles have been detailed in Ref. [5]. Based on Fig. 2.3 via tracking a round-trip passage through the slab, the phase delay between two partial waves (whether RCP or LCP mode) which is attributable to one additional round trip, may be derived to be:

$$\delta_{1,2} = \delta_{R,L} = \frac{4\pi(n_{p2} \mp \tilde{\kappa})d\cos\theta_{t(1,2)}}{\lambda_o}, \tag{2.6}$$

where the δ_R and δ_L, respectively, refer to the phase difference for RCP and LCP (with " $-$ " for RCP and " $+$ " for LCP) transmitted waves with λ_o the (monochromatic) free-space wavelength.

2.3.1 Total amplitude of the transmitted and reflected right- and left-circular modes

If the incident wave has the form $\overline{E}_{\perp i} = \mathrm{Re}[|E_0|e^{j(\omega t - kz)}\hat{a}]$, or, equivalently the phasor field $E_0 = |E_0|e^{-jkz}$, expressed as a scalar. Following the methodology established in

Refs. [6,7], the four complex field amplitudes (two each for the RCP ad LCP modes) may be obtained by summing the respective geometric series as follows:

$$(E_{r(\text{RCP})})_n = E_0 \left[\Gamma_1 + T_1 \tilde{T}_1 \tilde{\Gamma}_1 e^{-j\delta_R} \frac{\left(1 - \left(\tilde{\Gamma}_1^2\right)^n e^{-jn\delta_R}\right)}{\left(1 - \tilde{\Gamma}_1^2 e^{-j\delta_R}\right)} \right], \qquad (2.7)$$

where $(E_{r(\text{RCP})})_n$ is the reflected RCP-mode electric field summed over n passages within the slab. It must be noted that the series summed in Eq. (2.7) is finite to n terms, and is convergent. Under conditions whereby $|\tilde{\Gamma}_1|$ exceeds 1, the infinite series (say for large n) may not converge. For those cases where this happens, the analysis demonstrates singularities in the computation leading to nonphysical outcomes.

Likewise, for the transmitted RCP-mode electric field summed over n passages, one may obtain:

$$(E_{t(\text{RCP})})_n = E_0 T_1 \tilde{T}_1 \frac{\left(1 - \left(\tilde{\Gamma}_1^2\right)^n e^{-jn\delta_R}\right)}{\left(1 - \tilde{\Gamma}_1^2 e^{-j\delta_R}\right)}, \qquad (2.8)$$

where all the quantities in Eqs. (2.7) and (2.8) have been described earlier.

For the corresponding LCP modes, the reflected and transmitted fields following similar geometric series summations become:

$$(E_{r(\text{LCP})})_n = E_0 \left[\Gamma_2 + T_2 \tilde{T}_2 \tilde{\Gamma}_2 e^{-j\delta_L} \frac{\left(1 - \left(\tilde{\Gamma}_2^2\right)^n e^{-jn\delta_L}\right)}{\left(1 - \tilde{\Gamma}_2^2 e^{-j\delta_L}\right)} \right], \qquad (2.9)$$

and

$$(E_{t(\text{LCP})})_n = E_0 T_2 \tilde{T}_2 \frac{\left(1 - \left(\tilde{\Gamma}_2^2\right)^n e^{-jn\delta_L}\right)}{\left(1 - \tilde{\Gamma}_2^2 e^{-j\delta_L}\right)}. \qquad (2.10)$$

It must be noted the $\tilde{\Gamma}_2$ FCs for the LCP mode are found to be bounded in magnitude in the range $(0,1)$ for all parametric and system configurations. As a result, it is found that the series in Eqs. (2.9) and (2.10) actually converge in the limit $n \to \infty$. These convergent fields may be written as:

$$E_{r(\text{LCP})} = \frac{E_0 \left(\Gamma_2 - \Gamma_2 \tilde{\Gamma}_2^2 e^{-j\delta_L} + T_2 \tilde{T}_2 \tilde{\Gamma}_2 e^{-j\delta_L}\right)}{1 - \tilde{\Gamma}_2^2 e^{-j\delta_L}} = \frac{E_0 \Gamma_2 (1 - e^{-j\delta_L})}{1 - \tilde{\Gamma}_2^2 e^{-j\delta_L}}, \qquad (2.11)$$

and

$$E_{t(\text{LCP})} = \frac{E_0 T_2 \tilde{T}_2}{1 - \tilde{\Gamma}_2^2 e^{-j\delta_L}}. \qquad (2.12)$$

We note that the same procedure as was used for the RCP case is applied for the LCP mode except to designate the derived FCs with subscripts "2" (Γ_2, $\tilde{\Gamma}_2$, T_2, and $\tilde{T}_2$).

We observe next that the single-passage FCs for the RCP modes may sometimes exceed 1 in magnitude; for such a situation, therefore, the infinite series representation may not work. Hence, the finite series versions are used in the simulations subject to practical limits. In the finite series context, it is instructive to note that for certain angles of incidence and physical parameters, the transmitted RCP mode from the ACC interface may be evanescent (recall the inverse TIR phenomenon). For such a situation, the fact that the total transmitted field into the medium outside is always propagating, indicates that when one mode inside the slab becomes evanescent, the other must be propagating. Hence, for an evanescent RCP mode, the corresponding LCP mode must be propagating, and therefore is infinitely summable. The finite series summation (n being typically quite large for relatively thick slabs) for such a decaying RCP mode is therefore not necessary or relevant. For the LCP modes, however, the single-passage FCs are found to be consistently less than 1 in magnitude, so that the infinite series convergence formulas may be applied. It must be noted further that in some instances, especially those dealing with singularities (associated with the amplitude FCs), the power coefficients relative to the slab exhibit fluctuations around the expected sum of 1, indicating numerical artifacts.

As has been derived in Ref. [7] relative to thick F−P slabs, the effective reflectance and transmittance for the chiral F−P slab may be derived using the appropriate Poynting vectors relative to the two slab surfaces with power flow directed *normal* to the surface [6,16]. Thus, after some algebra, we may write the expression for the total *reflectance* of the slab as

$$\mathcal{R} = \left(\frac{I_{r(\text{Total})}}{I_i} \right)_{\text{normal}} = \frac{E_{r(\text{Total})} E^*_{r(\text{Total})}}{E_i E^*_i} \left(\frac{\eta_1}{\eta_1} \right) \frac{\cos \theta_r}{\cos \theta_i} = \frac{E_{r(\text{Total})} E^*_{r(\text{Total})}}{E_0^2}, \qquad (2.13)$$

so that

$$\mathcal{R} = \left| \left(\Gamma_1 + T_1 \tilde{T}_1 \tilde{\Gamma}_1 e^{-j\delta_R} \frac{\left(1 - \left(\tilde{\Gamma}_1^2 \right)^n e^{-jn\delta_R} \right)}{\left(1 - \tilde{\Gamma}_1^2 e^{-j\delta_R} \right)} \right) + \frac{\Gamma_2 \left(1 - e^{-j\delta_L} \right)}{1 - \tilde{\Gamma}_2^2 e^{-j\delta_L}} \right|^2, \qquad (2.14)$$

where the vertical bars represent absolute value. It must be noted that the cosine factors arise from resolving the Poynting vectors normally; also, since the incident and reflected fields are in the same medium, the wave impedances and the cosine terms ultimately cancel. Proceeding similarly, the total transmittance of the slab is found to be:

$$T = \left(\frac{I_{t(\text{Total})}}{I_i} \right)_{\text{normal}} = \frac{E_{t(\text{Total})} E^*_{t(\text{Total})}}{E_i E^*_i} \left(\frac{\eta_1}{\eta_3} \right) \frac{\cos \tilde{\theta}_t}{\cos \theta_i}, \qquad (2.15)$$

so that

$$T = \left| T_1 \tilde{T}_1 \frac{\left(1 - \left(\tilde{\Gamma}_1^2\right)^n e^{-jn\delta_R}\right)}{\left(1 - \tilde{\Gamma}_1^2 e^{-j\delta_R}\right)} + \frac{T_2 \tilde{T}_2}{1 - \tilde{\Gamma}_2^2 e^{-j\delta_L}}\right|^2 \times \left[\left(\frac{\eta_1}{\eta_3}\right) \frac{\cos \tilde{\theta}_t}{\cos \theta_i}\right], \qquad (2.16)$$

where we note that the cosine factors due to resolved Poynting vectors and the wave impedances *do not cancel* because the media on the right and left of the slab are no longer identical in the lithographic configuration. Note once again that the main symbols in Eqs. (2.13)−(2.16) are consistent with those indicated in Figs. 2.2 and 2.3.

2.4 Interpretive exposition of slab structures

In what follows, a variety of chiral slab structures (including uniform surface and substrate layers, and also lithographic tri-layers) are examined in some detail.

2.4.1 Standard achiral Fabry−Perot characteristics revisited

Case A
Lossless achiral Fabry−Perot characteristics

Shown in Fig. 2.4A and B are the typical intensity transmission modes for a standard F−P resonator corresponding to a slab of thickness 100 nm, thereby implying the upper limit of a thin film regime. We find that the two separate graphs in each case correspond to a different value of the single surface reflectance (based on $n_{p2} = 1.5, 3.5$) at the interface (with the equivalent resonator Q-factor increasing noticeably as the reflectance increases toward 1 for the higher slab index), as expected [11]. We also note the consistent raising of

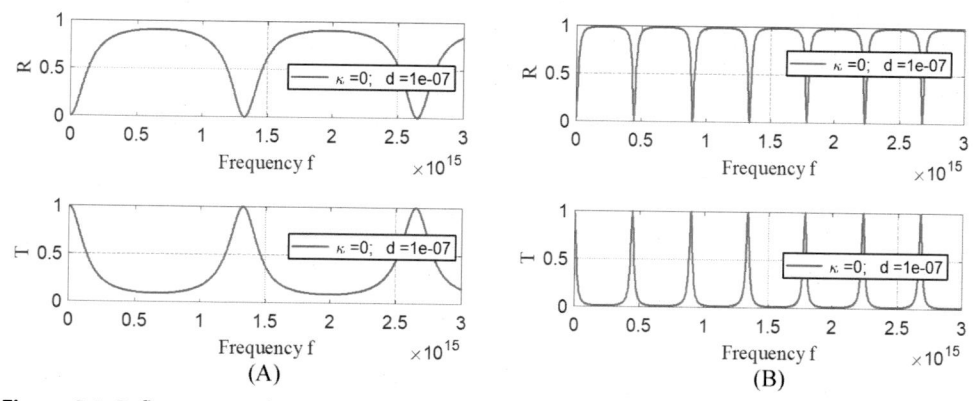

(A)　　　　　　　　　　　　　(B)

Figure 2.4 Reflectance and transmittance characteristics of a standard achiral Fabry−Perot resonator (A) $n_{p2} = 1.5$; (B) $n_{p2} = 3.5$.

the (periodic) resonance minima in the transmittance as the interface reflectance decreases. It is also noted that the mode separation decreases with the slab index; hence, there is a higher mode density in Fig. 2.4B corresponding to an n_{p2} of 3.5. Since the chiral F–P problem is considerably more complex, and in most cases lacks any analytical solution, an early test of the validity of any chiral analysis is based on deriving the solutions for the corresponding achiral problem directly out of the chiral analysis, and thereafter comparing the results. Accordingly, we consider the chiral F–P etalon in the limit of $\tilde{\kappa} = 0$ to reduce the problem to the achiral case (a process which generally results in achiral ($\tilde{\kappa} = 0$) FC graphs which are subject to factors 2 or 1/2 due the chirality implying dual modes in the algebra; the characteristics, however, are identical within a scale) [5,21]. Plotting the reflectance and transmittance characteristics of the slab by setting $\tilde{\kappa}$ to be zero (not shown here) for different chiral materials (including the lithographic case as presented in this work) reveals that the results are identical to those obtained by working with an achiral slab from first principles (as shown in Fig. 2.4). This confirmation allows further chiral analysis since the results for the achiral limit match the expected outcome.

Case B
Lossy achiral Fabry–Perot characteristics

By introducing lossy to the dielectric materials, permittivity (ε_2) becomes complex and may be written as [15,22]:

$$\varepsilon_2 = \varepsilon_0\left(\varepsilon'_{r2} - j\varepsilon''_{r2}\right), \tag{2.17}$$

where ε'_{r2} is the real relative permittivity corresponding to lossless behavior, and ε''_{r2} represents the dielectric loss.

The refractive index of the medium becomes complex and is given by

$$n_{p2c} = \sqrt{\left(\varepsilon'_{r2} - j\varepsilon''_{r2}\right)}, \tag{2.18}$$

where its impedance $\eta_{2_{\text{eff}}}$ may be written as

$$\eta_{2_{\text{eff}}} = \eta_0 \frac{1}{\sqrt{\left(\varepsilon'_{r2} - j\varepsilon''_{r2}\right)}} = \eta_0 \frac{1}{n_{p2c}}, \tag{2.19}$$

since the material is considered nonmagnetic.

By introducing the lossy the effective power reflection ($\mathcal{R}_O$) and transmission (T_O) coefficients for the ordinary slab written respectively as:

$$\mathcal{R}_O = \frac{2|\Gamma_\perp|^2(1 - \cos\delta)}{1 - \tilde{\Gamma}_\perp^{2*}e^{j\delta} - \tilde{\Gamma}_\perp^2 e^{-j\delta} + \left|\tilde{\Gamma}_\perp\right|^4}, \tag{2.20}$$

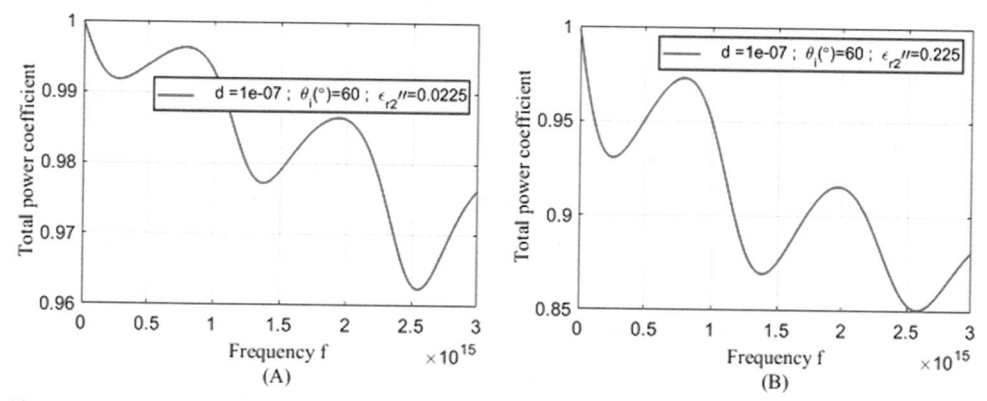

Figure 2.5 Total power characteristics of a standard achiral Fabry–Perot resonator with loss.

and

$$T_O = \frac{|T_\perp|^2 |\tilde{T}_\perp|^2}{1 - \tilde{\Gamma}_\perp^{2*} e^{j\delta} - \tilde{\Gamma}_\perp^2 e^{-j\delta} + |\tilde{\Gamma}_\perp|^4},$$

(2.21)

where the single-pass phase shift parameter δ given is by:

$$\delta = \frac{4\pi (n_{p2c}) d \cos \theta_t}{\lambda_o},$$

(2.22)

and is here complex due to the complex phase index n_{p2c}. Note that the reflection and transmission coefficients ($\Gamma_\perp$, $\tilde{\Gamma}_\perp$, $T_\perp$, $\tilde{T}_\perp$) are also complex. Following numerical computation, Fig. 2.5 shows the total power of the slab and the impact of dielectric loss on the characteristics of the slab. The advantage of carrying out the lossy $\mathcal{R}_O + T_O$ analysis is that for an achiral device, the result is analytically tractable. And expectedly, as seen from Fig. 2.5, the transverse resonant modes become increasingly lossy (total power coefficients less than 1) as the loss parameter (measured from ε_{r2}'') increases. Thus the sum of the power coefficients drops from a low of 0.962 to 0.85 as ε_{r2}'' is increased from 0.0225 to 0.225.

2.4.2 Lossless thin film chiral slab

Case A
Lossless thin film chiral slab immersed in the same medium

The chiral thin film for this section is immersed in the same dielectric material, leading to a uniform background (n_{p1}: n_{p2}: n_{p1}) configuration, with slab thickness in the 100−10 nm range.

Case A1
With $n_{p2} = 1.5$

When we make 1:1.5:1 index ratio with thickness 100 nm or 10 nm, as in Figs. 2.6 and 2.7, we get acceptable results for energy conservation only up to certain angles of incidence (limited by singularities). At angles higher than said singularities, the computation is internally aborted. In other words, there is a passband where the results are physical acceptable and a stop band (nonphysical) over the singularity angle. The passband is up to some angle of incidence approaching the singularity point (where the corresponding amplitude coefficient exceeds unity in magnitude) around which the total power coefficient hovers around $\mathcal{R} + T = 1$ within $\pm 10\%$. Thus cases as in Figs. 2.6B and 2.7B are within the 10% variability around 1.0. The fluctuations, of course, are not a violation of

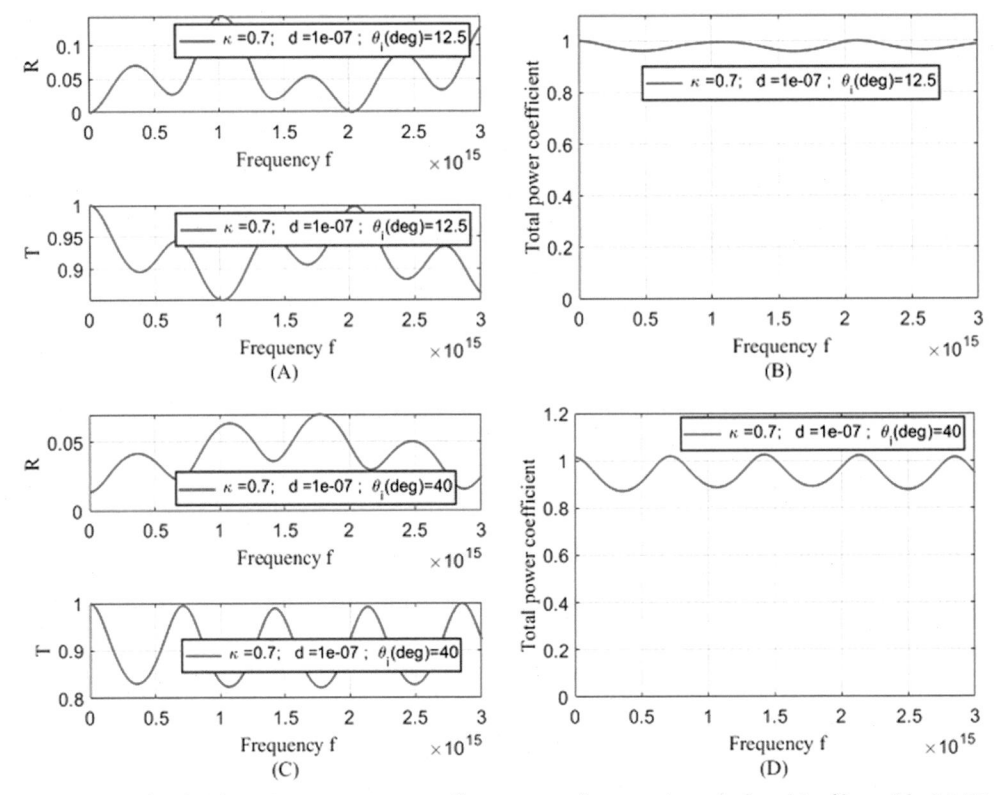

Figure 2.6 Chiral Fabry–Perot resonator reflectance and transmittance for thin film with 1:1.5:1 index ratio and $d = 100$ nm; T & $\mathcal{R}$ (A and C) and $\mathcal{R} + T$ (B and D).

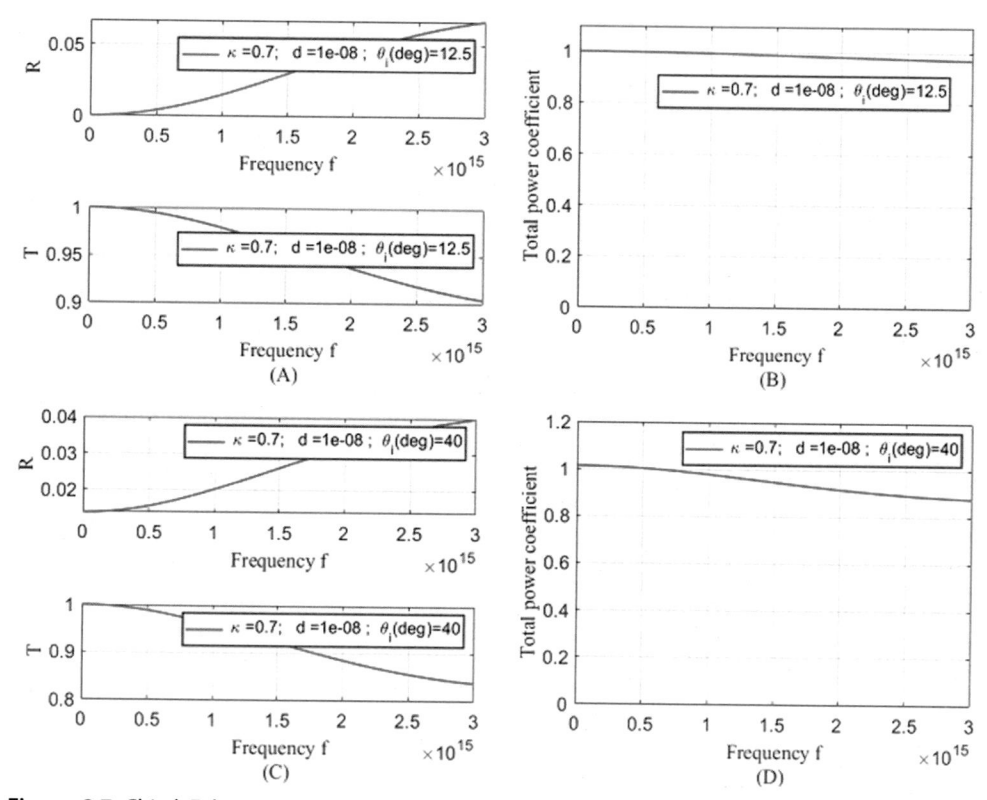

Figure 2.7 Chiral Fabry–Perot resonator reflectance and transmittance for thin film with 1:1.5:1 index ratio and $d = 10$ nm; T & $\mathcal{R}$ (A and C) and $\mathcal{R} + T$ (B and D).

energy conservation, but only indicative of computational instabilities around singularities.

Case A2
With $n_{p2} = 3.5$

Here in Fig. 2.8 when we choose 1:3.5:1 index ratio with thickness 100 nm, the results are consistently computationally unstable in terms of energy conservation. However; with a 1:3.5:1 index ratio and thickness 10nm as in Fig. 2.9, the results are computationally stable in terms of energy conservation even at higher angles of incidence (shown up to 60 degrees). An important case to be made is that even for the passband results, if the characteristics are plotted at sufficiently higher frequencies, it is found that $\mathcal{R} + T$ fluctuations are once again computationally unstable. These higher frequencies tend to enter into the X-ray and higher bands, and hence may be ignored.

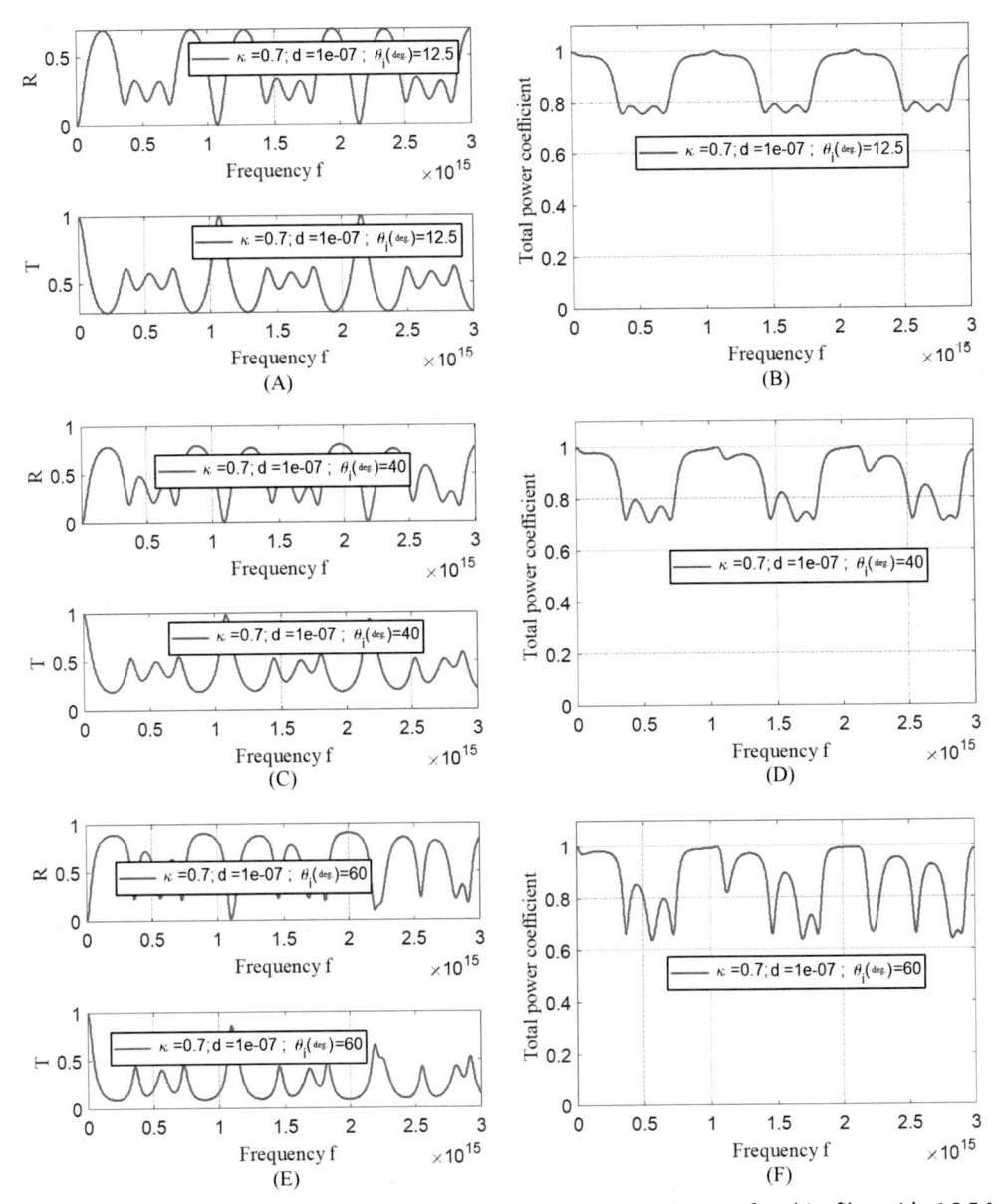

Figure 2.8 Chiral Fabry–Perot resonator reflectance and transmittance for thin film with 1:3.5:1 index ratio and $d = 100$ nm; T & $\mathcal{R}$ (A, C, and E) and $\mathcal{R} + T$ (B, D, and F).

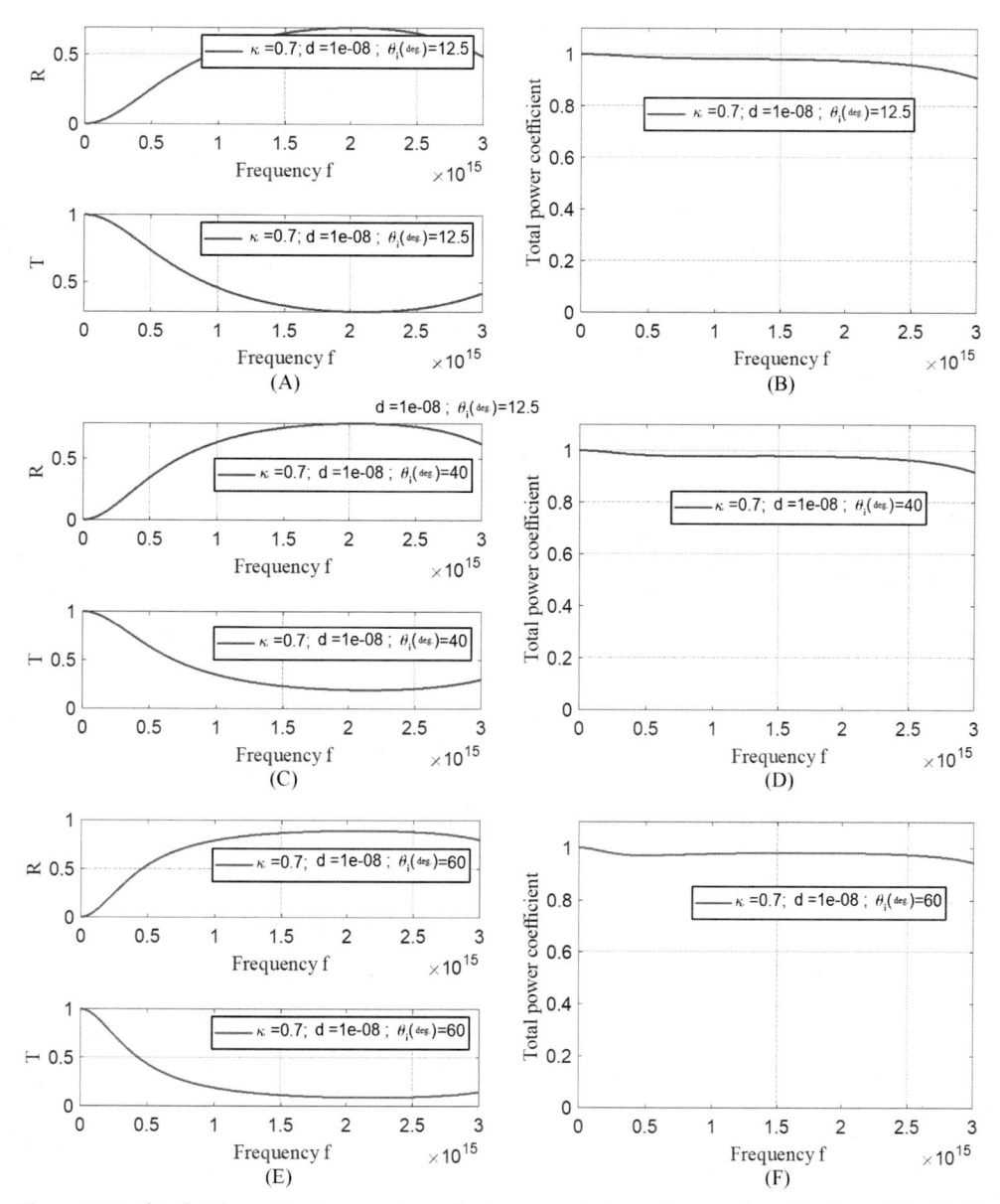

Figure 2.9 Chiral Fabry—Perot resonator reflectance and transmittance for thin film with 1:3.5:1 index ratio and $d = 10$ nm; T & $\mathcal{R}$ (A, C, and E) and $\mathcal{R} + T$ (B, D, and F).

Case B

Lossless thin film chiral slab immersed in dissimilar media (lithographic structure)

The chiral thin film is here configured in a geometry similar to a thin film lithography-based device, as in integrated circuits, leading to an $n_{p1}: n_{p2}: n_{p3}$ physical setup, with the n_{p3} ($\neq n_{p1}$) representing a *substrate*. Note that we present only the case for n_{p2} index of 1.5 in this presentation.

(1) Chiral F−P in thin film limit (100 nm, $n_{p2} = 1.5$)

(2) Lossless chiral F−P in the thin film limit (10 nm, $n_{p2} = 1.5$)

From Figs. 2.10 and 2.11, we find that Figs. 2.10B and 2.11B and D exhibit stable computational outcomes for the sum of the power coefficients; the rest of the cases are computationally unstable.

2.5 Mode density and linewidth versus slab thickness

In the graphical results which follow, impressions are generated regarding the changes in the transverse mode densities and also the effective mode linewidths as the slab

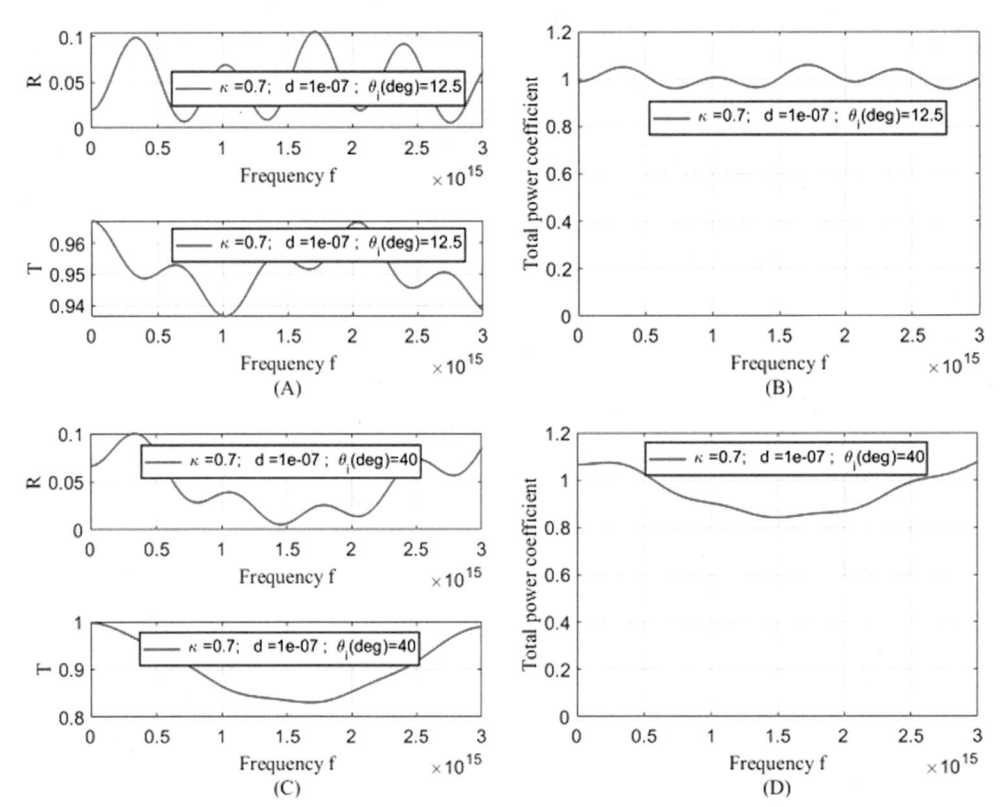

Figure 2.10 Chiral F−P resonator reflectance and transmittance for thin film with 1:1.5:1.772 index ratio and $d = 100$ nm; T & R (A,C) and $R + T$ (B and D).

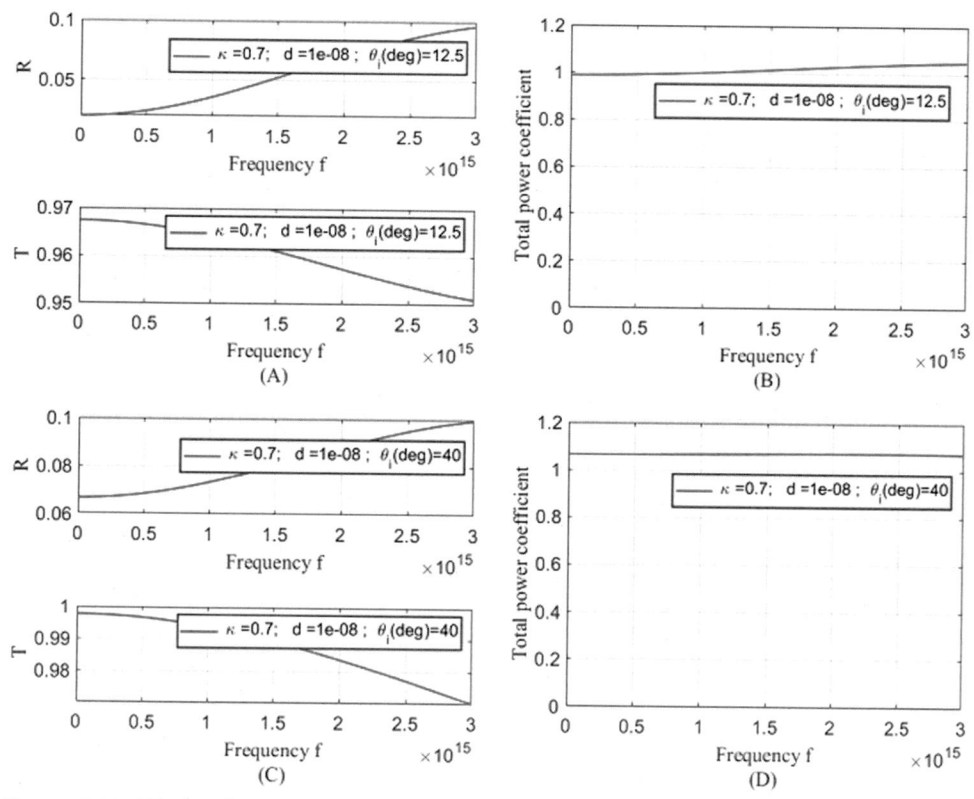

Figure 2.11 Chiral Fabry–Perot resonator reflectance and transmittance for thin film with 1:1.5:1.772 index ratio and $d = 10$ nm; T & R (A and C) and $R + T$ (B and D).

thickness changes in the range 1 μm to about 10 nm (about the lower thin film limit). This is to establish the general trends in resonator behavior transitioning from thick to thin films. The case for mode density (number of resonant modes within a specified optical band) as a function of slab thickness d is presented first, followed by the case for the effective mode linewidth, also as a function of the slab thickness d. We note also that a third sub-section introduces the lossy slab problem.

2.5.1 Mode density versus thickness

Shown in Fig. 2.12 are the mode density (number of resonant transverse modes within a specific optical band of $1-2 \times 10^{15}$ Hz, or a selected bandwidth of $\Delta f = 10^{15}$ Hz) versus the slab thickness d. Two cases for n_{p2} equal to 1.5 are shown, corresponding to the angles of incidence (θ_i) equal to 12.5 and 40 degrees. The mode counts were based on clearly discernable resonance peaks within the band, maintaining the same observation band in all cases of thickness. It is clear from both n_{p2} cases that the mode density increases with the slab thickness. Thus, as the slab thickness reduces toward the thin film limit, the mode density becomes considerably lower.

2.5.2 Resonance linewidth versus thickness

Shown in Fig. 2.13 are the effective mode linewidths versus the slab thickness corresponding to the same specified frequency band as in Section 2.5.1. It must be noted that since the transverse resonance modes become noticeably nonuniform within a chiral dielectric,

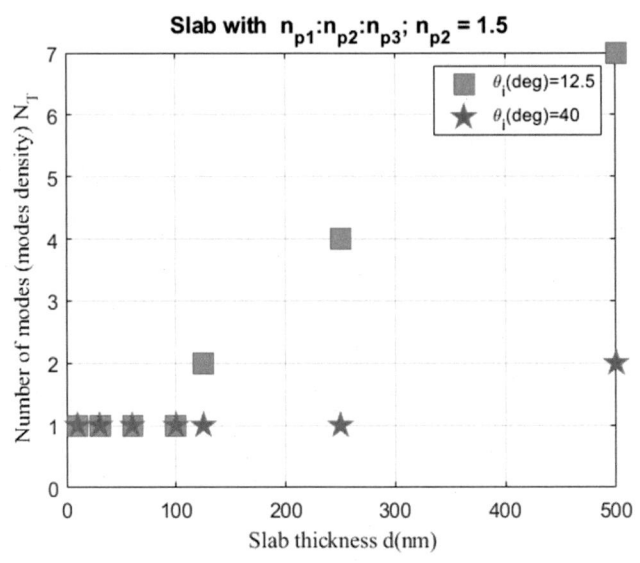

Figure 2.12 Number of modes N_T within specific frequency band versus slab thickness with $n_{p2} = 1.5$.

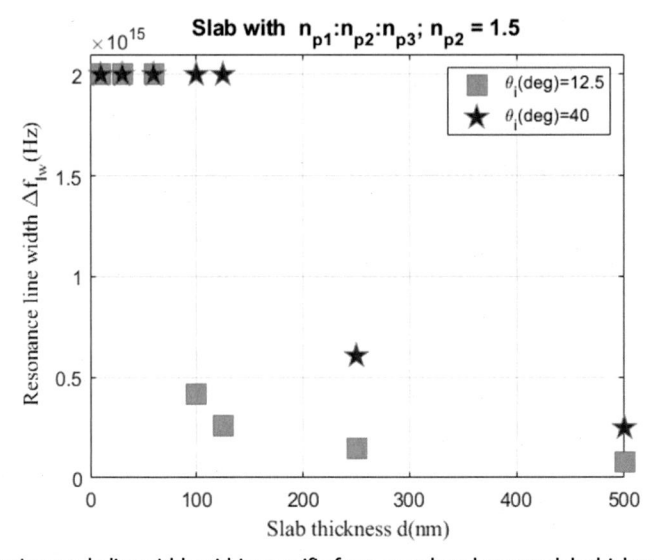

Figure 2.13 Effective mode linewidth within specific frequency band versus slab thickness with $n_{p2} = 1.5$.

approximate measures for the linewidths were determined between the peaks and troughs of each mode (with the troughs noticeably above zero), and averaged over the modes in the band. For sufficiently low thicknesses (typically below 100 nm), there was essentially a single propagating mode, for which the linewidth was effectively the entire frequency band under observation. Hence the linewidths are saturated on the top of the frequency axis. Overall, it is seen that the linewidths decrease noticeably as the slab thickness increases, and the drop is once again more pronounced corresponding to 40 degree for $n_{p2} = 1.5$.

2.5.3 Lossy thin film chiral slab

For this section we introduce the loss to the chiral thin film is immersed in the same/ dissimilar dielectric material. Comparing with the section on the achiral lossy F–P device with complex permittivity ε_2, we get

$$\left(n_{p2c} - \tilde{\kappa}\right)\sin\tilde{\theta}_{i1} = \sqrt{\varepsilon_{r3}}\,\sin\tilde{\theta}_{t1}, \quad \text{and} \tag{2.23a}$$

$$\left(n_{p2c} + \tilde{\kappa}\right)\sin\tilde{\theta}_{i2} = \sqrt{\varepsilon_{r3}}\,\sin\tilde{\theta}_{t2}. \text{ (Snell's laws under chirality)} \tag{2.23b}$$

and phase difference of lossy chiral slab given by

$$\delta_{RC,LC} = \frac{4\pi(n_{p2c} \mp \tilde{\kappa})d\cos\theta_{t(1,2)}}{\lambda_o} \tag{2.24}$$

Case A

Lossy thin film chiral slab immersed in the same medium

Shown in Figs. 2.14 and 2.15 are the total power coefficients for a lossy chiral slab resonator with thickness $d = 100$ nm and 10 nm, respectively which are numerically

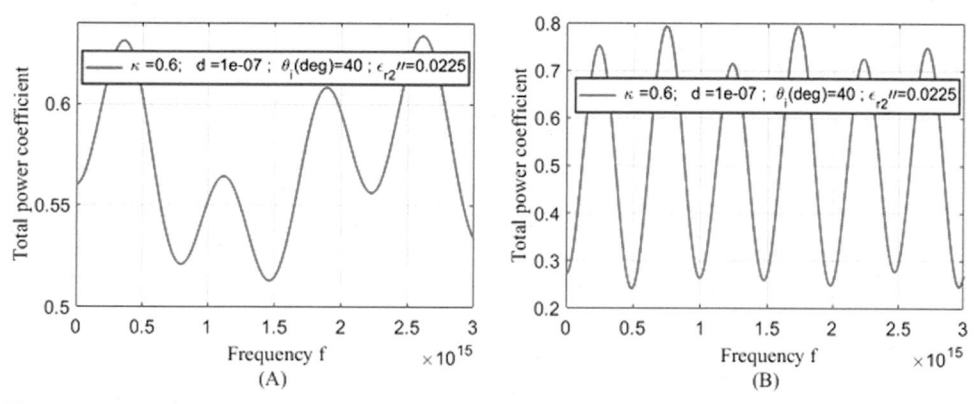

Figure 2.14 Total power characteristics of a lossy chiral Fabry–Perot resonator with $d = 100$ nm; (A) $n_{p2} = 1.5$; (B) $n_{p2} = 2.5$.

within the energy conservation bounds and consistently below unity for choices of n_{p2} of 1.5 and 2.5. The lossy problem is intrinsically complex, and is examined only numerically; further investigations of the resonator behavior under chiral losses are ongoing.

Case B
Lossy thin film chiral slab immersed in dissimilar media

Shown in Figs. 2.16 and 2.17 are the total power coefficients for a lossy chiral slab resonator with thickness $d = 100$ nm and 10 nm, respectively, which are also numerically within the energy conservation bounds and consistently below unity for choices of n_{p2} of 1.5 and 2.5. As mentioned before, the lossy problem is intrinsically complex and is examined only numerically; further investigations of the resonator behavior under chiral losses are ongoing.

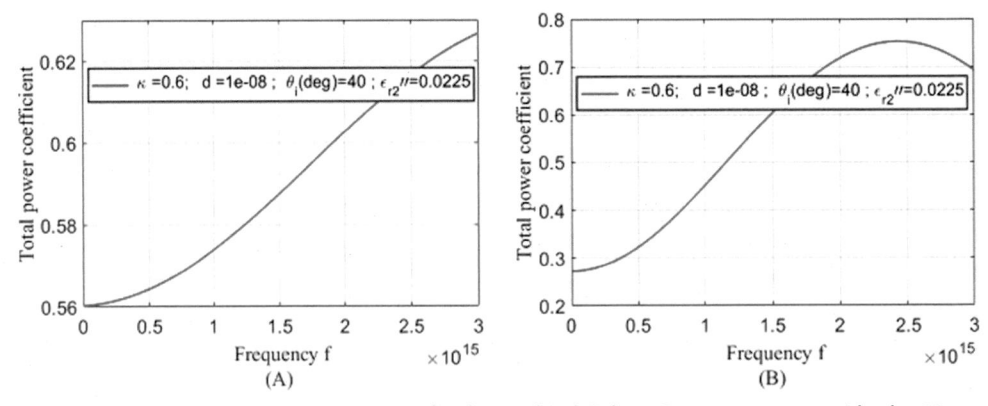

Figure 2.15 Total power characteristics of a lossy chiral Fabry–Perot resonator with $d = 10$ nm; (A) $n_{p2} = 1.5$; (B) $n_{p2} = 2.5$.

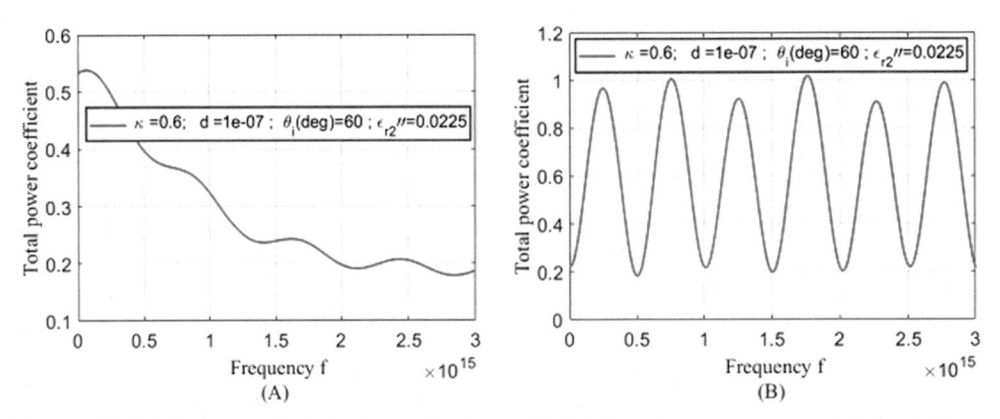

Figure 2.16 Total power characteristics of a lossy chiral Fabry–Perot resonator with $d = 100$ nm; (A) 1:1.5:2.5; (B) 1:2.5:1.5 index ratio.

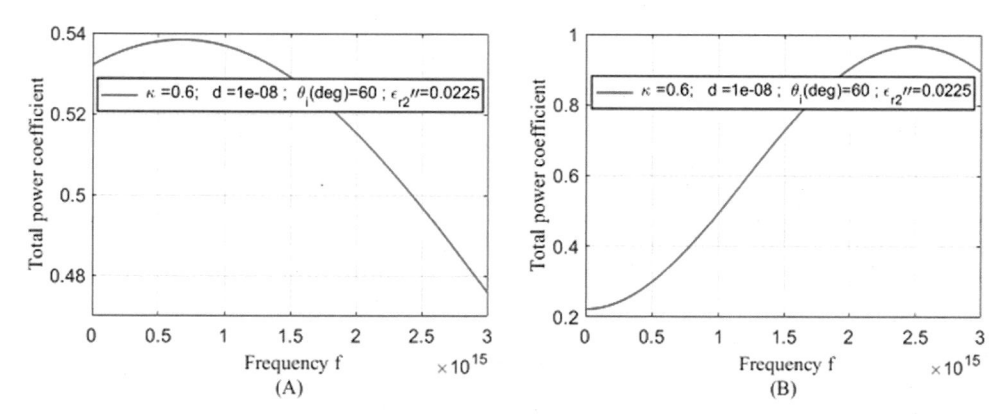

Figure 2.17 Total power characteristics of a lossy chiral Fabry−Perot resonator with $d = 10$ nm; (A) 1:1.5:2.5; (B) 1:2.5:1.5 index ratio.

2.6 Conclusion

Based on the extent of examining the behavior of chiral F−P resonators from the thick down to the thin film limits, with the cases involving 1 μm, 100 nm, and 10 nm (the ones involving thicknesses in the range 100 nm and 10 nm having been discussed in this work), it may be concluded that for different physical configurations, the overall behavior falls either in the passband or in the stopband regime (which have been defined as frequency bands over which the sum of the power coefficients is computationally stable, and energy conservation is demonstrated; in the stopbands, computation becomes unstable and is seen to be the consequence of the corresponding FC magnitudes exceeding unity, indicating singularities in the angles of incidence) when a spectral analysis is carried out. Overall, one standard trend is to note that even in the passband regimes, the devices exhibit much higher densities of transverse resonant modes in the passband for thicker films (in the 1 μm or thicker range), and considerably fewer in the thin film regime (100 nm or less), going down close to essentially two or three modes, and sometimes even a single mode. Also, as mode densities reduce, the corresponding *free spectral* ranges and mode linewidths become wider. Whether the propagation is realizable or not, depends on several different factors, including not only the thickness regime (which is somewhat less critical), but more importantly, on the dielectric properties (almost exclusively the permittivities, and hence the densities of the three layers). Thus whether the two chiral boundaries (ACC and CAC) are R/D or D/R greatly influences the passband and stopband behavior. As is usually the case, there are also specific *chirality* bands over which the behavior changes.

  In virtually all cases, singularities in the behavior leading to stopbands are a function of the angle of incidence (varying with different material parameters), and typically the device becomes propagationally nonrealizable once the singularity angle is approached—a property which is also dependent on the frequency band. The main difference between the thick and thin film regimes exists in the notable reduction of resonant modes as one approaches the thin film limit, wherein additional distinctions occur based on the phase index relations between the three layers. The main finding is concerned with computational instabilities for the chiral slabs (which are not analytically tractable) in certain parametric bands (choices of the chiral phase index, host phase index, and more significantly, the choice of the loss parameter in permittivity, n_{p2c}, and other parametric choices which are critical to the understanding of the device behavior).

Future work will involve further investigations of cases under dielectric losses, and also examining the resonator behavior under material dispersion affecting permittivity, permeability and chirality, including cases involving possible negative index in the chiral material corresponding to sidebands around the optical carrier.

Acknowledgments

One of us (R.Y.A.) gratefully acknowledges the Libyan Ministry of Education (LME) for financial support in enabling this work. The authors wish to sincerely thank Salaheddeen Bugoffa for assistance in developing portions of the lossy chiral material modeling.

References

[1] A. Lakhtakia, V.V. Varadan, V.K. Varadan, A parametric study of microwave reflection characteristics of a planar achiral—chiral interface, IEEE Trans. Electromagn. Compat. 28 (2) (1986) 90—95. Available from: https://doi.org/10.1109/TEMC.1986.4307254.

[2] S. Bassiri, C.H. Papas, N. Engheta, Electromagnetic wave propagation through a dielectric—chiral interface and through a chiral slab, J. Opt. Soc. Am. A 5 (9) (1988) 1450—1459. Available from: https://doi.org/10.1364/JOSAA.5.001450.

[3] S. Bassiri, C.H. Papas, N. Engheta, Electromagnetic wave propagation through a dielectric—chiral interface and through a chiral slab, J. Opt. Soc. Am. A 7 (11) (1990) 2154—2155. Available from: https://doi.org/10.1364/JOSAA.7.002154.

[4] H. Cory, I. Rosenhouse, Electromagnetic wave reflection and transmission at a chiral—dielectric interface, J. Mod. Opt. 38 (7) (1991) 1229—1241. Available from: https://doi.org/10.1080/09500349114551401.

[5] M.R. Chatterjee, R.Y. Ataai, Anomalous Brewster effect and evanescent modes at an achiral/chiral dielectric boundary under planar electromagnetic propagation, J. Mod. Opt. 66 (10) (2019) 1089—1105. Available from: https://doi.org/10.1080/09500340.2019.1601781.

[6] R.Y. Ataai, Analysis of Electromagnetic Signal Propagation Across Dispersive Achiral-Chiral Interfaces and Application to Resonators And Discrete Components (Ph.D. dissertation), University of Dayton, Dayton, OH, 2021, to appear.

[7] R.Y. Ataai, M.R. Chatterjee, Polarization, reflection, and transmission states of plane electromagnetic propagation through a reciprocal chiral slab, in: Light in Nature VII, in: Proceedings, 11099, SPIE Optical Engineering + Applications, San Diego, CA, 2019, p. 1109909. Available from: https://doi.org/10.1117/12.2528470.

[8] E. Georgieva, Reflection and refraction at the surface of an isotropic chiral medium: eigenvalue—eigenvector solution using a 4 x 4 matrix method, J. Opt. Soc. Am. A 12 (10) (1995) 2203—2211. Available from: https://doi.org/10.1364/JOSAA.12.002203.

[9] A. Lakhtakia, V.V. Varadan, V.K. Varadan, Reflection of plane waves at planar achiral—chiral interfaces: independence of the reflected polarization state from the incident polarization state, J. Opt. Soc. Am. A 7 (9) (1990) 1654—1656. Available from: https://doi.org/10.1364/JOSAA.7.001654.

[10] P.P. Banerjee, M.R. Chatterjee, Negative index in the presence of chirality and material dispersion, J. Opt. Soc. Am. B 26 (2) (2009) 194–202. Available from: https://doi.org/10.1364/JOSAB.26.000194.

[11] A. Yariv, Optical Electronics in Modern Communications, fifth ed., Oxford University Press, 1997, p. 768.

[12] F. Babaei, H. Savaloni, Reflection, transmission and circular dichroism in axially excited slab of a copper thin film helicoidal bianisotropic medium, Opt. Commun. 278 (2) (2007) 321–328. Available from: https://doi.org/10.1016/j.optcom.2007.06.029.

[13] A. Lakhtakia, On extending the Brewster law at planar interfaces, Optik 84 (5) (1990) 160–162.

[14] A. Lakhtakia, J.R. Diamond, Reciprocity and the concept of the Brewster wavenumber, Int. J. Infrared Millim. Waves 12 (10) (1991) 1167–1174.

[15] F. Mariotte, N. Engheta, Effect of chiral material loss on guided electromagnetic modes in parallel-plate chirowaveguides, J. Electromagn. Waves Appl. 7 (10) (2012) 1307–1321. Available from: https://doi.org/10.1163/156939393X00499.

[16] R.Y. Ataai, M.R. Chatterjee, Bi-modal resonance and spectral characteristics of a thin chiral slab resonator on an achiral substrate using Fresnel coefficients, In: International Workshop on Thin Films for Electronics, Electro-Optics, Energy, and Sensors, in: Proceedings, 11371, SPIE, Reykjavik, 2019, p. 1137105. Available from: https://doi.org/10.1117/12.2531858.

[17] M. Tanaka, H. Sueyoshi, Scattering of EM waves by a chiral slab, Electron. Commun. Jpn. (Part II: Electronics) 75 (9) (1992) 34–43. Available from: https://doi.org/10.1002/ecjb.4420750904.

[18] H. Cory, I. Rosenhouse, Reflection and transmission of EM waves by a chiral slab, J. Mod. Opt. 39 (6) (1992) 1321–1330. Available from: https://doi.org/10.1080/713823543.

[19] M.R. Chatterjee, R.Y. Ataai, Propagation across chiral interfaces and tunable slab resonators without and with dispersion, in: IEEE Photonics Conference, Reston, VA, (2018) 1–2. Available from: https://doi.org/10.1109/IPCon.2018.8527299.

[20] C.C. Davis, Laser and Electro-Optics: Fundamentals and Engineering, second ed., Cambridge University Press, 2014, p. 882.

[21] M.R. Chatterjee, R.Y. Ataai, Fresnel coefficients for electromagnetic propagation across a non-chiral and chiral dispersion interface with negative index, in: Frontiers in Optics / Laser Science Conference, Optical Society of America Technical Digest, Washington, DC, 2017, p. JW3A.121. Available from: https://doi.org/10.1364/FIO.2017.JW3A.121.

[22] S.G. Bugoffa, M.R. Chatterjee, Impact of losses on the negative index behavior in a chiral dielectric and tradeoffs between dielectric and magnetic losses toward zero effective attenuation, Opt Eng. 59 (11) (2020) 117106. Available from: https://doi.org/10.1117/1.OE.59.11.117106.

Appendix A

Symbols b_1 through b_5 using to calculate FCs of CAC

In this appendix, MATLAB-generated symbols which appear in the analytic solution discussed earlier in relation to the derivation of the Fresnel coefficients (FCs) corresponding to the chiral–achiral (CAC) interface, are expressed in terms of the physical system parameters.

The FCs for the CAC interface, as discussed, are composed of the following symbols (from MATLAB):

$$
\begin{aligned}
b_1 = {}& \eta_1^2 \cos\theta_{i1}\cos\theta_{t1} + \eta_1^2 \cos\theta_{i1}\cos\theta_{t2} + \eta_1^2 \cos\theta_{i2}\cos\theta_{t1} + \eta_2^2 \cos\theta_{i1}\cos\theta_{t1} \\
& + \eta_1^2 \cos\theta_{i2}\cos\theta_{t2} + \eta_2^2 \cos\theta_{i1}\cos\theta_{t2} + \eta_2^2 \cos\theta_{i2}\cos\theta_{t1} + \eta_2^2 \cos\theta_{i2}\cos\theta_{t2} \\
& + 4\eta_1\eta_2 \cos\theta_{i1}\cos\theta_{i2} - 2\eta_1\eta_2 \cos\theta_{i1}\cos\theta_{t1} + 2\eta_1\eta_2 \cos\theta_{i1}\cos\theta_{t2} \\
& + 2\eta_1\eta_2 \cos\theta_{i2}\cos\theta_{t1} - 2\eta_1\eta_2 \cos\theta_{i2}\cos\theta_{t2} + 4\eta_1\eta_2 \cos\theta_{t1}\cos\theta_{t2}.
\end{aligned}
$$

$$(A.1)$$

$$b_2 = E_{o2}\eta_2\cos\theta_{i1}\cos\theta_{i2}, \quad \text{and} \quad b_3 = E_{o2}\eta_1\cos\theta_{i1}\cos\theta_{i2} \tag{A.2}$$

$$b_4 = E_{o1}\eta_2\cos\theta_{i1}\cos\theta_{i2}, \quad \text{and} \quad b_5 = E_{o1}\eta_1\cos\theta_{i1}\cos\theta_{i2} \tag{A.3}$$

We next introduce the symbols:

$$
\begin{aligned}
A_1 =\ & E_{o1}\eta_1^2\cos\theta_{i1}\cos\theta_{t1} + E_{o1}\eta_1^2\cos\theta_{i1}\cos\theta_{t2} - E_{o1}\eta_1^2\cos\theta_{i2}\cos\theta_{t1} \\
& + E_{o1}\eta_2^2\cos\theta_{i1}\cos\theta_{t1} - E_{o1}\eta_1^2\cos\theta_{i2}\cos\theta_{t2} + E_{o1}\eta_2^2\cos\theta_{i1}\cos\theta_{t2} \\
& - E_{o1}\eta_2^2\cos\theta_{i2}\cos\theta_{t1} + 2E_{o2}\eta_1^2\cos\theta_{i2}\cos\theta_{t1} - E_{o1}\eta_2^2\cos\theta_{i2}\cos\theta_{t2} \\
& + 2E_{o2}\eta_1^2\cos\theta_{i2}\cos\theta_{t2} - 2E_{o2}\eta_2^2\cos\theta_{i2}\cos\theta_{t1} - 2E_{o2}\eta_2^2\cos\theta_{i2}\cos\theta_{t2} \\
& + 4E_{o1}\eta_1\eta_2\cos\theta_{i1}\cos\theta_{t2} - 2E_{o1}\eta_1\eta_2\cos\theta_{i1}\cos\theta_{t1} + 2E_{o1}\eta_1\eta_2\cos\theta_{i1}\cos\theta_{t2} \\
& - 2E_{o1}\eta_1\eta_2\cos\theta_{i2}\cos\theta_{t1} + 2E_{o1}\eta_1\eta_2\cos\theta_{i2}\cos\theta_{t2} - 4E_{o1}\eta_1\eta_2\cos\theta_{t1}\cos\theta_{t2}
\end{aligned}
\tag{A.4}
$$

$$
\begin{aligned}
A_2 =\ & 2E_{o1}\eta_1^2\cos\theta_{i1}\cos\theta_{t1} + 2E_{o1}\eta_1^2\cos\theta_{i1}\cos\theta_{t2} - 2E_{o1}\eta_2^2\cos\theta_{i1}\cos\theta_{t1} \\
& - E_{o2}\eta_1^2\cos\theta_{i1}\cos\theta_{t1} - 2E_{o1}\eta_2^2\cos\theta_{i1}\cos\theta_{t2} - E_{o2}\eta_1^2\cos\theta_{i1}\cos\theta_{t2} \\
& + E_{o2}\eta_1^2\cos\theta_{i2}\cos\theta_{t1} - E_{o2}\eta_2^2\cos\theta_{i1}\cos\theta_{t1} + E_{o2}\eta_1^2\cos\theta_{i2}\cos\theta_{t2} \\
& - E_{o2}\eta_2^2\cos\theta_{i1}\cos\theta_{t2} + E_{o2}\eta_2^2\cos\theta_{i2}\cos\theta_{t1} + E_{o2}\eta_2^2\cos\theta_{i2}\cos\theta_{t2} \\
& + 4E_{o2}\eta_1\eta_2\cos\theta_{i1}\cos\theta_{t2} + 2E_{o2}\eta_1\eta_2\cos\theta_{i1}\cos\theta_{t1} - 2E_{o2}\eta_1\eta_2\cos\theta_{i1}\cos\theta_{t2} \\
& + 2E_{o2}\eta_1\eta_2\cos\theta_{i2}\cos\theta_{t1} - 2E_{o2}\eta_1\eta_2\cos\theta_{i2}\cos\theta_{t2} - 4E_{o2}\eta_1\eta_2\cos\theta_{t1}\cos\theta_{t2}
\end{aligned}
\tag{A.5}
$$

$$B_1 = E_{o1}b_1, \quad B_2 = E_{o2}b_1 \tag{A.6}$$

$$
\begin{aligned}
C_1 = (& 4\eta_1(b_5 + b_4 - b_3 + b_2 + E_{o1}\eta_1\cos\theta_{i1}\cos\theta_{t2} + E_{o1}\eta_2\cos\theta_{i1}\cos\theta_{t2} \\
& + E_{o2}\eta_1\cos\theta_{i2}\cos\theta_{t2} - E_{o2}\eta_2\cos\theta_{i2}\cos\theta_{t2})
\end{aligned}
\tag{A.7}
$$

$$
\begin{aligned}
C_2 = (& 4\eta_1(b_4 - b_5 + b_3 + b_2 + E_{o1}\eta_1\cos\theta_{i1}\cos\theta_{t1} - E_{o1}\eta_2\cos\theta_{i1}\cos\theta_{t1} \\
& + E_{o2}\eta_1\cos\theta_{i2}\cos\theta_{t1} + E_{o2}\eta_2\cos\theta_{i2}\cos\theta_{t1})
\end{aligned}
\tag{A.8}
$$

Finally, the FCs are expressible as:

$$\tilde{\Gamma}_{\perp 1} = A_1/B_1; \tilde{\Gamma}_{\perp 2} = A_2/B_2; \tilde{T}_{\perp 1} = C_1/B_1; \text{ and } \tilde{T}_{\perp 2} = C_2/B_2. \tag{A.9}$$

CHAPTER 3

Structural, electrical, and electromagnetic properties of nanostructured vanadium dioxide thin films

Guru Subramanyam[1], Eunsung Shin[1], Prudhvi Ram Peri[2], Ram Katiyar[3], Golali Naziripour[4] and Sandwip Dey[2]

[1]Center of Excellence for Thin-Film Research and Surface Engineering (CETRASE), University of Dayton, Dayton, OH, United States
[2]School of Engineering of Matter, Transport and Energy, Arizona State University, Tempe, AZ, United States
[3]Department of Physics, University of Puerto Rico Rio Piedras, San Juan, Puerto Rico
[4]Indiana University Purdue University Indianapolis, Indianapolis, IN, United States

3.1 Introduction

Vanadium dioxide (VO_2) exhibits a first-order transition from the insulating monoclinic phase (M1) to the metallic tetragonal rutile phase (R) at a critical temperature (T_c) of about 341K (68°C) [1]. Typically, the transition occurs over a wider temperature range (58°C−72°C) where a mixed phase exists. The transition is reversible, but hysteretic. Due to the distinct electrical, optical, and electromagnetic properties exhibited by these crystallographic phases, applications of this binary oxide include reversible switching devices [2−4], switchable and tunable metamaterials [5−7], switchable and tunable antenna for microwave devices [8−10], sensors [11,12], and smart windows [13−16]. This oxide may be readily deposited in thin film form by sol−gel processing [17,18], sputtering [19,20], chemical vapor deposition [21,22], and pulsed laser deposition (PLD) [23,24]. Also, the T_c may be engineered to suit practical applications, for example, Cr, Fe, Al, and Ga doping increases T_c, whereas W, Ti, Mo, F, and Nb decreases T_c [13,14,24−28]. Specifically, the addition of a small amount of tungsten (W) not only causes a reduction in the T_c at the rate of about 50K per atomic percentage (at.%) of W [29,30], a broadening of the transition temperature range, coupled with increased conductivity of the insulating phase are observed [14,26,31−33]. Hence, W-doped VO_2 has gained much attention for application in smart devices including thermochromic and variable-reflectance mirrors [26,30,34], reconfigurable/tunable radio frequency (RF) devices (such as variable capacitors, variable inductors, and variable resistors), and reconfigurable antenna for portable and adaptive communication systems [4].

In this study, VO_2 thin films were deposited on sapphire substrates by PLD. This method was selected due to its simplicity, cleanliness, high deposition rate, and

Thin Film Nanophotonics
DOI: https://doi.org/10.1016/B978-0-12-822085-6.00005-4

reproducibility in fabrication of high-quality, oxide thin films with dopants on large-area (up to 4″ diameter) substrates. The overall objectives of this work were to (1) demonstrate the processing, validate the epitaxial quality, and characterize the physicochemical properties of nanostructured, undoped, and W-doped VO_2 (i.e., 0.34 at.%, 0.54 at.%, and 1.1 at. % W) thin films using X-ray diffraction (XRD), Rutherford backscattering (RBS), scanning electron microscopy (SEM), X-ray photoelectron spectroscopy (XPS), and high-angle annular dark-field images (HAADF) recorded in a scanning transmittance electron microscope (STEM); (2) compare the temperature dependences of the electrical resistivity of undoped and W-doped VO_2 thin films; and (3) validate the contrasting electromagnetic properties in the insulating and conducting states, including transmittance and reflectance of the VO_2 thin films at microwave and near-infrared (NIR) frequencies.

3.2 Experimental

3.2.1 Fabrication

Undoped and W-doped vanadium dioxide (VO_2) thin films were fabricated by a large-area PLD (Neocera Pioneer 180) system, capable of depositing up to 4″ diameter wafers. In the case of undoped films, a metallic vanadium disk (2″ diameter, 1/4″ thick) was used as the target. The separation distance between the target and substrate was 50 mm, and the output of a KrF excimer laser ($\lambda = 248$ nm) was used to ablate the target material. In the case of W-doped films, 0.5 mm diameter W wires were attached on the surface of the metallic V target. Different separation distances of the wires, used for controlling the W percentage in the film, were 15, 10, and 5 mm. The laser energy density at the target surface was set to approximately 2.5 J cm^{-2} and the laser repetition rate was maintained at 10 Hz. The ambient oxygen pressure and the substrate temperature were fixed at 25 mTorr and 500°C, respectively, during deposition. Three-inch, C-cut Sapphire wafers were used as the substrate for this study. Further details of processing and compositional analyses by XPS of the four VO_2 thin film samples used in this study are given elsewhere [4]. Table 3.1 depicts the films characterized in this study: one undoped and three W-doped VO_2 with 0.34, 0.54, and 1.1 at.% of W. The thicknesses of the films were determined on the micro-lithographically patterned test structures using a surface profilometer.

3.2.2 Structural characterization

XRD of the films was carried out using a PANalytical X-Pert diffractometer with a hybrid monochromator for Cu $K\alpha1$ radiation ($\lambda = 1.5406$ Å). The valence states and compositions of W-doped VO_2 films were determined by XPS, using Kratos Axis Ultra 165 Surface Analysis System. All XPS data were obtained using the

Table 3.1 Identification and thickness of undoped and W-doped VO_2 thin film samples characterized in this study; note, the atomic percentages (at.%) of W were calculated using: at.% W = [100 W/(W + V + O)].

Sample	Identification	Total film thickness (nm)
VO_2 undoped	VO_2	146.82
VO_2-0.34 at.% W	VO_2-0.34	132.13
VO_2-0.54 at.% W	VO_2-0.54	132.13
VO_2-1.1 at.% W	VO_2-1.1	117.45

monochromatic, Al $K\alpha$ X-ray source run at 300 W power (20 mA/15 keV) with the charge neutralizer on for charge compensation. The samples were loaded at the same time and pumped down to 2×10^{-6} Torr prior to entering the analysis chamber. For all analysis, the base pressure was maintained below 1×10^{-8} Torr. The analysis area in Hybrid Lens mode and Slot Aperture was limited to a 300 mm $\times$ 700 mm rectangular region. Low-resolution survey spectrum was recorded using 160 eV pass energy, and high-resolution data was collected from individual regions using 20 eV pass energy. All data fitting and quantitation were done using Kratos Vision2 Processing software and built-in relative sensitivity factors specific to this instrument configuration. All SEM images were taken on JEOL's JXA-8530F Electron Probe Microanalyzer (EPMA) with an electron probe, focused typically at 2–50 nm diameter onto the specimen. The STEM-HAADF imaging was taken on the ARM200F, which is an aberration-corrected STEM equipped with both an x-ray spectrometer and an electron spectrometer, with ultra-fast Gatan Enfinium EELS spectrometer. The ARM200F offers resolution of 0.8 Å at 200 kV and is equipped with a Schottky field-emission gun and a CEOS CESCOR hexapole aberration corrector. The compositional analyses and thin film depth profiles were carried out using RBS in the ion beam analysis facility equipped with a 1.7 MeV tandem accelerator.

3.2.3 Electrical and electromagnetic properties

The electrical resistivity of the VO_2 thin films was measured using a dc test structure made of a rectangular VO_2 thin film bar (2 mm $\times$ 0.1 mm) with two probe pads at the ends. A probe station with a temperature-controlled chuck was utilized to record resistivity changes as a function of temperature.

For microwave transmittance and reflectance properties, an X-band RF measurement system was designed and constructed. The system (Fig. 3.1) consisted of a HP 8510C vector network analyzer (VNA) equipped with a 45 MHz–26.5 GHz RF source and test set, a WR-90 X-band waveguide setup with a custom designed and fabricated variable-temperature sample platen, incorporating an Omega CN-142 PID temperature controller.

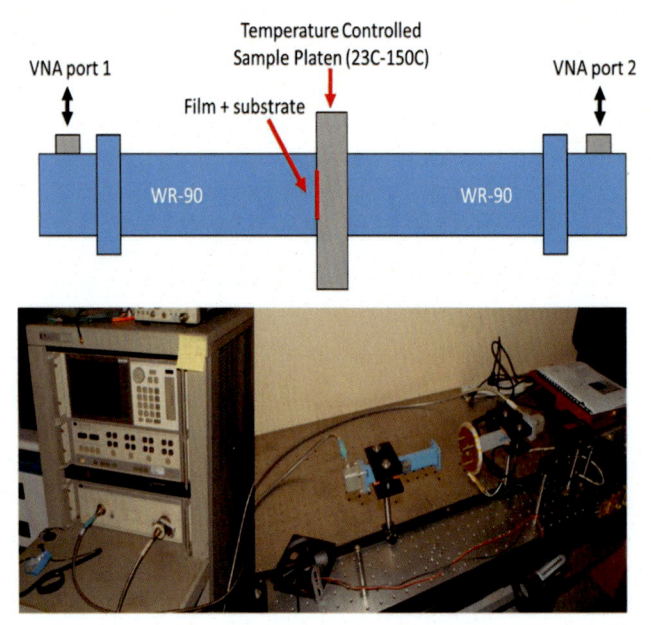

Figure 3.1 Measurement system for the microwave properties of VO$_2$ thin films as functions of temperature in the X-band (8−12 GHz) using WR-90 waveguides.

Using this system, a technique for systematically acquiring calibrated transmittance data as a function of temperature was developed. At room temperature, the VNA was run through its calibration procedure using X-band waveguide calibration standards. Once calibrated, the sample was loaded into the platen and temperature-dependent s-parameter measurements were taken using a custom-coded LabVIEW VI that automatically stepped the temperature up and down.

Optical transmittance and reflectance were measured for the films using a broadband optical source. The system, shown in Fig. 3.2, allows for calibrated, spectrally resolved measurements of reflectance and transmittance coefficients over a wide (500 nm−5 µm) spectral range with excellent (<1 nm full width at half maximum (FWHM)) spectral resolution and measurement repeatability.

For transmittance measurements, the sample was illuminated with a white-light source (100 W Oriel Optics Regulated Quartz Halogen Lamp) and focused through a Keplerian beam condenser (two lenses of equal numerical aperture, but different diameter), and the signal was modulated with an optical chopper (EG&G PAR 197) at the beam focus of the condenser. The collimated white-light was then passed through a 50/50 VIS-NIR Cube beam-splitter and onto a 50-mm focal length $f/2$ focusing lens pair and the sample is placed at the focal point in the heated stage. This configuration allowed for integration of many angles of incidence rather than just normal to the surface and compensated for scattering and beam divergence after propagation through

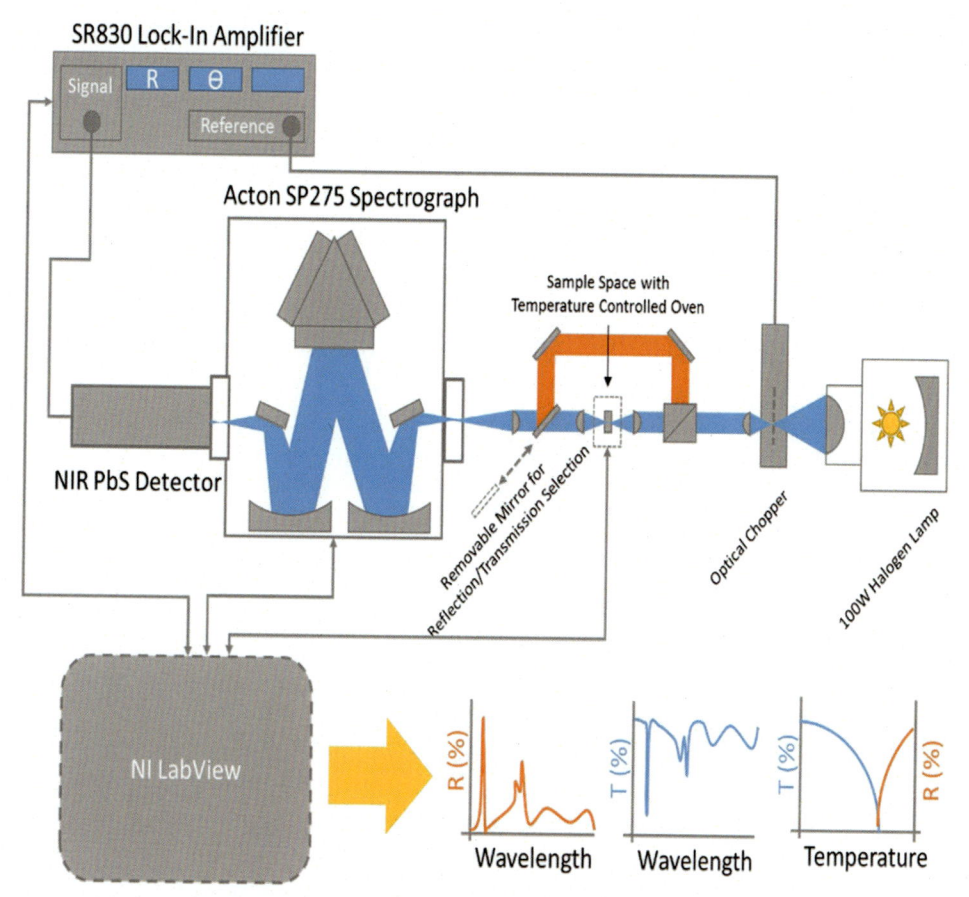

Figure 3.2 Schematic depiction of VIS-NIR reflectance/absorptance spectroscopy capability for measurements of phase-change materials with a variable-temperature oven sample space. *NIR,* Near infrared.

the sample film and substrate. The Spectrograph (Acton SP275) is a triple-grating, $f/3.8$ mm Czerny-Turner geometry, digitally controlled. Light is dispersed onto a 2-μm blaze, 300 g mm^{-1} diffraction grating and then coupled out through an exit slit onto the detector. As VO_2 has a bandgap of ~ 0.6 eV, most of the optical features occur in the 1−5-μm mid-wave infrared (MWIR) range. An amplified PbS photodiode (Oriel 70343) detector was suitable for this MWIR range. The system is integrated via a general purpose instrumentation bus (GPIB) control via National Instruments LabVIEW Software for measuring the sample heater power, spectrograph center wavelength utilizing a lock-in amplifier.

For reflectance measurements, an alternate path was employed that collected reflected light via a beam-splitter and then inserted into the spectrograph path via a retractable mirror on a translation stage. The ability to rapidly switch between reflectance and transmittance measurements allowed calibrated and reproducible measurements of the absorptance coefficient. This measurement consisted of systematically performing four measurements per temperature. First, a calibration reference mirror was measured in reflectance, and then the growth substrate transmittance spectrum was collected. The thin film sample was mounted, and temperature stabilized either at room temperature or elevated temperature and aligned in the reflection mode. The reflectance spectrum of the sample and finally the transmittance spectrum were collected. The mirror reference and substrate measurements formed the correction factors for the instrument response of the measurement system. Fig. 3.3 shows a photograph of the measurement system used for the NIR thermo–optic transparency modulation in VO_2 thin films.

For optical transmittance measurements, a bare substrate was utilized as a calibration standard for the spectral dependence of incident radiation, which allowed for the correction of film spectra for the instrument response of the system. The transmittance spectra, $T(\lambda)$, were generated using:

$$T(\lambda) = \frac{I(\lambda)}{I_0(\lambda)}, \tag{3.1}$$

where $I_0(\lambda)$ is the calibration standard spectrum, $I(\lambda)$ is the transmitted spectrum from the test sample, and $T(\lambda)$ is the transmittance.

For reflectance measurements a standard sample of known spectral reflectance, such as a calibrated mirror was first used for calibration. The calibration sample was placed in the sample space and a spectrum was collected by scanning the spectrograph over

Figure 3.3 VIS-NIR reflectance/absorptance spectroscopy system for measurements of phase-change materials. *NIR*, Near infrared VIS, visible light.

the wavelength range and measuring the lock-in voltage at each center wavelength. The reflectance spectra, $R(\lambda)$, were generated using:

$$R(\lambda) = \frac{I(\lambda)}{I_S(\lambda)R_S(\lambda)},$$

(3.2)

where $I(\lambda)$ is the measured reflectance spectrum of the unknown sample, $I_S(\lambda)$ is the measured reflectance spectrum of the known calibration standard, and $R_s(\lambda)$ is the known spectral reflectance of the calibration standard. Since the thickness, t, of the sample was known and both transmittance and reflectance spectra were generated, the absorptance coefficient $\alpha(\lambda)$ was calculated using the following relation derived from Beer–Lambert Law:

$$\alpha(\lambda) = -\frac{1}{t}\mathrm{Ln}(R(\lambda) + T(\lambda)).$$

(3.3)

Fig. 3.4 shows the measurement setup for dynamic illumination of VO_2 thin films using an optically shuttered high-power continuous wave (CW) laser (Coherent, 532 nm) to switch the film from insulating to the metallic state. The setup shows the triggered-actuation of the variable power CW laser illumination on the sample and detection of the transmitted and reflected near infrared (NIR) beam. A 40 mW 1550 nm vertical cavity surface emitting laser (VCSEL) was used for the NIR source and an InGaAs NIR photodiode was used as a detector.

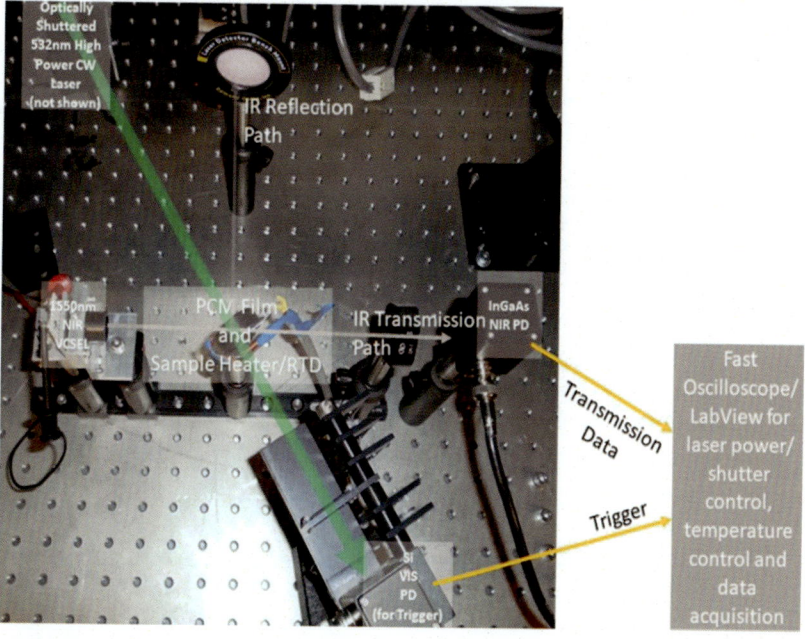

Figure 3.4 Experimental setup, for dynamic illumination testing with triggered acquisition, for switching speed measurement.

A spectroscopic ellipsometer (J.A. Woollam V-VASE UV-VIS-NIR), with a linearly polarized light, was incident on the test sample at an oblique angle of 70 degrees. The relative change in the amplitude, Ψ, and the phase difference, Δ, of the light reflected off the substrate were measured and related to the complex Fresnel reflection coefficient as follows:

$$\rho = \frac{R_p}{R_s} = \tan(\Psi)e^{j\Delta}, \tag{3.4}$$

where R_p and R_s are the Fresnel reflection coefficients for p-polarized and s-polarized light, respectively. The optical constants are determined by fitting the measured data utilizing the Drude−Lorentz model for the relative dielectric permittivity, which is a function of energy (E):

$$\varepsilon(E) = \varepsilon_\infty - \frac{A_0}{E^2 + jB_0 E} + \sum_1^n \frac{A_n}{E_n^2 - E^2 + jE}, \tag{3.5}$$

where A_n is the amplitude, E_n is the energy of the Lorentz oscillator, and B_0 is the bandwidth of the oscillator.

3.3 Experimental results and discussions

3.3.1 Thin film characterization

The room-temperature XRD patterns acquired from the undoped and W-doped VO_2 films deposited on C-cut sapphire wafers are presented in Fig. 3.5. Note, the critical metal-insulator transition temperature (T_c), the structure of insulating VO_2 is

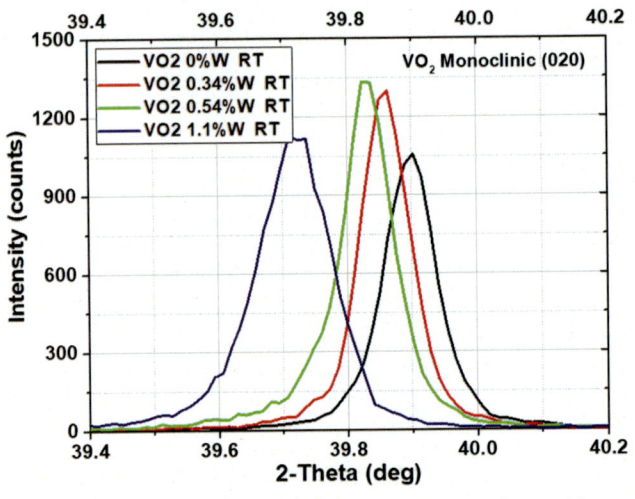

Figure 3.5 XRD plots of undoped and W-doped VO_2 films on Sapphire with peaks corresponding only to the (0 2 0) plane. *XRD*, X-ray diffraction.

monoclinic (space group No. 14, $P21/c$) with $a = 5.743$ Å, $b = 4.517$ Å, $c = 5.375$ Å, $\alpha = 90$ degrees, $\beta = 122.61$ degrees, and $\gamma = 90$ degrees [35,36]. The structure of sapphire is rhombohedral (space group No. 167, R-$3c$) with $a = b = 4.758$ Å, $c = 12.991$ Å, $\alpha = \beta = 90$ degrees, and $\gamma = 120$ degrees [35,36]. The presence of one diffraction peak shows that the monoclinic-structured VO_2 thin films only exhibit the (0 2 0) *out-of-plane* orientation. Additionally, the shift of this (0 2 0) peak to lower 2θ values with increasing W doping content is indicative of the increase in lattice constant, b, and ascertaining the incorporation of W into the VO_2 structure.

From analysis of the SEM images (Fig. 3.6) the average grain size of the films was approximately 125 nm, and the surface of the films became denser and smoother with increasing at.% of W, in agreement with a previous study [4]. From RBS analyses, composition of the films matched the expected values given in column 1 and estimated thickness of the films reported in column 3 of Table 3.1.

The XPS spectra of VO_2 films did not identify any W due to the lack of instrumental sensitivity for detecting W at or below 1% level. For near-surface (5 nm) stoichiometry, a finer scan of the region containing the O 1s and V 2p peaks (between 514 and 535 eV) was carried out on undoped and all W-doped VO_2 films. An XPS spectrum for VO_2-0.34 is illustrated in Fig. 3.7. A shoulder on the high binding energy end of the O 1s peak is indicative of at least two different forms of oxygen on the surface. Since the presence of C 1s peak (not shown in eV range of Fig. 3.7) is associated with adventitious carbon, the additional source of oxygen (in addition to O from VO_2) may be from adsorbed species. Both the V $2p_{1/2}$ and V $2p_{3/2}$ peaks were deconvoluted into two Gaussian profiles: one indicating a V oxidation state of 5+, consistent with the presence of V_2O_5, and the other a V oxidation state of 4+ for VO_2. Note, the unintended presence of surface V_2O_5 is likely due to postgrowth cooling under an oxygen-rich environment, which has been observed previously [37]. The relative percentages of each oxide in the surface layer of all samples are tabulated in Table 3.2 (column 2).

Although the growth rate of V_2O_5 is comparatively lower than VO_2, the V_2O_5 percentages within 5 nm of the surface are significantly high. This is attributed to the low electronegativity (1.63 on the Pauling scale) and electronic configuration ([Ar] $3d^3\ 4s^2$) of V, which facilitates the formation of ions with a higher oxidation state resulting in the ultrathin, passivation layer of V_2O_5 on VO_2 thin film surface [37]. Since VO_2 signals are observable and V_2O_5 layer thicknesses are less than the XPS electron escape depth ($\sim$5 nm), the peak areas (A) of the V 2p were used to estimate the surface V_2O_5 thicknesses, t, using the following relationship:

$$t = d\left[A_{V^{5+}}/(A_{V^{5+}} + A_{V^{4+}})\right], \qquad (3.6)$$

where d is the XPS sampling depth (5 nm), and $A_{V^{5+}}$ and $A_{V^{4+}}$ are the V 2p peak areas for the V^{5+} and V^{4+} oxidation states, respectively. The results are shown in

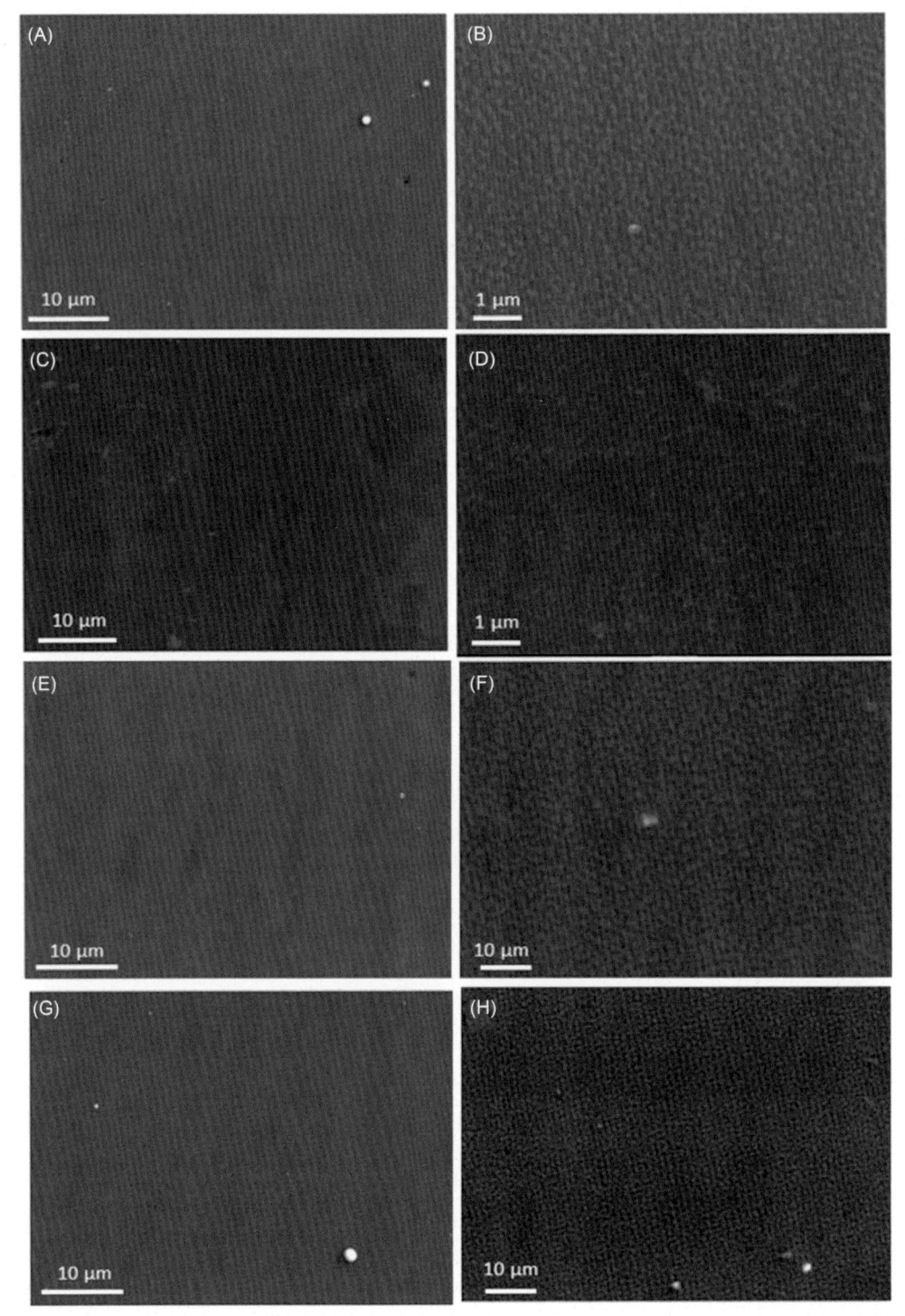

Figure 3.6 SEM images of VO$_2$ films: (A) undoped, (C) VO$_2$-0.34, (E) VO$_2$-0.54, and (G) VO$_2$-1.1 at low magnification; (B) undoped, (D) VO$_2$-0.34, (F) VO$_2$-0.54, and (H) VO$_2$-1.1 at high magnification. *SEM*, Scanning electron microscopy.

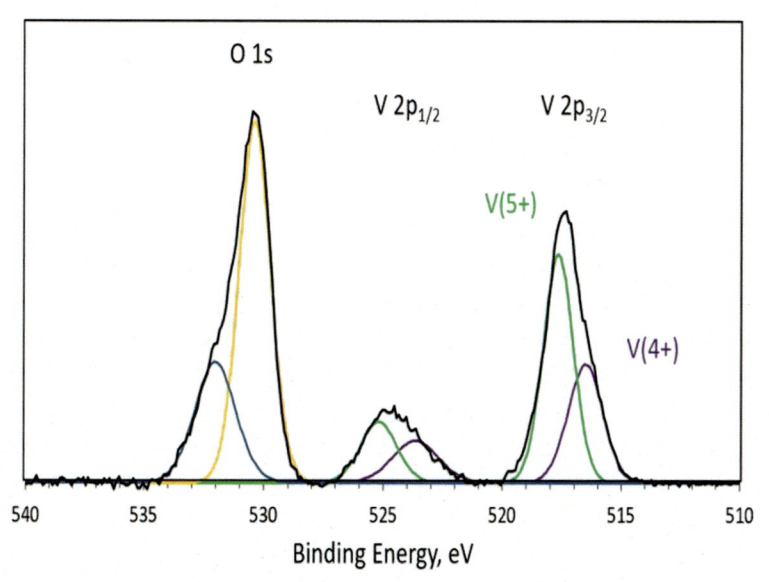

Figure 3.7 An XPS spectrum for VO$_2$-0.34 showing the deconvoluted O 1s and V 2p peaks. *XPS*, X-ray photoelectron spectroscopy.

Table 3.2 Relative percentages of VO$_2$ and V$_2$O$_5$ and calculated thicknesses of the surface layer in all samples determined by X-ray photoelectron spectroscopy.

Sample	% VO$_2$ (% V$_2$O$_5$)	Surface layer thickness (nm)
VO$_2$	37 (63)	2.98
VO$_2$–0.34	38 (62)	3.81
VO$_2$–0.54	39 (61)	4,02
VO$_2$–1.1	33 (67)	3.88

Table 3.2 (column 3). Since the bandgap of PLD V$_2$O$_5$ films grown at 500°C is greater than 2.1 eV [38], all electrical measurements and analyses of VO$_2$ films due to direct tunneling within the thin V$_2$O$_5$-rich surface layer, as well as all optical measurements (in the wavelength range over 590 nm that corresponds to 2.1 eV) and analyses of VO$_2$ films should not be affected by this thin (3–4 nm) surface layer.

High-resolution transmittance electron microscope images of all VO$_2$ films indicate the high quality of epitaxial films with sharp interfaces with the sapphire substrates. The fast Fourier transforms of atomic resolution images were indexed to VO$_2$ viewed along the [2 0 0] zone axis; typical data is illustrated in Fig. 3.8 for VO$_2$-0.34 and observations at different regions within the film indicate a phase-pure growth of the epitaxial VO$_2$ crystals on sapphire, which is also consistent with phase purity of the XRD data.

3.3.2 Electrical resistivity versus temperature characteristics

Fig. 3.9 shows the resistivity (ρ) versus temperature (T) plots of W–doped VO_2 films, measured from test structures of 2 mm length and 0.1 mm width. The measurements were performed, during heating and cooling, in the temperature range between 10°C and 90°C.

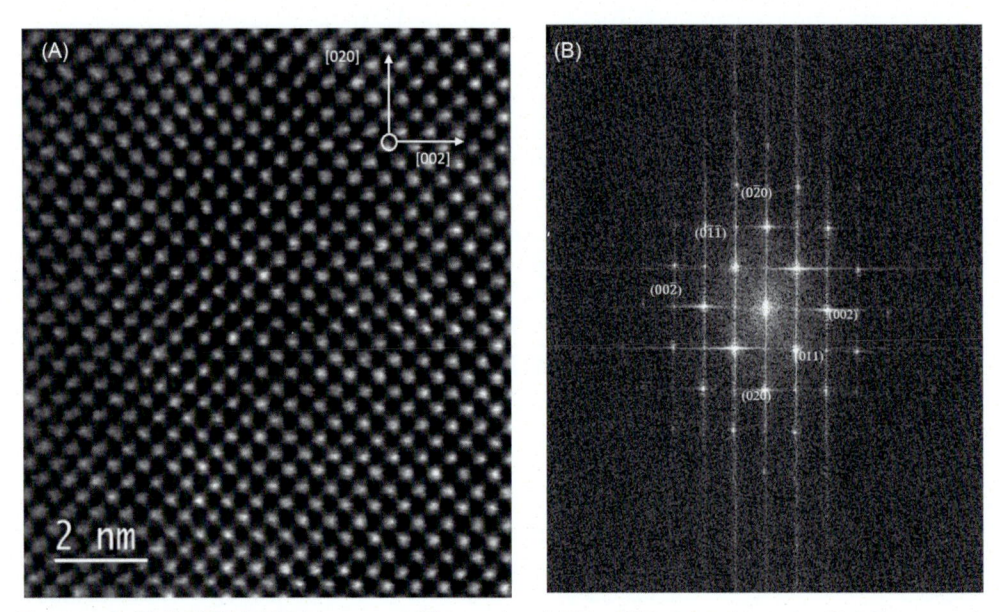

Figure 3.8 (A) HAADF STEM image and (B) associated *FFT* of VO_2-0.34 along the [2 0 0] zone axis. *FFT*, Fast Fourier transform; *HAADF STEM*, high-angle annular dark-field scanning transmittance electron microscope.

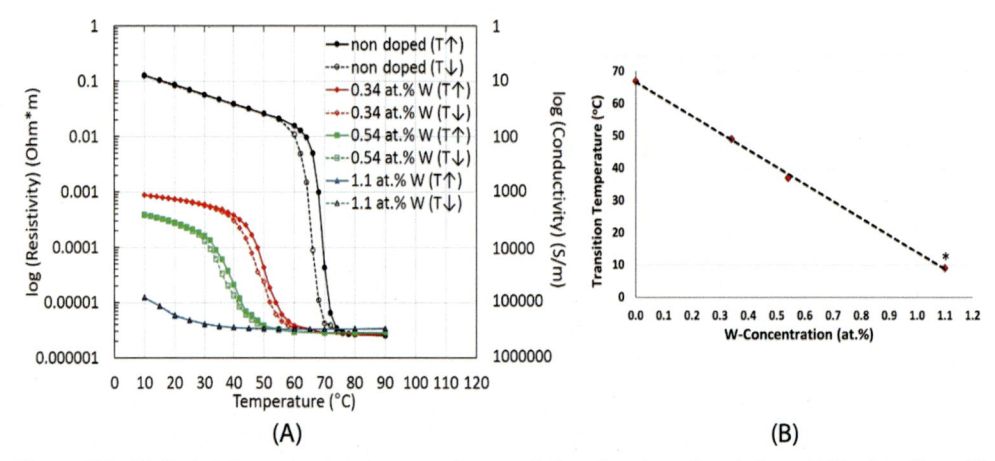

Figure 3.9 (A) Resistivity versus temperature characteristics of undoped and doped VO_2 thin films. (B) Plot of transition temperature as function of W concentration. (*): approximated value by Vegard's law.

The $\rho - T$ characteristics, exhibiting a small hysteresis, demonstrate the transition from an insulator (or semiconductor) state to a metallic conductor state for the undoped VO_2 thin film. The measurement reveals that the undoped VO_2 thin film is of high quality due to the resistivity change of four orders of magnitude between insulating and conducting states. The transition temperature (T_c) decreases, and the width of transition behavior is broadened as the at.% of W in the film structure increases. Additionally, the resistivity value of the film at low-temperature (below the phase-transition temperature) also decreases as at.% of W increases. These phenomena are consistent with previous reports [29,30,33,34]. The results also allow us to determine the W concentration-dependence of the T_c in the films as illustrated in Fig. 3.9B. This linear dependence, which also supports Vegard's law, allows us to extrapolate a T_c of $\sim 9°C$ for 1.1 at.% W-doped VO_2 film.

Also, a model was developed to verify the nature of the insulator to metal transition in the film. The $\rho - T$ characteristics show two distinct patterns: (1) semiconductor state below the T_c, and (2) a drastic drop in resistivity due to phase change from the monoclinic to the rutile structure in the transition zone. The electrical resistivity in the semiconductor state can be modeled as follows:

$$R(T) = R_0 \cdot e^{E_a/kT}, \qquad (3.7)$$

where R_0, E_a, and k are preexponential factors, activation energy, and Boltzmann constant, respectively. Therefore a regression analysis on resistivities below T_c resulted in an activation energy of 0.33 eV, which is in the range of other experimental reports of activation energies in the range of $0.1 - 0.65$ eV [37]. The scatter in experimental data may be attributed to microstructural effects including defects in the thin film. The semiconductor state is followed by a drastic drop in resistivity in the vicinity of T_c, as shown in Fig. 3.10. Note, the electrical resistivity of the metallic state is significantly lower and in agreement with reported data [37]. In the transition region the Bruggeman effective medium approximation (EMA) is used to model the electrical conductivity of the heterogeneous two-phase mixture and observe the phase-transition pattern [36]. In EMA the metallic particles are randomly dispersed in the host material, and the electrical conductivity may be expressed as:

$$\sigma(f, \sigma_i, \sigma_m) = \frac{1}{4}\left[2\sigma_p - \sigma'_p + \sqrt{\left(2\sigma_p - \sigma'_p\right)^2 + 8\sigma_i\sigma_m}\right], \qquad (3.8)$$

where $\sigma_p = (1-f)\sigma_i + f\sigma_m, \sigma^I_p = f\sigma_i + (1-f)\sigma_m$, and σ_i, σ_m, and f are electrical conductivity of insulator, electrical conductivity of metal, and volume fraction of metallic particles, respectively. The volume fraction, f, varies from 0 (fully insulator state) to

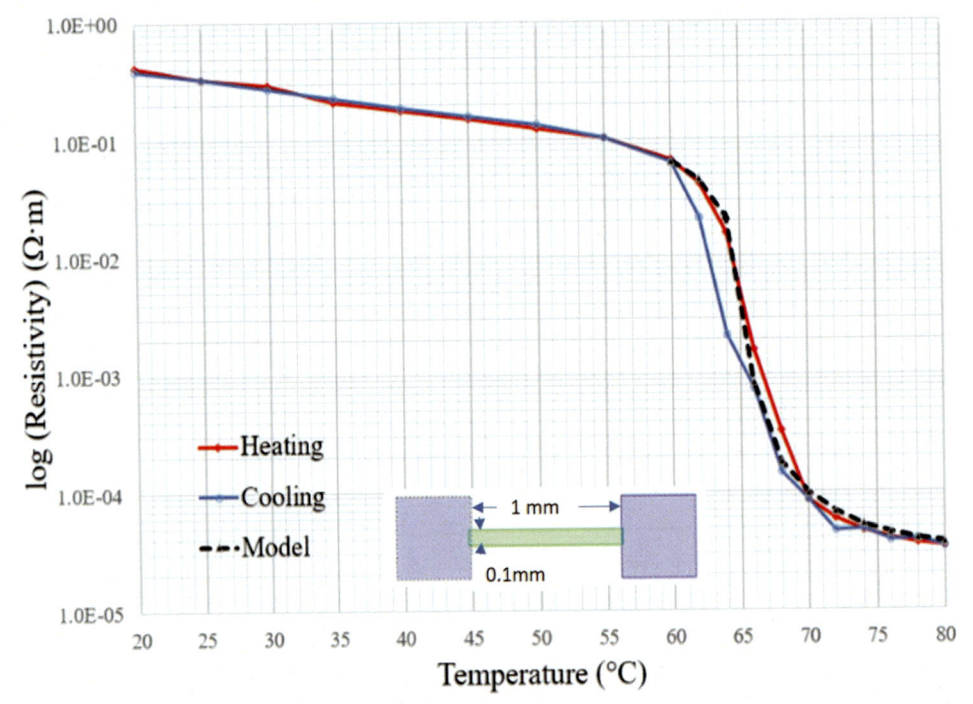

Figure 3.10 Measured resistivity of an undoped VO$_2$ thin film on a sapphire substrate. Measurements were made on the 1 mm long, 0.1 mm wide, and 150-nm-thick film test structure shown. The modeled resistivity matches the experimental data reasonably well.

1 (fully metallic state) with increasing temperature. The particle distribution as a function of time is approximated with a Gaussian distribution [36]:

$$N(T) = Ae^{-[(T-T_C)/b]^2}. \tag{3.9}$$

where b and T_c are model fitting parameters; $b = 10.5°C$, $T_c = 70°C$, and $A = 1/b\sqrt{\pi}$. Finally, the evolution of volume fraction may be found as:

$$f = \int_{T_o}^{T} N(T)dT \tag{3.10}$$

where T_o is the initial temperature in the transition zone. Eqs. (3.7)−(3.10) are used to simulate the electrical resistivity, $\rho = 1/\sigma$, in the transition zone. Simulation results plotted in Fig. 3.10 show an excellent agreement with experimental data. This model verifies that the electronic (Mott) transition takes place first before the structural (Peierls) transition, which indicates that the phase change starts with the formation of the metallic rutile phase at one or more random sites and then gradually spreads within the entire thin film under the influence of temperature [39].

3.3.3 Microwave transmission and reflection properties

The microwave frequency-dependent scattering parameter representing the forward transmission coefficient (S_{21}) measurements at the X-band (8–12 GHz) are shown in Fig. 3.11. The parameter S_{21} is the ratio of the power transmitted to the output port to the power available at the input port. Fig. 3.11 shows swept frequency S_{21} above and below the transition temperature for a 225-nm-thick VO_2 as shown. These materials exhibit low insertion loss across the X-band in the insulating phase, as most of the RF power is transmitted to the output port. Above the phase-transition temperature, transmission loss (also known as isolation) increases to >25 dB with isolation peaking at 8.3 GHz. This shows that the film is reflecting the incident microwave signals above the transition temperature due to its metallic nature.

Fig. 3.12 shows the measured thermal hysteresis curves of the RF transmittance coefficient for a series of VO_2 samples prepared by identical growth conditions but varying film thickness. As can be readily observed, the 225 and 300 nm VO_2 films exhibit thermally driven transmittance modulation, whereas the 150 nm film shows no such modulation. The hypothesis is that the film thickness of 150 nm is much smaller than the skin depth at 10 GHz such that insufficient RF coupling occurs.

Also of note is the nearly invariant attenuation level of >25 dB for both the 225- and 300-nm-thick films. This effect is seen even though there are differences in the

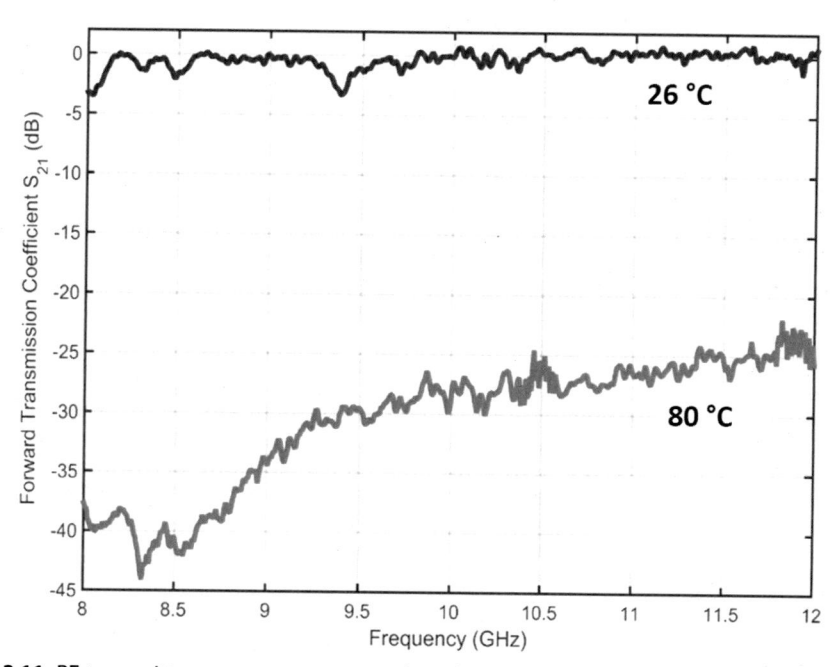

Figure 3.11 RF transmittance spectrum across the X-band frequency range for a 225-nm-thick VO_2 thin film on a sapphire substrate. *RF*, Radio frequency.

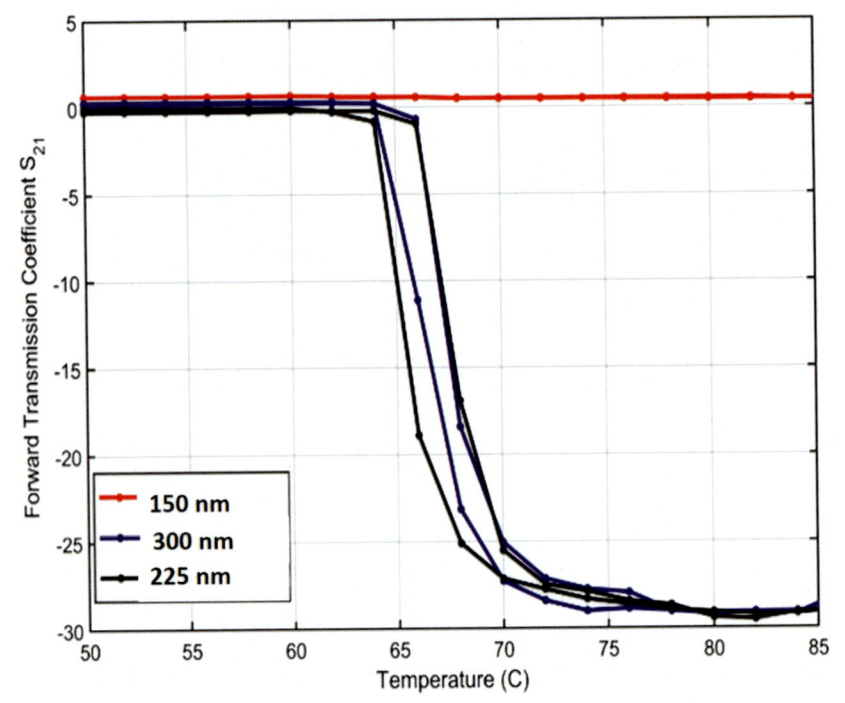

Figure 3.12 Comparison of forward transmission coefficient with temperature for three VO₂ films of different thicknesses.

widths of their hysteresis loops. Such a disparate effect may be attributed to microstructural differences in grain size and surface morphology. Additionally, hysteresis curves were constructed for all measured frequencies and shown in Fig. 3.13, with the red traces for lower and blue traces for higher frequencies. These plots show that most of the frequency-dependent variability occurs in the metallic phase above T_c. The isolation is as high as 50 dB at the lower X-band frequency of 8 GHz and decreases to around 23 dB at the highest frequency of 12 GHz.

3.3.4 Optical transmittance, reflectance, and absorptance measurements

The previously described measurement system was used for the optical transmittance and reflectance spectra of three different thickness VO₂ films grown on sapphire substrates. From the measured transmittance and reflectance spectra, the absorptance was calculated over the entire spectrum. Fig. 3.14 shows the measured transmittance, reflectance, and absorptance spectra for 150, 225, and 300-nm-thick VO₂ films collected at two temperatures. The blue curve is measured at room temperature (~ 295K), which is lower than the T_c of VO₂ (341K). The red curve is collected at an

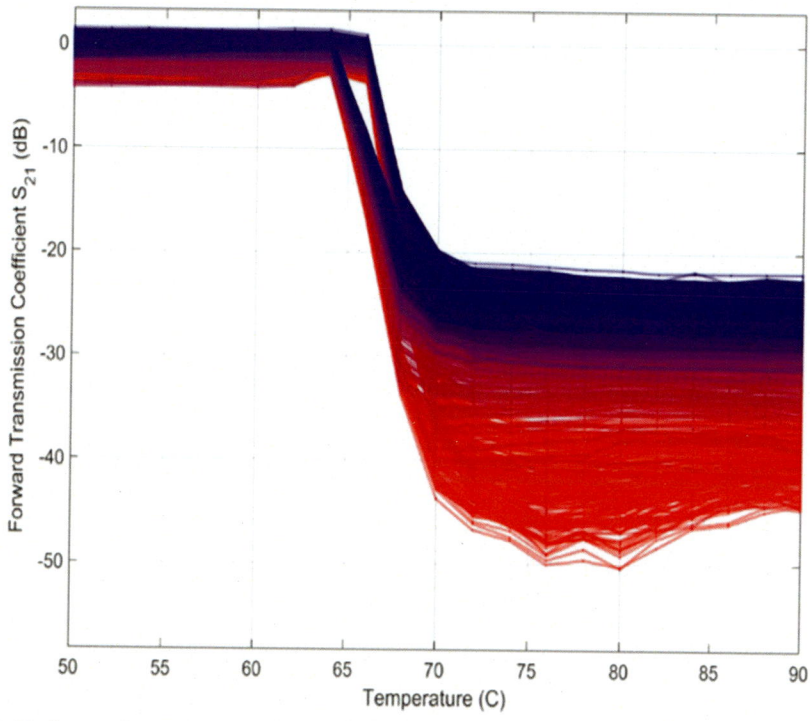

Figure 3.13 Forward transmission coefficient measurements of a 225-nm-thick VO_2 film, at temperatures above and below the phase-transition temperature plotted for multiple X-band frequencies, displaying hysteresis loops.

elevated temperature of $384K \pm 5K$, controlled by constant power heating of the sample stage, is above T_c of VO_2 and characteristic of the conducting phase.

The transmittance spectra show that the thicker films have some similarities compared to the thinner VO_2 film. The thicker films show increasing transmittance in the UV–VIS and near IR range up to 1.5 μm, with the 225–nm–thick film showing higher transmittance than the 300 nm film. The highest transmittance for the 225- and 300–nm-thick films is observed at 1.75 and 2.3 μm wavelength, respectively. The optical transition efficiency (or transmittance contrast) between the insulating and metallic states is also higher for the thicker films, especially in the MWIR range, compared to the thinner VO_2 film. The highest transition efficiency of over 55% observed in the 225 nm film at 1.75 μm wavelength is higher than recently reported values [38]. The 150–nm-thick film only showed a transition efficiency of 20% in the range of 1.5−5 μm range. The increase in contrast for the thicker films can be attributed to the increased sample thickness and therefore interaction volume and effective coupling to MWIR light. Note that the transmittance is almost close to zero in the entire MWIR range for the thicker VO_2 films, which indicates that the films are of excellent

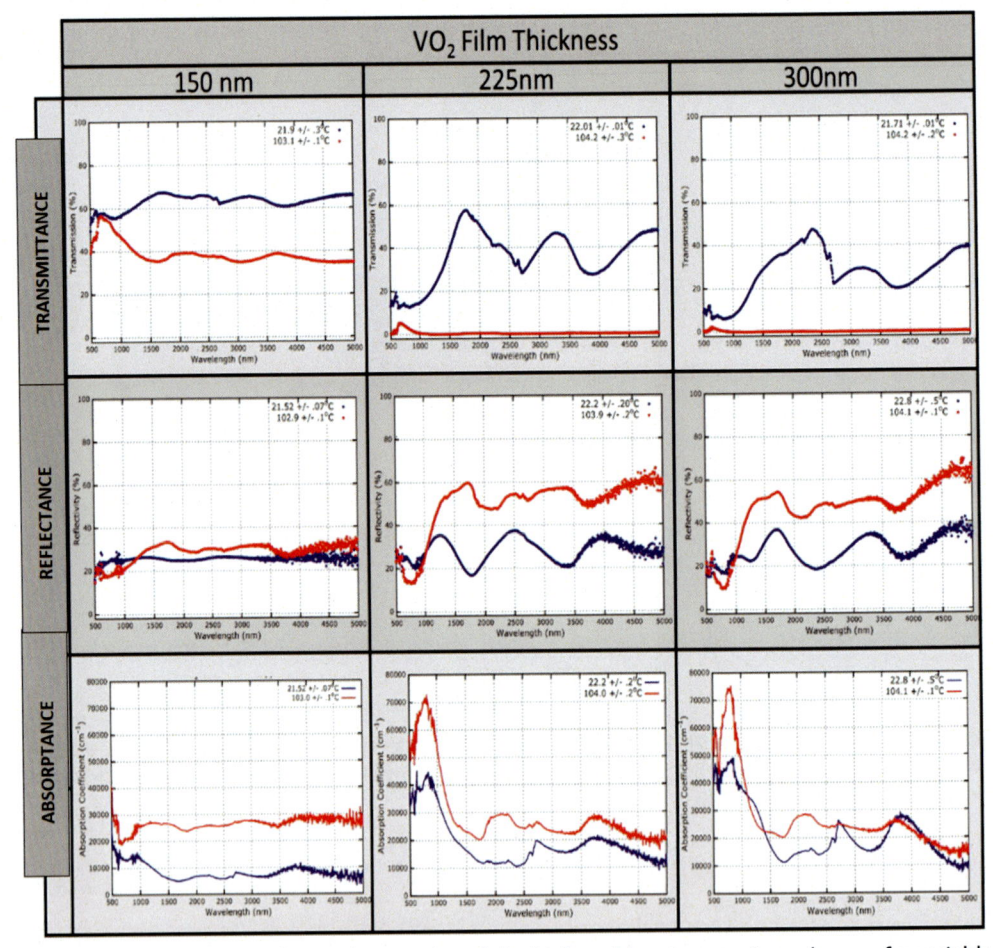

Figure 3.14 Comparison of optical properties of the high and low-temperature phases of a variable thickness VO$_2$ sample set. The red and blue curves represent measurements of the materials in the high and low-temperature phases, respectively.

crystalline quality. Fabry—Perot oscillations, due to constructive and destructive interferences in VO$_2$ epilayer of thickness in the order of incident radiation wavelength, are evident in both the transmittance and reflectance spectra.

The reflectance spectra of the thicker films also show similarities with increased reflectance contrast in the 225-nm-thick VO$_2$ film compared to the others. This increase in reflectance would account for the remarkable contrast seen in the transmittance spectra of the high and low-temperature phases of the 225 nm VO$_2$ film compared to the others. The absorptance spectra show higher absorptance at higher temperature metallic phase compared to the insulating state for the three samples. These data clearly demonstrate temperature-dependent absorptance behavior occurring in the NIR/SWIR range. The low-and-high-temperature measurements are self-consistent in the visible portion of the spectrum, where there is little difference in absorption behavior between the low-temperature and high-temperature phases of the thicker samples. The increased absorptance in the visible spectrum up to 750 nm highlights the region of interband transitions between the oxygen 2p band and vanadium 3d band with a Fermi level of approximately 2.5 eV. The effects of this increased absorptance are observable in both the transmittance and reflectance spectra.

The temperature dependence of transmittance and reflectance for the same samples were studied at the optical wavelength of 1.55 μm to assess their transition efficiency and hence the switching performance. These data allow us to understand the onset of the phase transition, measure the critical temperature, and understand its manifestation in terms of optical properties. Fig. 3.15 shows a comparison between three samples of varying thickness and indicates that optimum switching performance is achieved for a film thickness of 225 nm at a wavelength of 1.55 μm. The optical transition efficiency is over 20% for the 150 nm film, 30% for the 300 nm film, and >40% for the 225 nm film. On taking the temperature at which the transmittance is half the maximum, the T_c is determined as 72°C for the 150 nm film, 66°C for the 225 nm film, and 72°C for the 300 nm film, respectively. The deviations in the transition temperature are related to the imperfections such as defects, grain boundaries, strain, and nonstoichiometry [39].

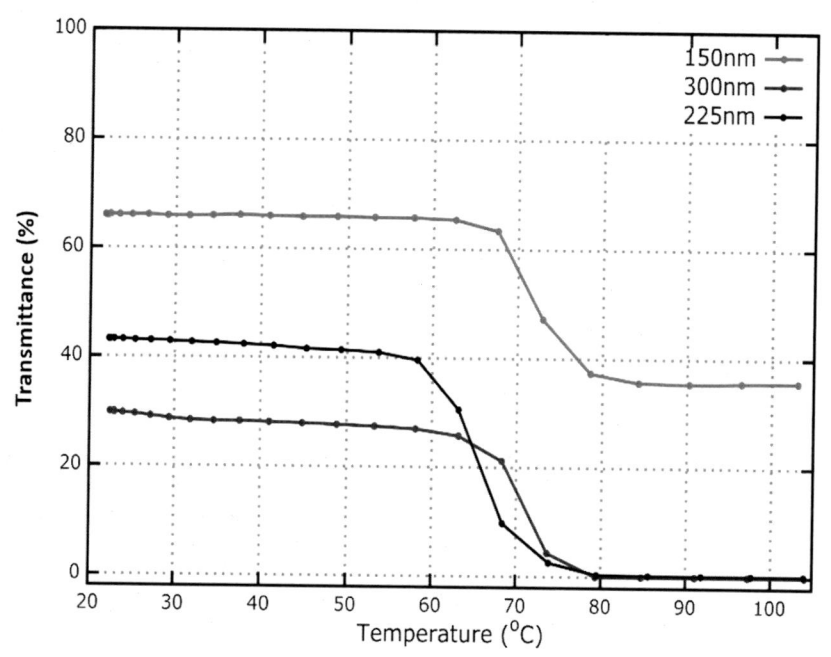

Figure 3.15 Temperature-dependent IR transmittance in VO$_2$ films of increasing thickness. *IR, Infrared.*

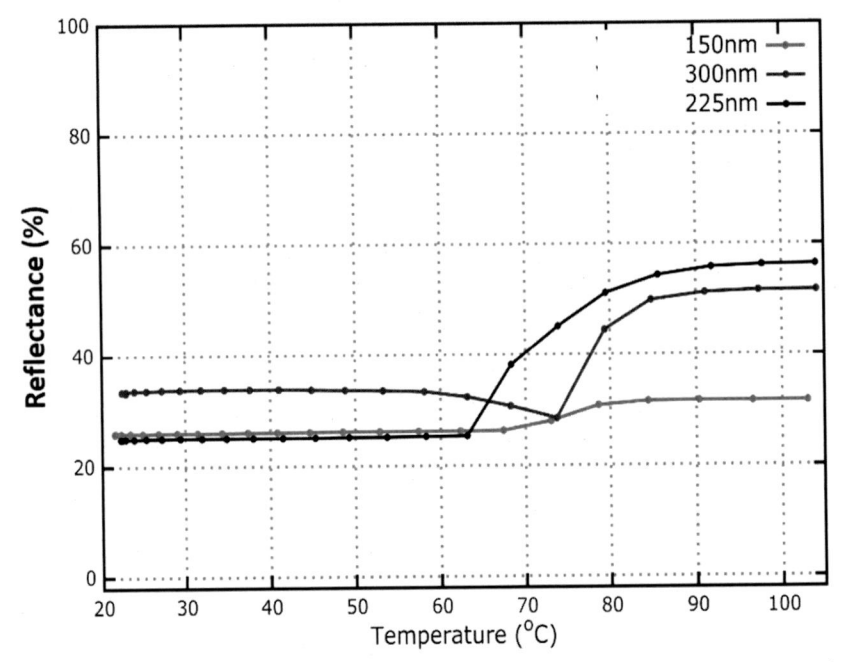

Figure 3.16 Temperature-dependent IR reflectance modulation in VO_2 films of increasing thickness. *IR*, infrared.

The reflectance mode thermal measurements for the same samples are shown in Fig. 3.16. The behavior is like the one shown in previous figure with the exception that the onset of transition appears to be slightly more gradual in reflectance. In this case, the 300-nm thick-film seems to show the highest transition efficiency in reflectance. These data suggest that the phase-transition traverses an intermediate state, with high absorptance but similar refractive index to the low-temperature phase before the material fully transitions to the high-temperature phase.

3.3.5 Optical switching

Fig. 3.17 shows the optical switching measurements performed using a high-power 1550 nm CW laser on a 225-nm-thick VO_2 film. The sample was thermally biased in the vicinity of T_c to impart high-speed switching using the CW laser. Fig. 3.17 shows that the high-power laser can impart a fast (<50 Ms) onset of the semiconductor to metallic phase transition. As the 532 nm CW laser's intensity is varied from 19.58 to 97.9 W cm^{-2}, the VO_2 transitions from fully transmissive at 1550 nm to 50% transmissive within tens of milliseconds (ms), which is reasonably rapid for a thermal transition. Ultra-fast photo-induced transitions have been reported in the order of hundreds of picoseconds to nanoseconds [40,41]. The ultra-fast photo-induced phase transition is attributed to the electron excitation in VO_2.

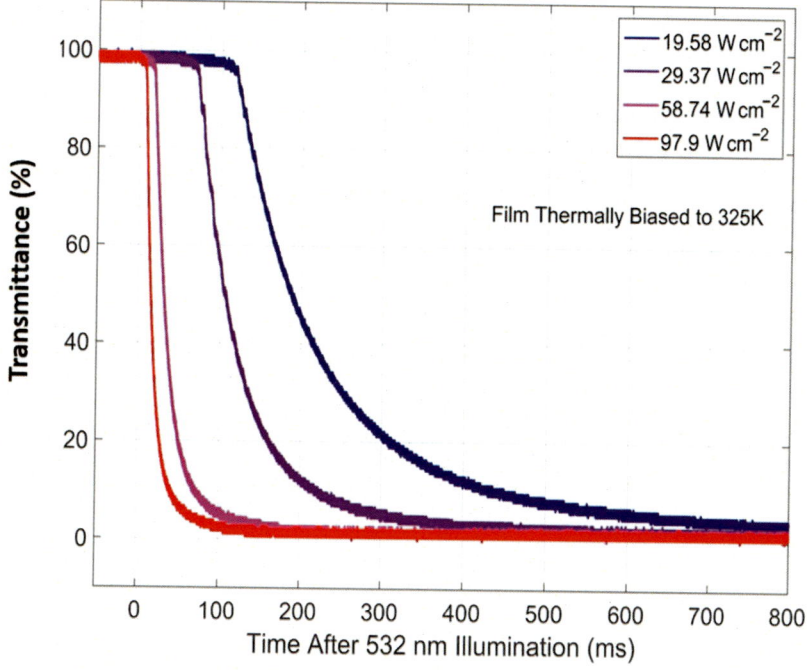

Figure 3.17 Dynamic illumination of the VO_2 thin film for optically induced phase transition. This measurement was performed with the film biased to 50°C. The optical intensity of the *CW* laser influences the speed of switching.

3.3.6 Optical constants for the VO_2 thin films

The plots of the complex refractive index of a 180-nm-undoped VO_2 thin film and a 0.34 at.% W–doped VO_2 on sapphire substrate, obtained from ellipsometry at room temperature and at 80°C, are shown in Figs. 3.18 and 3.19, respectively. In the MWIR region, the increase in both the real (n) and imaginary parts (k) of the refractive index is clear. While the n is slightly lower in the MWIR region, the k (also known as the decay constant) is higher for the W–doped VO_2 thin film, which results in higher loss-tangent and higher absorptance compared to the undoped film of the same thickness. The n and k for the undoped film is comparable to the reported values in the literature [42−44].

The attenuation coefficient, α, is calculated using the relationship $4\pi k/\lambda_o$, where λ_o is the wavelength of light in free-space. The indirect and direct bandgap of undoped and doped VO_2 thin films were determined from the attenuation coefficient plots; note, the former bandgap is associated with the splitting of the V_{3d} states. Fig. 3.20 shows the plots of $(\alpha E)^{1/2}$ versus photon energy E with the indirect bandgap approximated by linear extrapolation of the data onto the x-axis. The indirect bandgap at room temperature is ∼0.35 eV for the undoped VO_2 and 0.22 eV for the 0.34 at.%

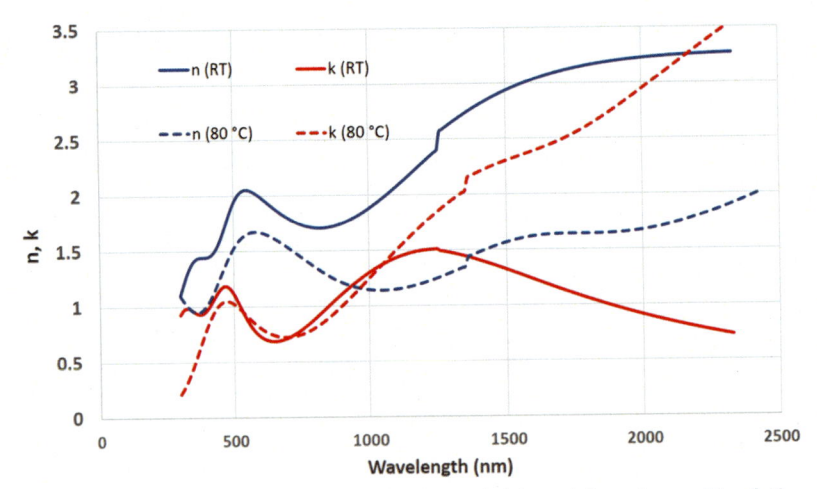

Figure 3.18 The wavelength dependence of the real (*n*) and imaginary (*k*) of the complex refractive index of a 180 nm thick undoped VO_2 film on sapphire substrate at room temperature and at 80°C.

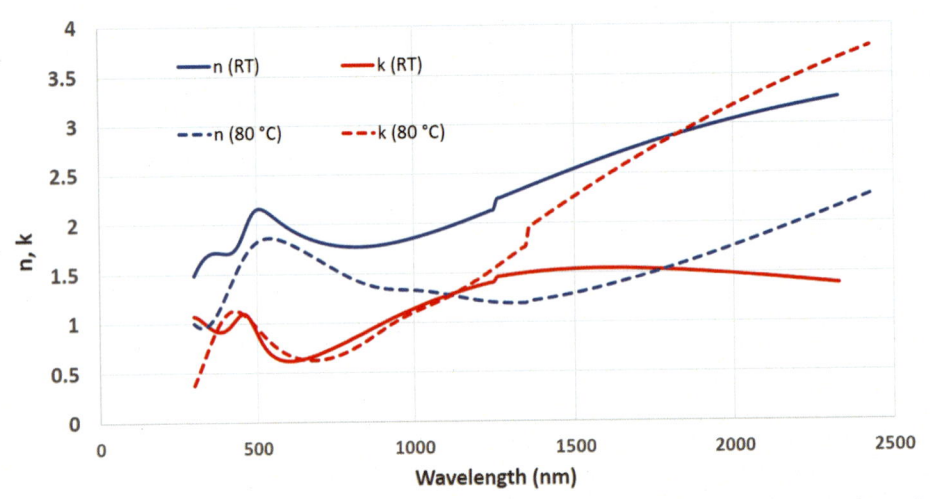

Figure 3.19 The wavelength dependence of the real (*n*) and imaginary (*k*) of the complex refractive index of a 180 nm thick 0.34 at.% W-doped VO_2 film on sapphire substrate at room temperature and at 80°C.

W-doped VO_2. After a sharp increase, there is a relatively flat band between 1 and 1.5 eV range, followed by a sharp increase above 1.5 eV. The flat band could be a result of inhomogeneity in the film consisting of both tetragonal and monoclinic forms resulting in multiband absorption. The rapid increase above 2 eV is attributed to a direct interband O_{2p} to V3d transition [44].

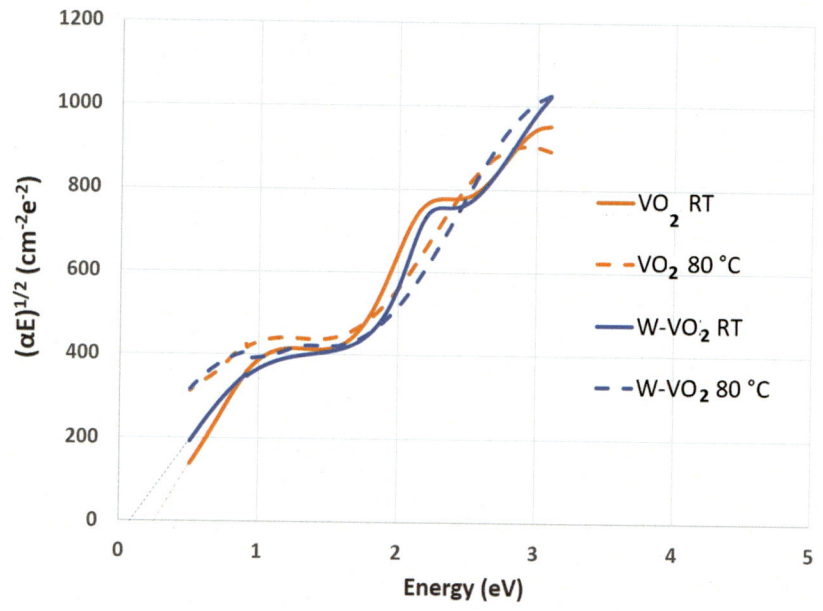

Figure 3.20 The indirect bandgap determination for the undoped and 0.34 at.% W-doped VO$_2$ thin films at room temperature and at 80°C.

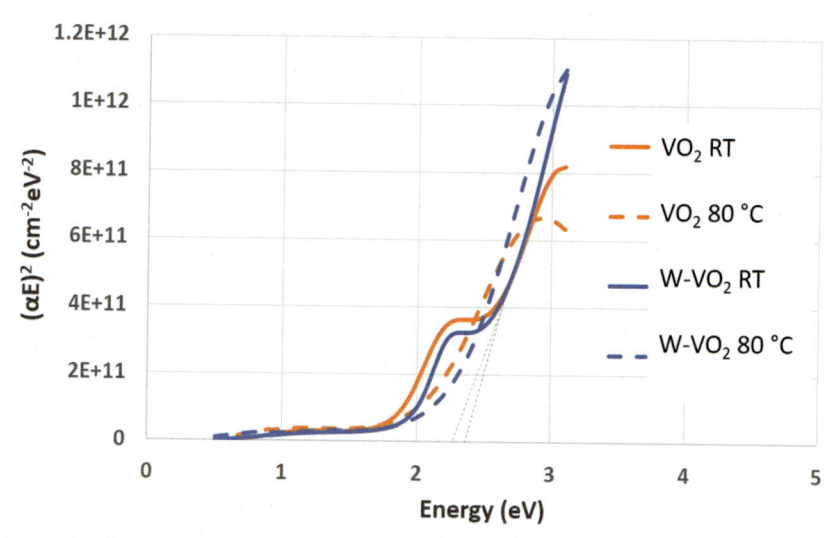

Figure 3.21 The direct bandgap determination for the undoped and 0.34 at.% W-doped VO$_2$ thin films at room temperature and at 80°C.

As shown in Fig. 3.21, the direct bandgap for the VO$_2$ thin films was determined from plots of $(\alpha E)^{1/2}$ versus E [44]. The effective bandgap can be obtained by extrapolating the slope of the curve with the sharpest rise in the vicinity of 2 eV. The effective bandgap is approximately 2.25 eV and 2.4 eV for the insulating state of the undoped and the 0.34 at.% W–doped VO_2 thin films, respectively. These are comparable to the published data in the literature [44].

3.4 Summary and conclusions

High-quality, nanostructured VO_2 and W–doped VO_2 thin films were synthesized using a PLD system, and comprehensively characterized for their structural, electrical, microwave, and optical properties. Notably, the temperature-dependent electrical resistivity, microwave transmission characteristics, and wavelength-dependent optical transmittance and reflectance were measured and analyzed. The electrical resistivity ratio of the undoped VO_2 between the insulating and the metallic states was in the order of 10^4. The microwave transmission measurements of the films at X-band showed very high contrast between the insulating and the metallic state of the films with almost no insertion loss between 8 and 12 GHz in the insulating state (i.e., allowing microwave transmission) and insertion loss higher than 25 dB in the metallic state (i.e., reflecting the microwave signal). The 225 nm VO_2 thin film showed the highest optical transition efficiency >55% and films thicker than 200 nm exhibited complete reflectance in the MWIR region. The VO_2 thin films were switched from the insulating state to the metallic state in tens of milliseconds using a high-power CW laser, with the sample heated to 50°C. The measured complex refractive indices are comparable to the values reported in the literature. The direct and indirect bandgaps, obtained from the complex refractive index measurements and attenuation coefficient, are also comparable to the values reported in the literature.

Acknowledgments

The authors acknowledge Dr. Kevin Leedy, Air Force Research Laboratory, for his collaboration on process development for large-area VO_2 thin films (up to 4″ diameter) using PLD. The authors acknowledge collaboration with Dr. Thomas Kent, and Dr. Jeff Berry, of Berryhill Corporation, for an Air Force SBIR Phase I research on VO_2 thin films. The authors also acknowledge collaborations with Dr. David Look, Wright State University and AFRL, Professor Zhihong Chen and his graduate research assistant Mr. Chincheng Chiang, from Birck Nanotechnology Center, Purdue University for the Ellipsometry measurements. Also, the use of facilities for all physicochemical characterization including atomic resolution imaging and electron diffraction at the *John M. Cowley Center* for High Resolution Electron Microscopy at Arizona State University (ASU) is gratefully acknowledged.

References

[1] M. Imada, A. Fujimori, Y. Tokura, Metal-Insul. Transit. 70 (1998) 1039−1263.
[2] S. Chen, H. Ma, X. Yi, H. Wang, X. Tao, M. Chen, et al., Optical switch based on vanadium dioxide thin films, Infrared Phys. Technol. 45 (2004) 239−242.
[3] Y. Zhou, X. Chen, C. Ko, Z. Yang, C. Mouli, S. Ramanathan, Voltage-triggered ultrafast phase transition in vanadium dioxide switches, IEEE Electron. Device Lett. 34 (2013) 220−222.

[4] E. Shin, K.C. Pan, W. Wang, G. Subramanyam, V. Vasilyev, K. Leedy, et al., Tungsten-doped vanadium dioxide thin film based tunable antenna, Mater. Res. Bull. 101 (2018) 287−290.

[5] H. Wang, Y. Yang, L. Wang, Wavelength-tunable infrared metamaterial by tailoring magnetic resonance condition with VO_2 phase transition, J. Appl. Phys. 116 (2014) 17−21.

[6] M.J. Dicken, K. Aydin, I.M. Pryce, L.A. Sweatlock, E.M. Boyd, S. Walavalkar, et al., Frequency tunable near-infrared metamaterials based on VO_2 phase transition, Opt. Express 17 (2009) 18330.

[7] T. Paik, S.H. Hong, E.A. Gaulding, H. Caglayan, T.R. Gordon, N. Engheta, et al., Solution-processed phase-change VO_2 metamaterials from colloidal vanadium oxide (VO_x) nanocrystals, ACS Nano 8 (2014) 797−806.

[8] J. Givernaud, A. Crunteanu, J.C. Orlianges, A. Pothier, C. Champeaux, A. Catherinot, et al., Microwave power limiting devices based on the semiconductor-metal transition in vanadium-dioxide thin films, IEEE Trans. Microw. Theory Tech. 58 (2010) 2352−2361.

[9] G. Subramanyam, E. Shin, D. Brown, H. Yue, Thermally controlled vanadium dioxide thin film microwave devices, Proc. Midwest. Symp. Circuits Syst. (2013) 73−76.

[10] E. Rahimi, K. Şendur, Temperature-driven switchable-beam Yagi-Uda antenna using VO_2 semiconductor-metal phase transitions, Opt. Commun. 392 (2017) 109−113.

[11] A.K. Prasad, S. Amirthapandian, S. Dhara, S. Dash, N. Murali, A.K. Tyagi, Novel single phase vanadium dioxide nanostructured films for methane sensing near room temperature, Sens. Actuators B Chem. 191 (2014) 252−256.

[12] L.A.L. De Almeida, G.S. Deep, A.M. Nogueira Lima, H.F. Neff, R.C.S. Freire, A hysteresis model for a vanadium dioxide transition-edge microbolometer, IEEE Trans. Instrum. Meas. 50 (2001) 1030−1035.

[13] P. Kiri, M.E.A. Warwick, I. Ridley, R. Binions, Fluorine doped vanadium dioxide thin films for smart windows, Thin Solid Films 520 (2011) 1363−1366.

[14] Z. Huang, C. Chen, C. Lv, S. Chen, Tungsten-doped vanadium dioxide thin films on borosilicate glass for smart window application, J. Alloys Compd. 564 (2013) 158−161.

[15] F. Guinneton, L. Sauques, J.C. Valmalette, F. Cros, J.R. Gavarri, Comparative study between nanocrystalline powder and thin film of vanadium dioxide VO_2: electrical and infrared properties, J. Phys. Chem. Solids 62 (2001) 1229−1238.

[16] J. Liu, Q. Li, T. Wang, D. Yu, Y. Li, Metastable vanadium dioxide nanobelts: hydrothermal synthesis, electrical transport, and magnetic properties, Angew. Chem. Int. Ed. 43 (2004) 5048−5052.

[17] G. Guzman, R. Morineau, J. Livage, Synthesis of vanadium dioxide thin films from vanadium alkoxides, Mater. Res. Bull. 29 (1994) 509−515.

[18] K.R. Speck, H.S. Hu, M.E. Sherwin, R.S. Potember, Vanadium dioxide films grown from vanadium tetra-isopropoxide by the sol-gel process, Thin Solid Films 165 (1988) 317−322.

[19] R.T. Kivaisi, M. Samiji, Optical and electrical properties of vanadium dioxide films prepared under optimized RF sputtering conditions, Sol. Energy Mater. Sol. Cell 57 (1999) 141−152.

[20] X. Yi, C. Chen, L. Liu, Y. Wang, B. Xiong, H. Wang, et al., A new fabrication method for vanadium dioxide thin films deposited by ion beam sputtering, Infrared Phys. Technol. 44 (2003) 137−141.

[21] T.D. Manning, I.P. Parkin, R.J.H. Clark, D. Sheel, M.E. Pemble, D. Vernadou, Intelligent window coatings: atmospheric pressure chemical vapour deposition of vanadium oxides, J. Mater. Chem. 12 (2002) 2936−2939.

[22] D. Vernardou, M.E. Pemble, D.W. Sheel, In-situ FTIR studies of the growth of vanadium dioxide coatings on glass by atmospheric pressure chemical vapour deposition for VCl_4 and H_2O system, Thin Solid Films 515 (2007) 8768−8770.

[23] Z.P. Wu, A. Miyashita, S. Yamamoto, H. Abe, I. Nashiyama, K. Narumi, et al., Molybdenum substitutional doping and its effects on phase transition properties in single crystalline vanadium dioxide thin film, J. Appl. Phys. 86 (1999) 5311−5313.

[24] C.S. Blackman, C. Piccirillo, R. Binions, I.P. Parkin, Atmospheric pressure chemical vapour deposition of thermochromic tungsten doped vanadium dioxide thin films for use in architectural glazing, Thin Solid Films 517 (2009) 4565−4570.

[25] M. Soltani, M. Chaker, E. Haddad, R.V. Kruzelecky, D. Nikanpour, Optical switching of vanadium dioxide thin films deposited by reactive pulsed laser deposition, J. Vac. Sci. Technol. A Vac. Surf. Films 22 (2004) 859.

[26] X. Tan, T. Yao, R. Long, Z. Sun, Y. Feng, H. Cheng, et al., Unraveling metal-insulator transition mechanism of VO_2 triggered by tungsten doping, Sci. Rep. 2 (2012) 1−6.

[27] B. Chen, D. Yang, P.A. Charpentier, M. Zeman, Al^{3+}-doped vanadium dioxide thin films deposited by PLD, Sol. Energy Mater. Sol. Cell 93 (2009) 1550−1554.

[28] T.D. Manning, I.P. Parkin, C. Blackman, U. Qureshi, APCVD of thermochromic vanadium dioxide thin films-solid solutions $V_{(2-x)}M_xO_2$ (M = Mo, Nb) or composites VO_2: SnO_2, J. Mater. Chem. 15 (2005) 4560−4566.

[29] K. Shibuya, M. Kawasaki, Y. Tokura, Metal-insulator transition in epitaxial $V_{(1-x)}W_xO_2$ ($0 \leq x \leq 0.33$) thin films, Appl. Phys. Lett. 96 (2010) 022102.

[30] T.L. Wu, L. Whittaker, S. Banerjee, G. Sambandamurthy, Temperature and voltage driven tunable metal-insulator transition in individual $W_xV_{(1-x)}O_2$ nanowires, Phys. Rev. B Condens. Matter Mater. Phys. 83 (2011) 073101.

[31] N. Émond, A. Hendaoui, M. Chaker, Low resistivity $W_xV_{(1-x)}O_2$-based multilayer structure with high temperature coefficient of resistance for microbolometer applications, Appl. Phys. Lett. 107 (2015) 140537.

[32] B.G. Chae, H.T. Kim, D.H. Youn, K.Y. Kang, Abrupt metal-insulator transition observed in VO_2 thin films induced by a switching voltage pulse, Phys. B: Condens. Matter 369 (2005) 76−80.

[33] J.H. Cho, Y.J. Byun, J.H. Kim, Y.J. Lee, Y.H. Jeong, M.P. Chun, et al., Thermochromic characteristics of WO_3-doped vanadium dioxide thin films prepared by sol-gel method, Ceram. Int. 38 (2012) S589−S593.

[34] Q. Gu, A. Falk, J. Wu, L. Ouyang, H. Park, Current-driven phase oscillation and domain-wall propagation in $W_xV_{(1-x)}O_2$ nanobeams,", Nano Lett. 7 (2007) 363−366.

[35] K. Okimura, J. Sakai, Changes in lattice parameters of VO_2 films grown on c-Plane Al_2O_3 substrates across metal-insulator transition, Jpn. J. Appl. Phys. 48 (2009) 1−6.

[36] V. Eyert, The metal-insulator transitions of VO_2: a band theoretical approach, Ann. Phys. (Leipz.) 11 (2002) 650−704.

[37] A.J. Littlejohn, Y. Yang, Z. Lu, E. Shin, K.C. Pan, G. Subramanyam, et al., Naturally formed ultrathin V_2O_5 heteroepitaxial layer on VO_2/sapphire (001) film, Appl. Surf. Sci. 419 (2017) 365−372.

[38] C.V. Ramana, R.J. Smith, O.M. Hussain, C.C. Chusuei, C.M. Julien, Correlation between growth conditions, microstructure, and optical properties in pulsed-laser-deposited V_2O_5 thin films, Chem. Mater. 17 (5) (2005) 1213−1219.

[39] S. Kumar, J.P. Strachan, A.L. David Kilcoyne, T. Tyliszczak, M.D. Pickett, C. Santori, et al., The phase transition in VO_2 probed using x-ray, visible and infrared radiations, Appl. Phys. Lett. 108 (2016) 073102.

[40] Z. Shao, X. Cao, H. Luo, P. Jin, Recent progress in the phase-transition mechanism and modulation of vanadium dioxide materials, Nat. Asia Mater. 10 (2018) 581−605.

[41] T.V. Son, V.V. Truong, J.-F. Bisson, A. Haché, Nanosecond polarization modulation in vanadium dioxide thin films, Appl. Phys. Lett. 111 (4) (2017) 041103.

[42] D. Fu, K. Liu, T. Tao, K. Lo, C. Cheng, B. Liu, et al., Comprehensive study of the metal-insulator transition in pulsed laser deposited epitaxial VO_2 thin films, J. Appl. Phys. 113 (2013) 1−7.

[43] C.N. Berglund, H.J. Guggenheim, Electronic properties of VO_2 near the semiconductor-metal transition,", Phys. Rev. 185 (1969) 1022.

[44] C. Zhang, C. Koughia, O. Güneş, J. Luo, N. Hossain, Y. Li, et al., Synthesis, structure and optical properties of high-quality VO_2 thin films grown on silicon, quartz and sapphire substrates by high temperature magnetron sputtering: properties through the transition temperature, J. Alloys Compd. 848 (2020) 1−13.

CHAPTER 4

Photoactive ZnO nanostructured thin films modified with TiO$_2$, and reduced graphene oxide

Pierre G. Ramos, Luis A. Sánchez and Juan M. Rodriguez
Center for the Development of Advanced Materials and Nanotechnology, National University of Engineering, Lima, Perú

4.1 Introduction

The unusual properties of transition metal oxides, which distinguish them from metallic elements, covalent semiconductors, and ionic insulators are due to the unique nature of the outer d-electrons. The bonding between the metal-d and oxygen-p orbitals varies anywhere from nearly ionic to metallic. Correlation of the structure and physical properties of transition metal oxides requires a description of the valence electrons that bind the atoms in the solid-state with both localized and itinerant d-electron behavior, which can be described either by the ligand field theory or the band theory. The band theory assumes that the electrons in a solid are distributed among a set of available stationary states following the Fermic Dirac statistics. In this, the number of allowed electron states per unit cell in an energy range is called the density of states, all states lower in energy than the Fermi energy E_F are occupied by electrons, while that above E_F unoccupied at $T = 0K$. In a metal E_F lies within a ban, whereas in an insulator or semiconductor E_F lies between bands with an energy gap E_g separating the highest-lying filled band (valence band, VB) and the lowest-lying empty band (conduction band, CB). Adding impurities or defects rise a perfectly stoichiometric compound with intrinsic properties to shown extrinsic ones, being then acceptors or donor of electrons, p-type or n-type, respectively.

The most important way of generating minority carriers in semiconductors with a wide bandgap is by light absorption, its energy must exceed the bandgap E_g of the semiconductor. The photogenerated electron in the CB and hole in the VB migrate under the influence of the electric field toward the bulk and the surface, by the effect of strongly charged surface states, counting as electric current, and if an electric circuit is built it could be ready to perform the incidence photon-to-current efficiency (IPCE) characterization. It could be possible that the electron—hole pairs can recombine and lost by heat. If the charges are localized by trapping at surface states, their

Thin Film Nanophotonics
DOI: https://doi.org/10.1016/B978-0-12-822085-6.00002-9

mean lifetime can be long enough to allow their transfer to adsorbed electron donors or acceptors, associated with pollutants following redox reactions is also called photocatalytic water purification.

Photocatalysis that is a promising, simple, and eco-friendly technology has been attractive to treat low-concentration organic contaminants in recent years [1]. So far, various semiconductor photocatalysts, such as ZnO, TiO_2, CuO, Fe_2O_3, and WO_3, have demonstrated promising photocatalytic activity [2−4], highlighting particularly zinc oxide due to its lower cost, nontoxicity, and size-tunable physicochemical properties [4,5].

Currently, the key challenge for improving the photocatalytic dye degradation efficiency of ZnO or TiO_2 is to inhibit the recombination of photogenerated charge carriers (electron and holes). Thus strategies like nanostructuring, doping, and formation of nanocomposites have been adopted to achieve this improvement [6,7]; likewise, the photocatalytic activity of the photocatalyst can also be improved by enlarging its specific surface area, through morphological modification [8].

Among the different types of nanostructures [9], 0D, 1D, 2D, and 3D nanostructures like nanoparticles, nanorods (NRs), nanofibers, films, within others have received much attention due to their photoactivity and potential optimization for photocatalysis [10,11].

In particular, these materials would be fabricated following up several techniques, from the point of view of soft chemistry routes, this chapter is focused on the sol−gel, spray pyrolysis (SP), spin coating, electrospinning, and hydrothermal bath techniques. Additionally, the doping and formation of nanocomposites with graphene-based materials will be discussed from the point of view of the photoactivity to get an improvement of the photocatalytic water purification.

4.2 Fabrication methods of modified ZnO nanostructures

The modified zinc oxide nanostructures focused here were synthesized through different methods, including: electrospinning, SP, spray-gel (SG), and wet chemical method. The processes carried out will be described below.

4.2.1 Electrospinning

The ZnO and ZnO/TiO_2 nanostructures were obtained from the calcination of precursor nanofibers fabricated by electrospinning using spinning solutions with different Zinc acetate/PVA (polyvinyl alcohol) mass ratios (1:2, 2:3, 1:2, and 3:2). The electrospinning parameters were optimized after running a series of experiments. The syringe supplied the feeding solution at a speed of $2 \, mL \, h^{-1}$ and the electrical potential applied was approximately 62 kV. The precursor nanofibers were deposited onto fluorine-doped tin oxide (FTO) glass that was placed 10 cm away from the tip of the needle as seen in Fig. 4.1.

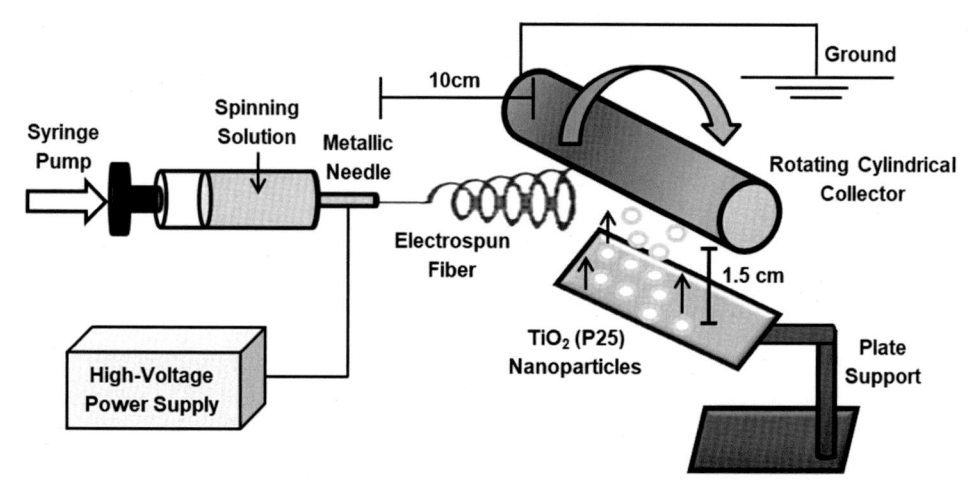

Figure 4.1 Schematic view of electrostatically modified electrospinning process applied to the formation of precursor fibers of ZnO and ZnO/TiO$_2$ nanostructures.

On the other hand, for the fabrication of ZnO/TiO$_2$ nanofibers, TiO$_2$-P25 nanoparticles were adhered in situ during the electrospinning process by electrostatic induction, as shown in Fig. 4.1. During 2 h of electrospinning, 25 mg of TiO$_2$-P25 nanoparticles was distributed every 30 min on a metal plate at a distance of 1.5 cm from the collector substrate. To remove the polymer after the electrospinning process and to obtain ZnO and ZnO/TiO$_2$ nanofibers, the deposited precursor nanofibers were calcined at 600°C for 3 h.

The ZnO and ZnO-reduced graphene oxide (rGO) nanorods were deposited and grown onto FTO glass plates by wet chemical method from ZnO and ZnO-rGO seeds layers fabricated by electrospinning, respectively. To obtain a pure ZnO seed layer, a spinning solution containing zinc acetate and polyvinylpyrrolidone dissolved in N−N dimethylformamide was used. Two different series of ZnO-rGO seed layers were obtained under different conditions. The first series of ZnO-rGO seed layers were fabricated applying 20 kV of voltage and using the same spinning solution seen above, but adding this time 0.1, 0.2, and 0.3 wt.% rGO. Meanwhile, for the second series we used the same spinning solution for ZnO seed layer with 0.2 wt.% rGO and three different spinning voltages was applied (20, 30, and 40 kV). Then, the ZnO-rGO seed layers were obtained by calcination in a muffle furnace at 400°C of the coated substrates. The solution medium used for the growth of the ZnO and ZnO-rGO NRs and the steps performed to obtain them were carried out based on Rodriguez et al. [12].

ZnO nanorods grown from ZnO seeds prepared with 0 wt.% rGO and 20 kV of spinning voltage were labeled as Z20, whereas the ZnO-rGO nanorods grown from ZnO-rGO seeds prepared with 20 kV, 30 kV, and 40 kV of spinning voltages were labeled as ZG20, ZG30, and ZG40, respectively. Likewise, ZnO-rGO nanorods grown from ZnO-rGO seeds prepared with 0.1, 0.2, and 0.3 wt.% rGO were labeled as ZG1, ZG2, and ZG3.

4.2.2 Spray pyrolysis

The ZnO nanorods were deposited and grown onto FTO conductive glass by wet chemical method from ZnO seed layers fabricated by SP technique. To obtain a ZnO seed layers, 0.10 M zinc acetate solution in a mixture of deionized water and ethanol was used. The values of molar ethanol/water ratio (Γ) were varied in three values: 0, 0.06, and 0.92. In this experiment, 10 layers of ZnO seed films were deposited using the SP system. ZnO nanorods were grown, from seed layers obtained by SP, according to the method seen above [12].

4.2.3 Spray-gel

ZnO nanorods were grown by a hydrothermal process using ZnO seed layers deposited onto FTO substrates by SG. A methanol-based sol of ZnO nanoparticles [13] was used as a spray solution. In these experiments, 10−50 seed layers were successively deposited and the growth solution used was the same as that used to obtain ZnO nanorods grown from seed layers fabricated by SP. ZnO grown from seed layers fabricated by SG with 10, 30, and 50 layers were labeled as P10, P30, and P50, respectively.

4.3 Characterization methods

4.3.1 Morphological characterization

The morphology of ZnO and ZnO-rGO NRs was visualized using a scanning electron microscopy HITACHI SU8230, and a Supra 40VP field emission scanning electron microscope (ZEISS).

4.3.2 Photoactivity

4.3.2.1 Incident photon-to-current efficiency

The IPCE of the ZnO and ZnO/TiO$_2$ fabricated nanostructures was evaluated in a three-electrode cell using the nanofibers as a working electrode, a platinum wire as a counter electrode and Ag/AgCl electrode immersed in 0.1 M KCl solution as a reference electrode, 0.1 N NaOH as an electrolytic solution, and a 400−1000 W Xe lamp as a light source.

4.3.2.2 Photoluminescence

The photoluminescence (PL) measurements were done at room temperature using a PTI Quantamaster QM-1 luminescence spectrometer using 370 nm as the excitation wavelength, whereas, Cathodoluminescence (CL) was carried out at room temperature in a FEI Quanta SEM, while the PL spectra were also obtained on the Renishaw equipment measured over the wavelength range of 350−950 nm using a He−Cd laser source with a wavelength of 325 nm.

4.3.2.3 Photocatalysis

The photocatalytic activities of the fabricated nanostructures were measured through degradation of an aqueous solution of methyl orange (MO) using a 300 W OSRAM Ultravitalux lamp placed approximately at 20 cm from the system, where 70 W m^{-2} in the UV-A range of intensity was measured. Blank experiment, as a reference, was carried out by applying only light irradiation without catalysts; no degradation of dye was observed, corroborating that the degradation was effectively driven by a photocatalytic process. The initial concentration of MO employed for the photocatalytic experiment using ZnO nanofibers, ZnO/TiO₂ nanofibers, ZnO nanorods, was 3 ppm, while the initial concentration used for ZnO-rGO nanorods was 5 ppm. The influence of fabrication parameters on photocatalytic performance was studied.

4.4 Surface-modified ZnO-based nanostructures

4.4.1 Nanofibers

4.4.1.1 ZnO and ZnO/TiO₂ nanofibers

Fig. 4.2A−D shows the FE-SEM (Field Emission Scanning Electron Microscope) images of the as-spun precursors fibers were obtained by electrospinning using spinning solutions with different zinc acetate/PVA mass ratios. As shown, the increase of the amount of zinc acetate loaded in the spinning solution produces an enlarged of the precursor fibers from 45 to 270 nm. This size increment is a result of the gelation of zinc acetate, due to the increase in the viscosity of the spinning solution with increasing the amount of zinc acetate [14].

The FE-SEM images for the ZnO nanofibers are shown in Fig. 4.2E−H. The results reveal a linear dependence between the average size and the amount of zinc acetate [15]. The average size of the ZnO nanofibers was: 45 ± 8 nm for 1:2, 47 ± 9 nm for 2:3, 55 ± 12 nm for 1:1, and 73 ± 18 nm for 3:2. Higher amounts of precursor (zinc acetate) produce nanofibers with larger nanoparticles connected along the fiber axis (Fig. 4.2F−H). However, ZnO nanostructures obtained from the lower zinc acetate/PVA mass ratio (1:2) show large area of ordered atomic arrangement (Fig. 4.2E) due to that ZnO nuclei are well separated and are unable to form primary particles during the calcination process [14].

Fig. 4.3A−D shows the FE-SEM images of the precursor nanofibers composed of the zinc acetate/polyvinyl alcohol/titanium dioxide (ZnAc/PVA/TiO₂). The regions enclosed by white dashed circles indicated in Fig. 4.3A−D confirm that TiO₂-P25 nanoparticles were coated or attached to all precursor fibers obtained by electrospinning. The amount of zinc acetate increase produced the diameters of the precursor fibers enlarged from 85 to 228 nm. Comparing the precursor fibers of ZnO with the precursor

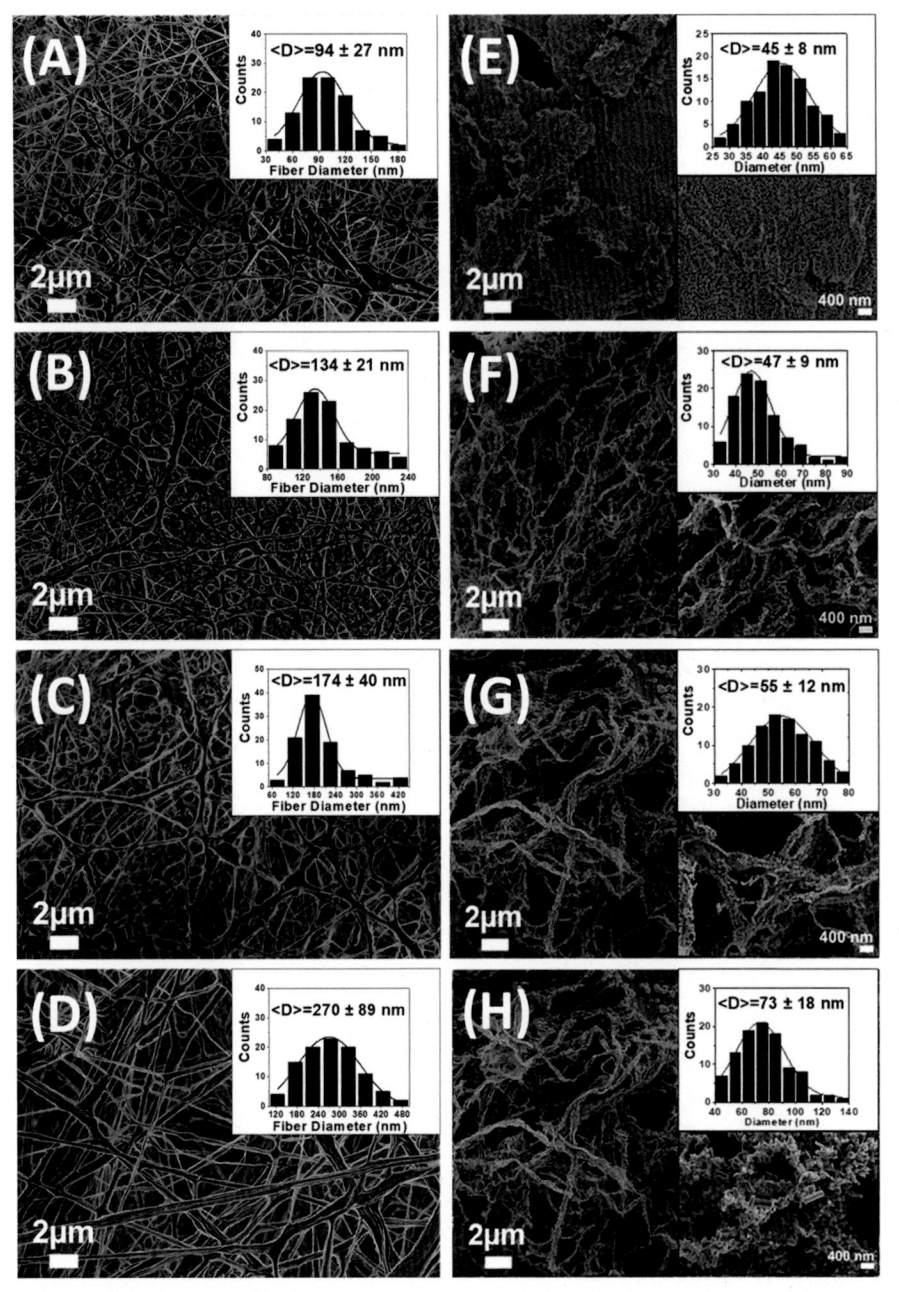

Figure 4.2 FE-SEM images of precursor nanofibers (A−D) and ZnO nanofibers (E−H) fabricated using spinning solutions with different zinc acetate/PVA ratios: 1:2 (A, E), 2:3 (B, F), 1:1 (C, G), and 3:2 (D, H). The size distributions are shown as insets. *Reprinted with permission from P.G. Ramos, N.J. Morales, R.J. Candal, M. Hojamberdiev, J. Rodriguez, Influence of zinc acetate content on the photoelectrochemical performance of zinc oxide nanostructures fabricated by electrospinning technique, Nanomater. Nanotechnol. 6 (2016) 1−7. Copyright 2016, SAGE journals.*

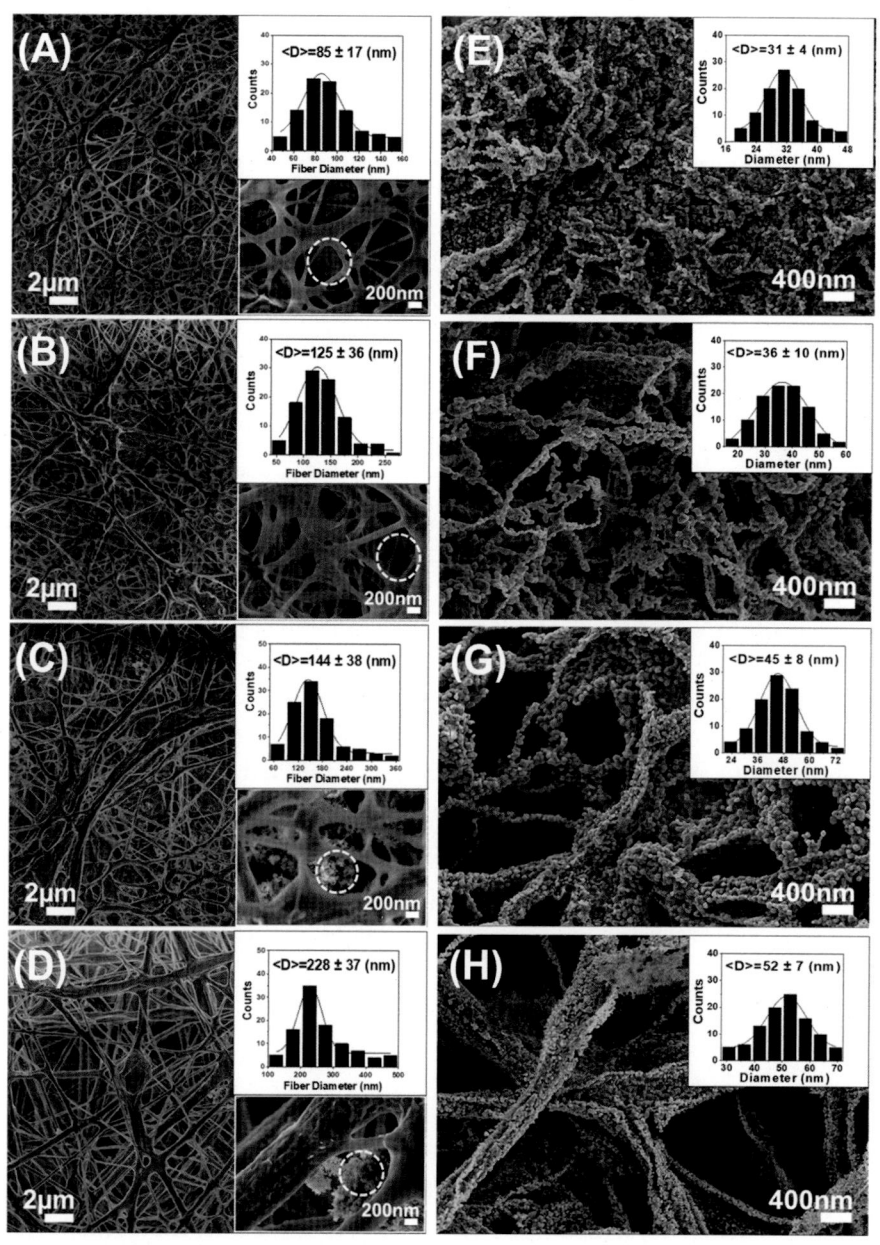

Figure 4.3 FE-SEM images of precursor nanofibers (A–D) and ZnO/TiO$_2$ nanofibers (E–H) fabricated using spinning solutions with different zinc acetate/PVA ratios: 1:2 (A, E), 2:3 (B, F), 1:1 (C, G), and 3:2 (D, H). The size distributions are shown as insets. *Reprinted with permission from P.G. Ramos, E. Flores, L.A. Sánchez, R.J. Candal, M. Hojamberdiev, et al., Enhanced photoelectrochemical performance and photocatalytic activity of ZnO/TiO$_2$ nanostructures fabricated by an electrostatically modified electrospinning, Appl. Surf. Sci. 426 (2017) 844–851. Copyright 2017, Elsevier.*

fibers composed of ZnAc/PVA/TiO$_2$ a decrease in the fiber diameters was noted, due to the presence of the collector/plate system (shown in Fig. 4.1). This new system generates a slight distortion in the field electric lines thereby causing further stretching of the polymer jet which reduced the fiber diameters [16,17].

The FE-SEM images of the ZnO/TiO$_2$ nanofibers are shown in Fig. 4.3E–H. The average size of the nanoparticles forming the ZnO/TiO$_2$ nanofibers was: 31 ± 4 nm for 1:2, 36 ± 10 nm for 2:3, 45 ± 8 nm for 1:1, and 52 ± 7 nm for 3:2, finding a linear dependence between the average size and the amount of zinc acetate. Furthermore, the mean diameter of the nanoparticles from the ZnO/TiO$_2$ nanofibers is smaller than that of the nanoparticles from the ZnO nanofibers even though both of them were fabricated using the same spinning solutions. This is due to the decrease in the fiber diameter caused by the presence of the plate, where the TiO$_2$ nanoparticles were placed.

The IPCE spectra of ZnO nanofibers and ZnO/TiO$_2$ nanofibers are plotted in Fig. 4.4A and B as a function of wavelength. As shown, depending on zinc acetate/PVA mass ratio, the IPCE values of the ZnO and ZnO/TiO$_2$ nanofibers increases in the following order: 4% for 3:2 <7% for 1:2 <28% for 1:1 <31% for 2:3 for ZnO and 7% for 3:2 <9% for 1:2 <31% for 1:1 <43% for 2:3 for ZnO/TiO$_2$ at about 350 nm. The ZnO and ZnO/TiO$_2$ nanofibers fabricated using the spinning solution with a zinc acetate/PVA mass ratio of 2:3 exhibited the highest IPCE value in the wavelength range from 330 to 420 nm compared with other nanostructures fabricated using the spinning solution with zinc acetate/PVA mass ratios of 1:2, 1:1 and 3:2. Although the ZnO and ZnO/TiO$_2$ nanostructures fabricated with a zinc acetate/PVA mass ratio of 3:2 contained the highest amount of zinc acetate, they showed the lowest

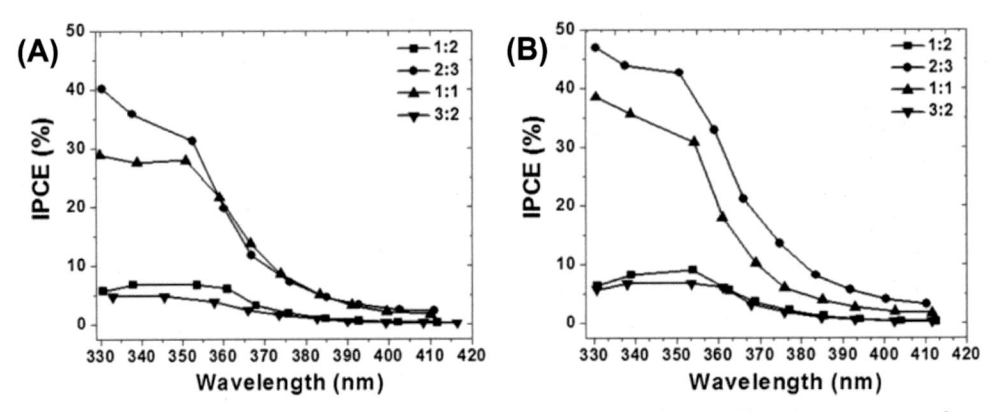

Figure 4.4 IPCE curves of (A) ZnO and (B) ZnO/TiO$_2$ nanofibers fabricated by electrospinning from spinning solutions with different Zinc acetate/PVA ratios: 1:2, 2:3, 1:1, and 3:2. *IPCE*, Incidence photon-to-current efficiency. *Reprinted with permission from P.G. Ramos, E. Flores, L.A. Sánchez, R.J. Candal, M. Hojamberdiev, et al., Enhanced photoelectrochemical performance and photocatalytic activity of ZnO/TiO$_2$ nanostructures fabricated by an electrostatically modified electrospinning, Appl. Surf. Sci. 426 (2017) 844–851. Copyright 2017, Elsevier.*

IPCE values in the wavelength range from 330 to 420 nm. Furthermore, the main difference in the IPCE values of all the fabricated nanostructures is presumably attributed to the enhanced light absorption, electron transfer rate, and delayed charge recombination [18,19], which depend on the morphology, adhesion of TiO$_2$-P25 nanoparticles to the substrates, and specific surface area of the fabricated nanostructures [18]. Therefore we believe that ZnO and ZnO/TiO$_2$ nanofibers fabricated with zinc acetate/PVA mass ratio of 2:3 showed the highest IPCE due to good adhesion of TiO$_2$-P25 nanoparticles, small crystal size resulting in a large specific surface area, and enhanced light absorption [18,19]. Finally, the results show an increase in the IPCE values of the ZnO/TiO$_2$ nanofibers compared with the IPCE values obtained for the ZnO nanofibers (see Fig. 4.4A). This can be due to the in situ adhesion of TiO$_2$-P25 nanoparticles within the ZnO nanostructures during the electrospinning process. According to previous studies [18−20], this can assist to improve the efficiency of light absorption, electron injection, and efficiency of collection, thus enhancing the IPCE values. The results of this study showed that the zinc acetate/PVA mass ratio was critical and must be controlled to achieve the highest IPCE value.

4.4.2 Nanorods

4.4.2.1 ZnO nanorods

Fig. 4.5 shows FE-SEM images of ZnO NRs grown over seeds deposited by SP formed for three representative ethanol/water ratios (Γ): (1) 0, (2) 0.06, and (3) 0.92. Nanorods exhibit hexagonal cross-section and grow in average perpendicularly oriented to the surface. The results indicate that the distribution, density, and size of the seeds depend on the molar ratio ethanol/water in the seeding solution, which also affects the characteristics of the ZnO nanorods [13].

FE-SEM images of ZnO seeds deposited on FTO by SG after 10 and 50 layers are shown in Fig. 4.6A and D, respectively. The main feature of SG seeding is the presence of rings with a high concentration of ZnO particles, with approximately 40 μm diameter and 2 μm thickness originated during drying. When a drop impacts the substrate heated it becomes pinned at the surface; further evaporation of the solvent produces transport of matter by convection from the center to the edge of the drop producing accumulation of material at the boundaries [21]. On the rings as shown in Fig. 4.6B and E, there is a higher density of better-oriented and thinner growth nanorods, which their diameter decreases as the number of seed layers increases. In contrast, outside the rings the nanorods are multioriented (*white dashed lines* in Fig. 4.6C and F) and not follow a clear trend with the number of seed layer as seen in the inset in Fig. 4.6E. Table 4.1 summarizes the thickness and average rod diameter of ZnO NRs films synthesized. ZnO nanorods fabricated exhibit notable differences when they grow from SP or SG seeds.

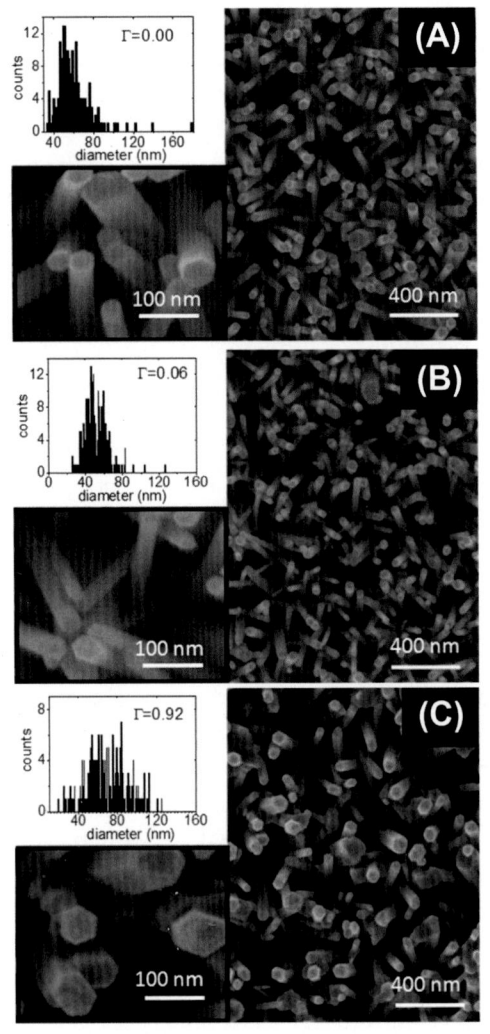

Figure 4.5 FE-SEM images of ZnO nanorods grown from seed layers obtained by SP at different ethanol/water molar ratios Γ of 0 (A), 0.06 (B), and 0.92 (C). The size distributions are shown as insets. *SP*, Spray pyrolysis. *Reprinted with permission from J. Rodríguez, D. Onna, L. Sánchez, M.C. Marchi, R. Candal, S. Ponce, et al., The role of seeding in the morphology and wettability of ZnO nanorods films on different substrates, Appl. Surf. Sci. 279 (2013) 197–203.Copyright 2013, Elsevier.*

Specular transmittance spectra for ZnO nanorods arrays were recorded in the $200 < \lambda < 1000$ nm range. In general, the transmittance is below 68% for wavelengths in the near-infrared (Fig. 4.7A). From the transmittance measurements and using the film thicknesses obtained (Table 4.1), and considering optical transitions, the direct bandgap, E_g can be estimated from the following empirical formula [22]: $\alpha h\nu = A\sqrt{E - E_g}$. In this formula, E_g is the optical bandgap, α is the absorption

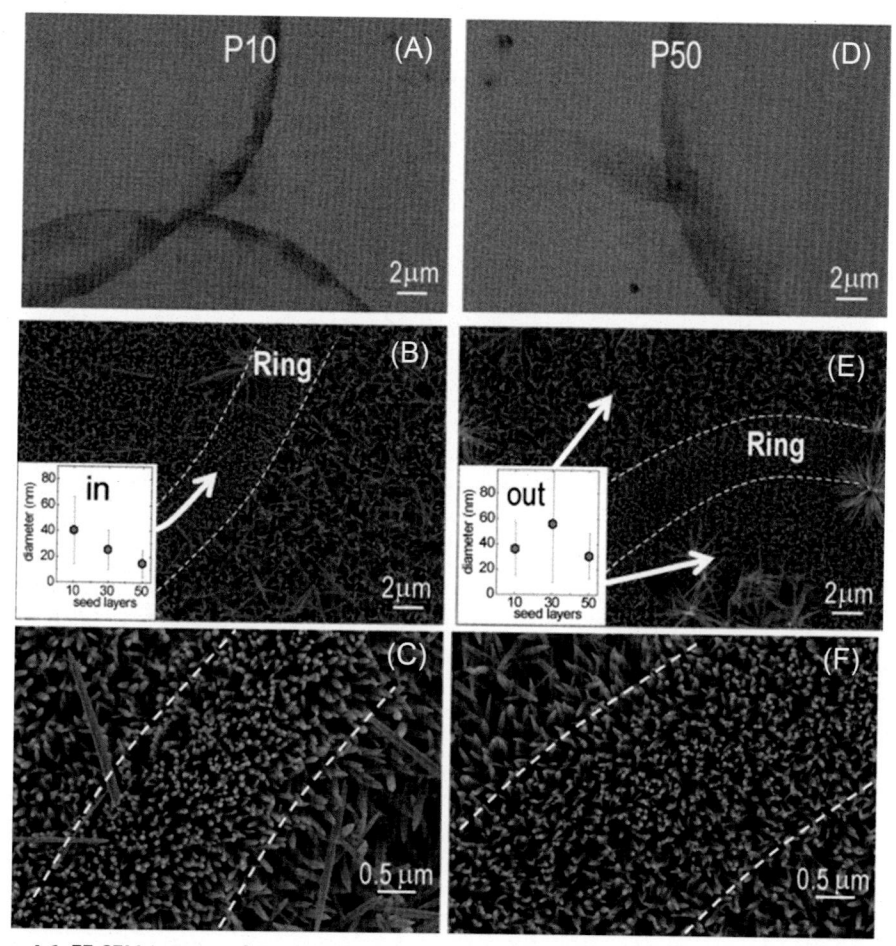

Figure 4.6 FE-SEM images of ZnO seeds deposited by SG applying 10 (A) and 50 (E) layers of ZnO sol; and top view of nanorods labeled as P10 (B, C) and P50 (E, F). The white lines indicate the location of rings with vertically aligned thin NRs. The rod diameter distributions on the rings or out the rings versus the number of layers deposited are shown as insets in (B) and (E), respectively. *SG, Spray-gel. Reprinted with permission from J. Rodríguez, D. Onna, L. Sánchez, M.C. Marchi, R. Candal, S. Ponce, et al., The role of seeding in the morphology and wettability of ZnO nanorods films on different substrates, Appl. Surf. Sci. 279 (2013) 197–203. Copyright 2013, Elsevier.*

coefficient, E is the photon energy which can be converted to a wavelength using the relation $E(\text{eV}) = 1240/\lambda(\text{nm})$ and A is a constant. The absorption coefficient α can be estimated from the normalized transmittance T using the equation: $\alpha = -\frac{1}{d}\ln T$ where d is the thickness of the samples. The extrapolation of a linear part of the curve $(\alpha E)^2$ versus E to 0 as shown in Fig. 4.7B allows us to obtain the optical bandgap (E_g) of the sample. The optical bandgap for the different ZnO nanorods grown at different seeding conditions $\Gamma = 0$, 0.06, and 0.92 were 3.10 ± 0.05, 3.20 ± 0.05, and

Table 4.1 Summary of parameters obtained for ZnO nanorods.

Sample		Thickness (nm)	Mean diameter (nm)	
SP on FTO	FTO	346 ± 5	–	
	$\Gamma = 0.0$	1600 ± 40	54 ± 17	
	$\Gamma = 0.06$	1730 ± 80	49 ± 19	
	$\Gamma = 0.92$	1530 ± 50	67 ± 37	
SG on FTO			On the drop	Outside the drop
	P10	2200 ± 50	40 ± 26	36 ± 21
	P20	2270 ± 50	26 ± 15	56 ± 46
	P30	2000 ± 50	15 ± 11	31 ± 18

FTO, Fluorine-doped tin oxide; *SP*, spray-gel.

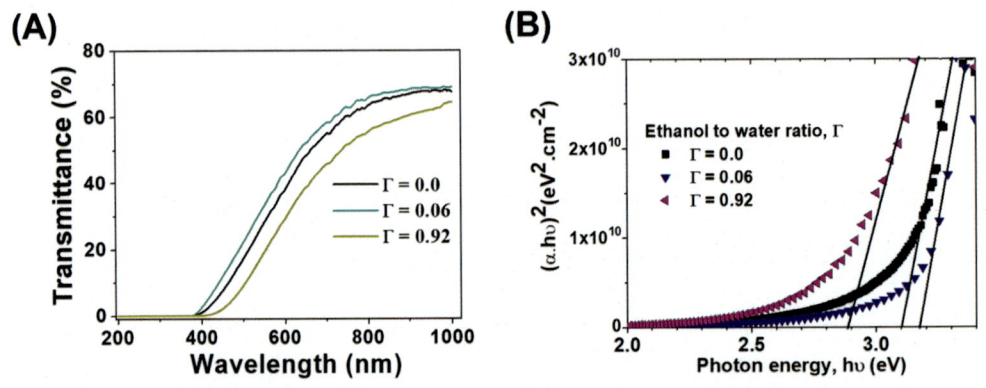

Figure 4.7 (A) Transmittance spectra at room temperature for ZnO NRs samples grown using different ethanol—water molar ratios Γ. (B) $(\alpha \cdot h\nu)^2$ versus $h\nu$ plots derived from transmittance spectra for ZnO NRs films grown from seeds deposited with different ethanol—water molar ratios Γ. The straight lines correspond to extrapolation of the linear parts of the spectra to extract the E_g. *Reprinted with permission from J. Rodríguez, G. Feuillet, F. Donatini, D. Onna, L. Sanchez, R. Candal, et al., Influence of the spray pyrolysis seeding and growth parameters on the structure and optical properties of ZnO nanorod arrays, Mater. Chem. Phys. 151 (2015) 378–384. Copyright 2015, Elsevier.*

2.90 ± 0.05, respectively, with a noticeable change as a function of Γ and a maximum obtained for $\Gamma = 0.06$, which correlates with the NR samples having a thinner diameter (see Fig. 4.5). Intrinsic values for E_g of ZnO at room temperature are reported to be around 3.3 eV [12,13], that is, higher than those obtained in the present case. However, it is necessary to point out that below the bandgap there is an absorption tail, called Urbach tail (Fig. 4.7B), extending toward longer wavelengths, which is mainly ascribed to structural disorder [23] and that calculated absorption spectra could be influenced by waveguide properties of the ZnO NRs [22]. As discussed extensively in Ref. [23], this method is valid for disordered semiconductors. In the present case, it allows to quantify the onset of the absorption edge of the spectra.

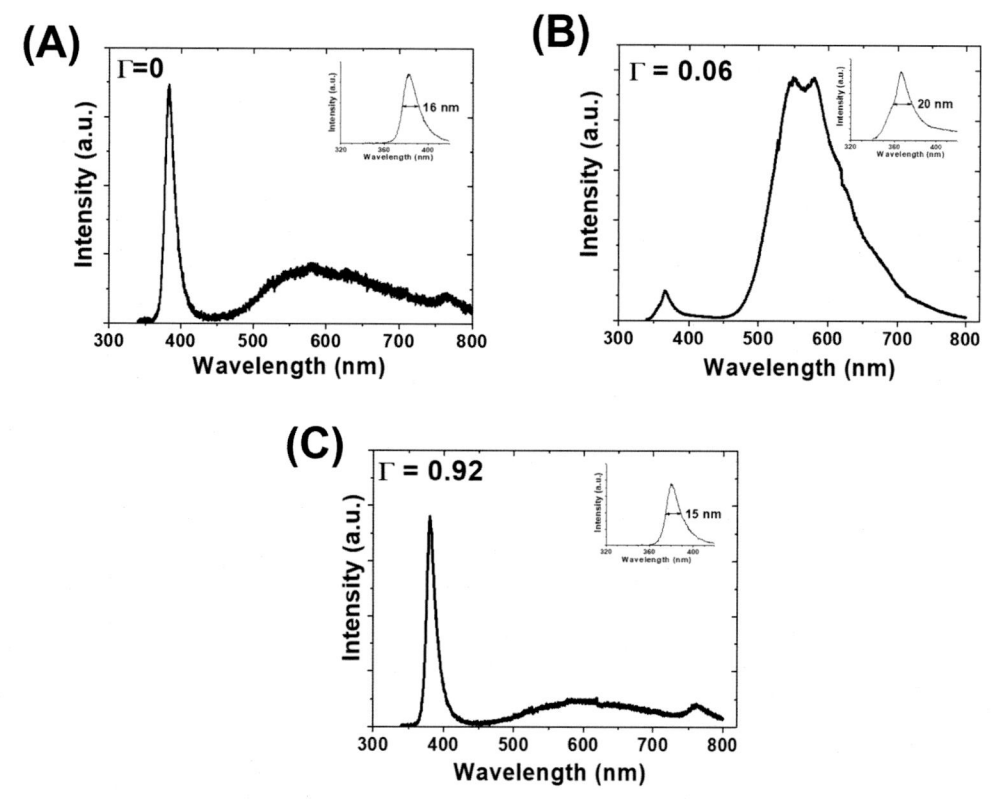

Figure 4.8 Room temperature CL spectra of ZnO nanorods grown from seed layers fabricated by SP at different ethanol/water molar ratios Γ of 0 (A), 0.06 (B), and 0.92 (C). An enlargement of the near band edge (*NBE*) peak of the ZnO nanorods are shown as insets in each image. *CL*, Cathodoluminescence; *SP*, spray-gel. *Reprinted with permission from J. Rodríguez, G. Feuillet, F. Donatini, D. Onna, L. Sanchez, R. Candal, et al., Influence of the spray pyrolysis seeding and growth parameters on the structure and optical properties of ZnO nanorod arrays, Mater. Chem. Phys. 151 (2015) 378–384. Copyright 2015, Elsevier.*

Fig. 4.8 shows the CL spectra measured at room temperature (RT) for ZnO nanorods grown on seeds deposited on FTO by SP using ethanol/water ratios of (A) $\Gamma = 0$, (B) $\Gamma = 0.06$, and (C) $\Gamma = 0.92$. The CL spectra show two major peaks: one at wavelengths around 370 nm and a much broader one above 450 nm. The peak at shorter wavelengths was attributed to near band edge (NBE) emission of the ZnO NRs, whereas the other one corresponds to deep level (DL) emission [24]. The NBE emission lines at RT for the ZnO nanorods fabricated with seeding conditions Γ of 0.0, 0.06, and 0.92 have their maximum at around 3.23, 3.36, and 3.27 eV, respectively. A value of 3.36 eV for the NBE peak position is in the range of the reported values for the gap of bulk ZnO [25], while the other values obtained may arise from

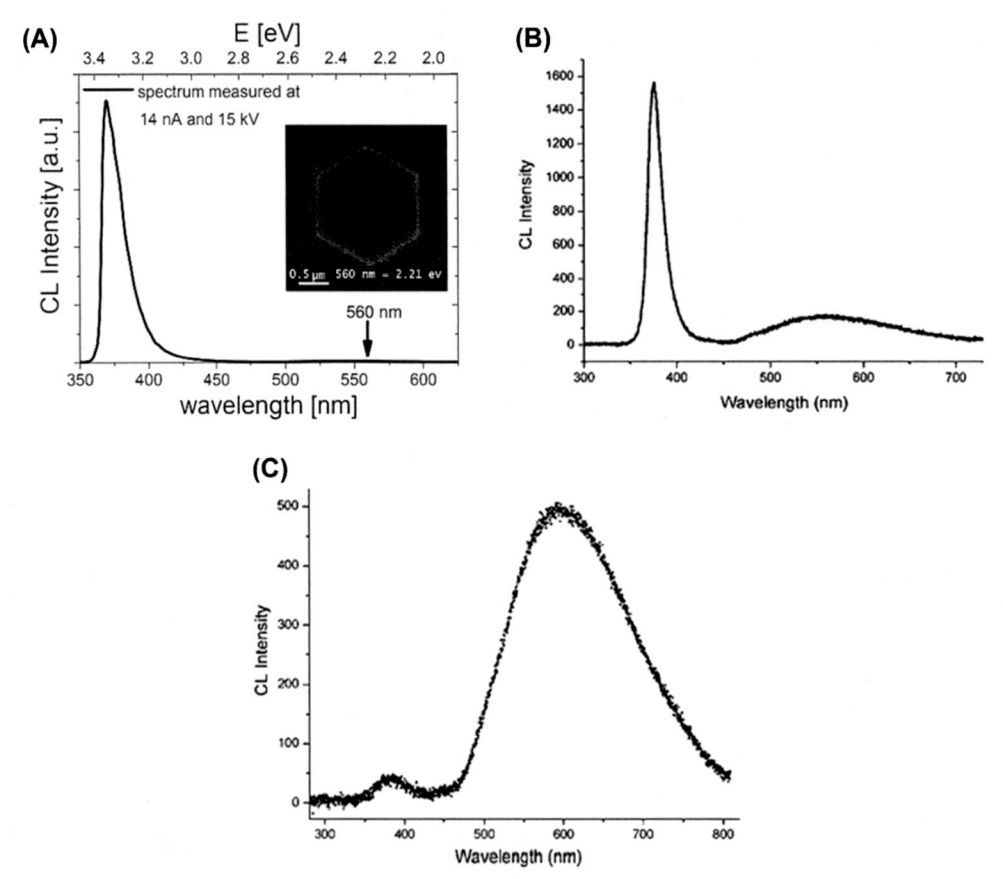

Figure 4.9 (A) Cathodoluminescence (*CL*) spectrum taken from ZnO microrods. The inset in (A) is the top-view monochromatic *CL* map of the individual ZnO microrod taken at 560 nm. (B) and (C) CL spectra for ZnO nanorods with average diameter of 120 nm and 30 nm, respectively. *(A) Reprinted with permission from A. Pieniążek, B.S. Witkowski, A. Reszka, M. Godlewski, B.J. Kowalski, Optical properties of ZnO microrods grown by a hydrothermal method — a cathodoluminescence study, Opt. Mater. Express 6 (2016) 3741–3750. Copyright 2016, The Optical Society; (B) and (C) Reprinted with permission from Y. Jiao, H.J. Zhu, X.F. Wang, L. Shi, Y. Liu, L.M. Peng, et al., A simple route to controllable growth of ZnO nanorod arrays on conducting substrates, CrystEngComm 12 (2010) 940–946. Copyright 2010, The Royal Society of Chemistry.*

other physical effects. The relative height of the DL emission increases dramatically for $\Gamma = 0.06$, where it becomes predominant compared to the NBE line. The CL image of the ZnO rod for $\Gamma = 0.06$ reveals inhomogeneities in the emission of the nanorods, as a consequence of two processes: the cluster mechanism and simultaneous dissolution [26], which produced thinner nanorods with higher defect density.

Other studies on the CL of ZnO nanorods [27,28] are shown in Fig. 4.9. CL spectrum and the CL monochromatic map taken at 560 nm are shown in Fig. 4.9A and in its inset, respectively. The results exhibit a lack of defect-related emission and the presence of a main peak of NBE emission at 3.35 eV (around 370 nm), which was originated from neutral donor bound exciton recombination [27]. CL spectra taken from ZnO nanorods with mean diameter around 120 nm and 30 nm are shown in Fig. 4.9B and C, respectively. It was found that depending on the diameter of the ZnO nanorod a strong band edge emission is observed for large diameters, whereas for nanorods with smaller diameters, band edge emission becomes weak and defect emission dominates in the spectra. Larger-diameter nanorods are found to possess higher band edge to defect emission ratio, indicating their better electronic structural quality, which is determined by both the growth conditions and the nanorod surface-to-volume ratio [28]. The defect emission from the thin nanorods covers a wide wavelength range starting from 450 to 800 nm, with maximum at centered at 600 nm. A similar broad defect emission but with very low intensity at 560 nm can also be observed for the wide nanorods.

Typical PL spectra of ZnO nanorods exhibit mainly of a near-UV excitonic emission peak (3.3 eV), which has been ascribed to the recombination of excitons through an exciton−exciton collision process [29,30]. In addition, ZnO usually exhibits visible luminescence which has been assigned to defects-related emissions [30−32]. The mechanisms behind the visible-light emission of ZnO nanomaterials have been widely investigated. Specially, the green PL emission with energies of 2.2−2.4 eV have been most commonly observed in ZnO and was generally attributed to originating from oxygen vacancies (Vo) into ZnO [30]. The emission peaks that appeared in the red region are due to the native defects like interstitial zinc atoms [31]. Blue emission is related to the electronic transmission from zinc interstitial donor energy level to the recipient of zinc vacancies, whereas yellow/orange emission is associated with excess oxygen [32]. Many studies on PL for ZnO nanorods [32−36], which were fabricated by distinct methods and varying different parameters, have been developed. For instance, Efafi et al. [33] investigated the optical properties of ZnO nanorods prepared with different growth solutions volumes: 25% (S1), 50% (S2), and 75% (S3) of autoclave volume. The results, illustrated in Fig. 4.10A, revealed that two main peaks were distinguishable in the PL spectrums, a strong peak in the UV and another broad one near the red region. Furthermore, the PL spectra of the samples denoted that the energy bandgap of the samples decreases as the volume of the growth solution increases in the autoclave. Another study [34] found that PL intensities of ZnO nanorods were directly related to the size of the rod due to the surface-related defects and that increasing the nanorod diameter, the intensity ratio of UV to visible emission increases. Meanwhile, the effect of annealing temperature and annealing time on PL in ZnO nanorods was investigated by Kurudirek et al. [35]. An optical study conducted for ZnO nanorods grown onto FTO from seed layers fabricated by SP ($\Gamma = 0$) and SG (P10) is shown in Fig. 4.10B. The spectra for the ZnO nanorods reveal two

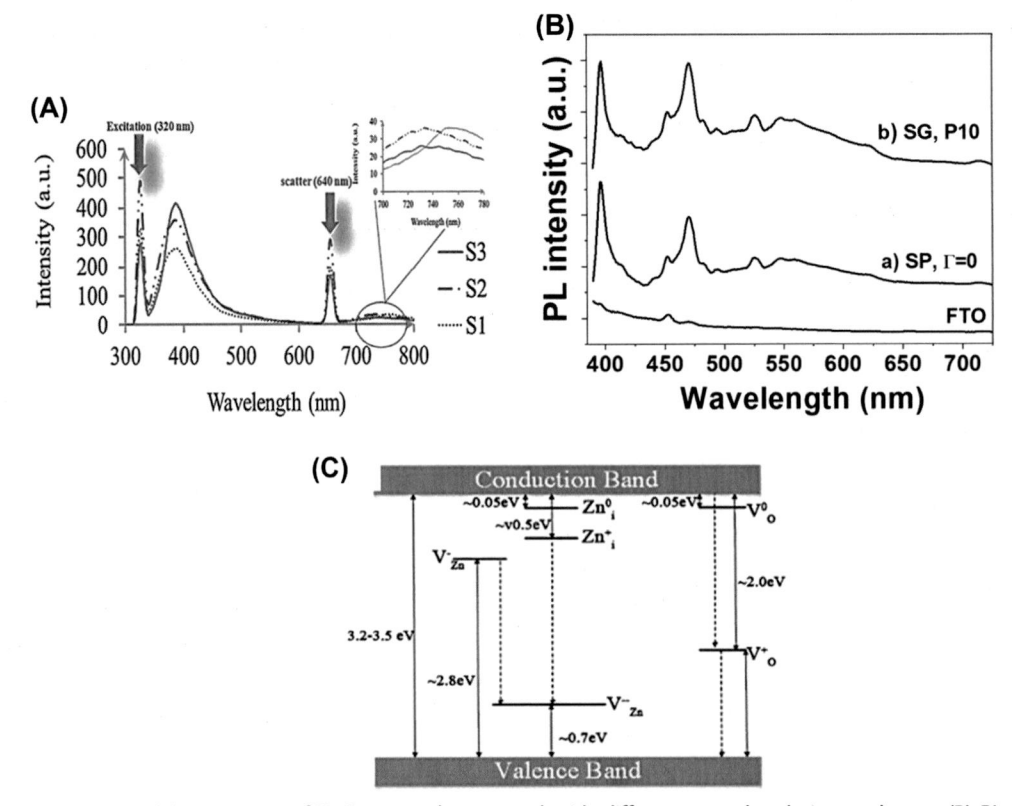

Figure 4.10 (A) PL spectra of ZnO nanorods prepared with different growth solutions volumes. (B) PL spectra of ZnO nanorods grown onto FTO from seed layers fabricated by SP-$\Gamma = 0$ and SG-P10. (C) Schematic electronic level diagram of ZnO. *FTO*, Fluorine-doped tin oxide; *PL*, photoluminescence; *SP*, spray-gel. *(A) Reprinted with permission from B. Efafi, H. Mazandarani, M.H.M. Ara, B. Ghafary, Improvement in photoluminescence behavior of well-aligned ZnO nanorods by optimization of thermodynamic parameters, Physica B Condens. Matter 579 (2020) 411915. Copyright 2020, Elsevier; (B) Reprinted with permission from J. Rodríguez, D. Onna, L. Sánchez, M.C. Marchi, R. Candal, S. Ponce, et al., The role of seeding in the morphology and wettability of ZnO nanorods films on different substrates, Appl. Surf. Sci. 279 (2013) 197−203. Copyright 2013, Elsevier; (C) Reprinted with permission from Y. Lv, Z. Zhang, J. Yan, W. Zhao, C. Zhai, J. Liu, Growth mechanism and photoluminescence property of hydrothermal oriented ZnO nanostructures evolving from nanorods to nanoplates, J. Alloys Compd. 718 (2017) 161−169. Copyright 2017, Elsevier.*

strong peaks at 397 and 470 nm with additional weak peaks around at 483, 494, and 526 nm, which are ascribed to oxygen vacancies and oxygen antisites [36,37]. PL spectra for SP and SG seeded films are similar, despite the different direction and organization of the grown NRs. The emission at 397 nm is assigned to shallow donor-related UV emission [38] and the unusually peak at 470 nm to DL emission related to intrinsic defects of the ZnO NRs, such as Zn vacancies or interstitial Zn or O [39].

A schematic electronic level diagram of ZnO was illustrated in Fig. 4.10C by Lv et al. [40]. According to the energy difference of the levels, and the different levels into the forbidden band can include the following transitions: BC $\rightarrow V^+$ at 1.75 eV (708 nm) and/or $V^+_0 \rightarrow$ BV at 1.75 eV (708 nm), and/or $V^{--}_{Zn} \rightarrow V^-_{Zn}$ at 2.1 eV (590 nm) [40].

4.4.2.2 ZnO-rGO nanorods

An example of ZnO and ZnO-rGO nanorods obtained by electrospinning assisted hydrothermal method is shown below. The FE-SEM image of pure ZnO nanorods is shown in Fig. 4.11A, whereas Fig. 4.11B–D shows the FE-SEM images of ZnO-rGO nanorods grown from seed layers fabricated by electrospinning using a spinning solution with 0.2 wt.% rGO and applying different spinning voltages. The FE-SEM images of ZnO-rGO NRs obtained from seed layers fabricated using different spinning solutions with 0.1, 0.2, and 0.3 wt.% rGO are shown in Fig. 4.11E–G. The higher magnification FE-SEM images showed as an inset in the lower-left corner of Fig. 4.11B–G and indicate that nanorods come together in a flower-like architecture [41,42]. When only ZnO layers were deposited on FTO, the diameters of the pure ZnO nanorods obtained (Z20) were in the range 42–58 nm. The adhesion of the rGO and the spinning voltage increase produce a decrease in the dispersion of diameter size and the mean diameter of the ZnO-rGO nanorods, with almost similar sizes. The average size of the ZnO-rGO NRs was: 34 ± 6 nm, 32 ± 3 nm, and 29 ± 3 nm for ZG20, ZG30, and ZG40, respectively. Meanwhile, the mean diameter of pure ZnO nanorods (50 ± 8 nm) decreases compared to the sizes obtained for ZnO-rGO nanorods fabricated from spinning solutions with 0.1 wt.% rGO (37 ± 5 nm) and 0.2 wt.% rGO (34 ± 6 nm). Unlike the average size of ZnO-rGO nanorods fabricated from spinning solutions with 0.3 wt.% rGO (49 ± 9 nm), which has a similar size value compared to pristine ZnO nanorods. It should be noted that the average size of ZnO-rGO decreases compared to the average size of pure ZnO nanorods, due to two reasons: (1) the presence of rGO in the spinning solution, which increased its electrical conductivity [43] and (2) the increase of the spinning voltage [44].

Measuring PL is an effective means of investigating the separation process of e$^-$ and h$^+$ pairs in semiconductors. It is demonstrated by several studies [45–48] that the use of rGO improves the separation efficiency of photoexcited electron–hole pairs generated from ZnO and boosts the efficient transfer of charge carriers. For example, Ding et al. [45] found a quench in PL spectra caused by the coating with rGO sheets in the ZnO nanostructures (see Fig. 4.12A). The quenching behavior could be explained by interfacial charge transfer from ZnO to graphene [49]. Likewise, in other study [47] when ZnO was adhered to rGO, the photoemission intensity of samples fell down, which was caused by the quenching of photoemission, as seen in Fig. 4.12B. What is more, with increasing the content of rGO, the quenching

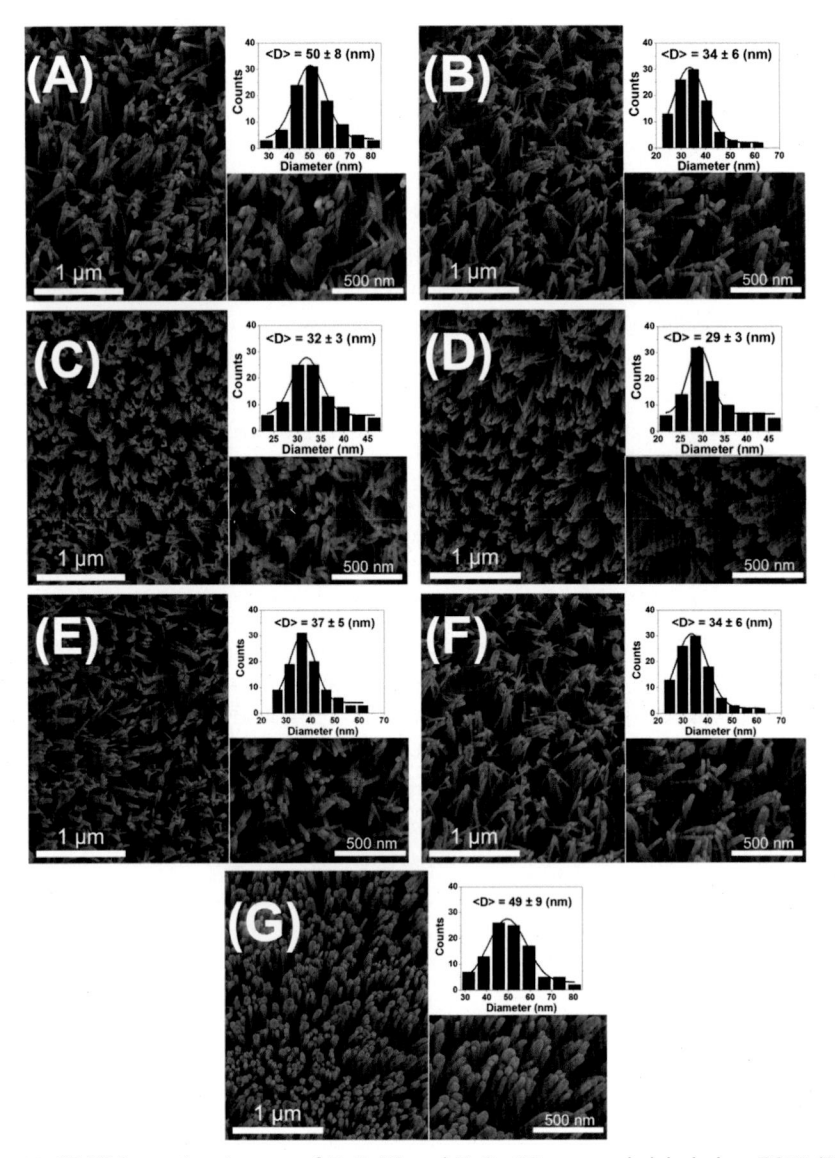

Figure 4.11 FE-SEM top-view images of ZnO (A) and ZnO-rGO nanorods labeled as ZG20 (B), ZG30 (C), ZG40 (D), ZG1 (E), ZG2 (F) and ZG3 (G). The size distributions are shown as insets. *Reprinted with permission from P.G. Ramos, C. Luyo, L.A. Sánchez, E.D. Gomez, J.M. Rodriguez, The spinning voltage influence on the growth of ZnO-rGO nanorods for photocatalytic degradation of methyl orange dye, Catalysts 10 (2020) 660. Copyright 2020, MDPI and with permission from P.G. Ramos, E. Flores, C. Luyo, L.A. Sánchez, J. Rodriguez, Fabrication of ZnO-RGO nanorods by electrospinning assisted hydrothermal method with enhanced photocatalytic activity, Mater. Today Commun. 19 (2019) 407–412. Copyright 2019, Elsevier.*

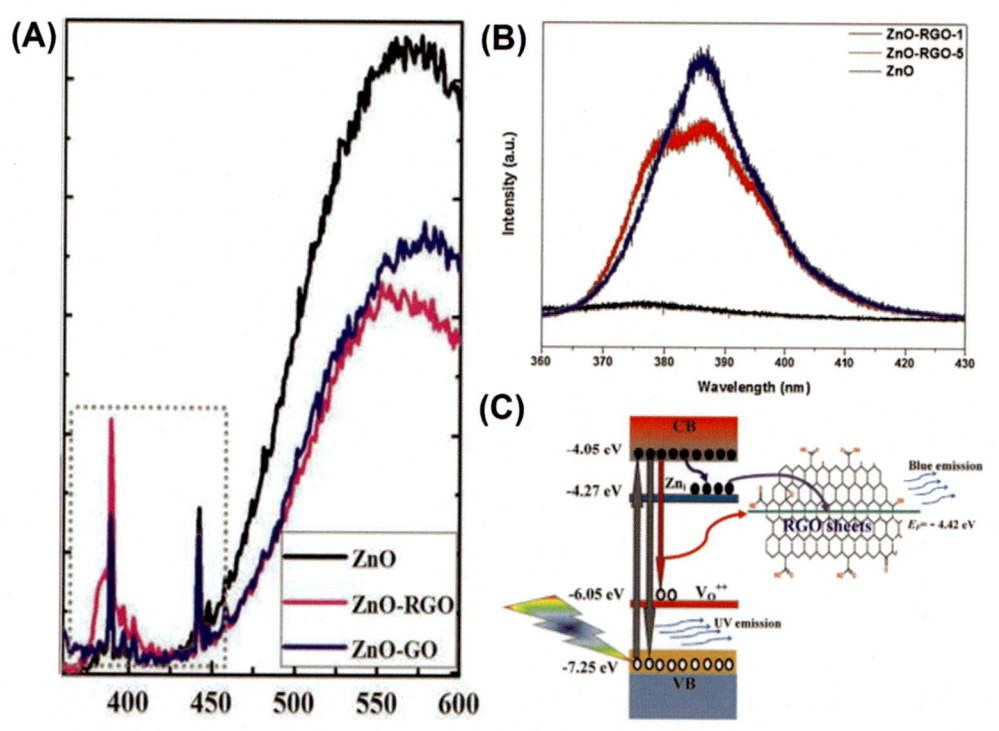

Figure 4.12 (A) Comparison of relative photoluminescence intensity of ZnO, ZnO-GO, ZnO-rGO nanorods. (B) Photoluminescence spectra of as-prepared ZnO-rGO-1, ZnO-rGO-5 and ZnO. (C) The band diagram demonstrating interfacial charge transfer that takes place from the trapping states (Zni) of ZnO to rGO. *(A) Reprinted with permission from J. Ding, M. Wang, J. Deng, W. Gao, Z. Yang, C. Ran, et al., A comparison study between ZnO nanorods coated with graphene oxide and reduced graphene oxide, J. Alloys Compd. 582 (2014) 29–32. Copyright 2014, Elsevier. (B) Reprinted with permission from X. Zhou, T. Shi, H. Zhou, Hydrothermal preparation of ZnO-reduced graphene oxide hybrid with high performance in photocatalytic degradation, Appl. Surf. Sci. 258 (2012) 6204–6211. Copyright 2012, Elsevier. (C) Reprinted with permission from G. Jayalakshmi, K. Saravanan, J. Pradhan, P. Magudapathy, B.K. Panigrahi, Facile synthesis and enhanced luminescence behavior of ZnO:reduced graphene oxide(rGO) hybrid nanostructures, J. Lumin. 203 (2018) 1–6. Copyright 2018, Elsevier.*

phenomenon of ZnO–rGO-1 became much more different than that of ZnO–rGO-5, which told photoelectron transfer from ZnO to rGO more effective to impede the combination of photoinduced electrons and holes [47]. The work concludes that the rGO sheets played an important role in accepting the photoelectrons from ZnO.

Other cases of the influence of rGO in the PL spectra of ZnO nanorods were demonstrated by Ramos et al. [41,42]. The results showed in Fig. 4.13 reveal that all samples have the same emission shape with three emission peaks located around 385, 600, and 760 nm. NBE emission around 385 nm (Fig. 4.13B and D) is attributed to the free exciton recombination process [41]. Furthermore, a blue-shift in this peak was

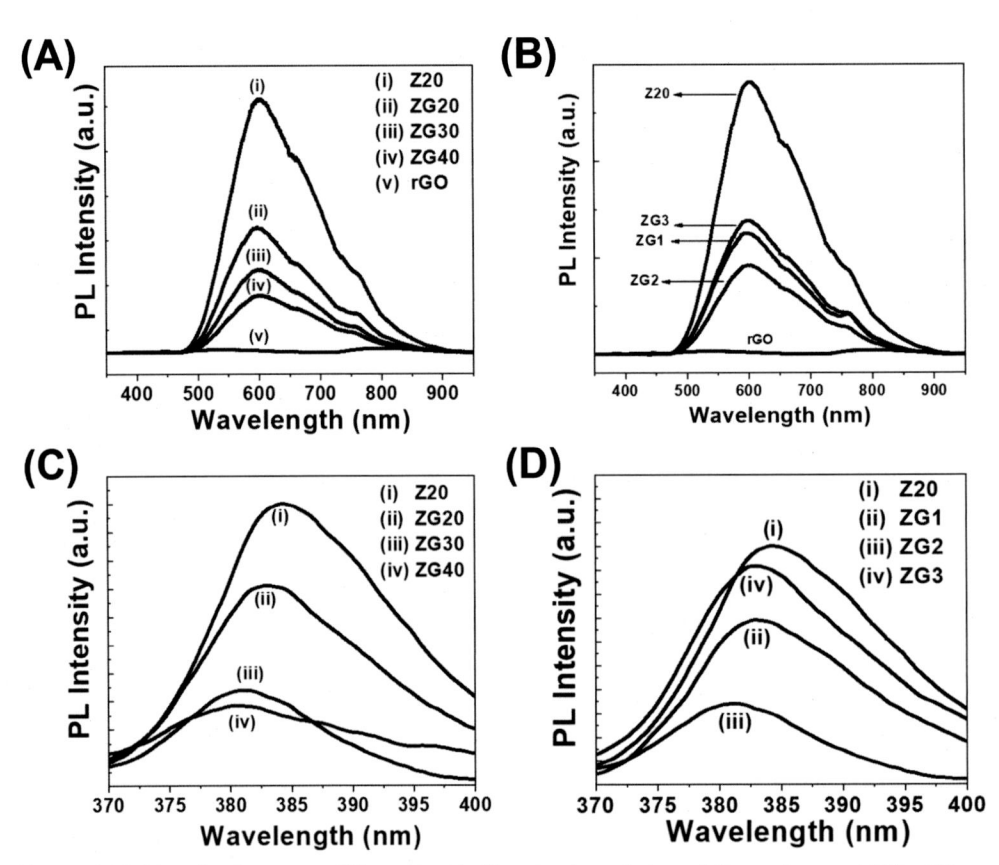

Figure 4.13 Photoluminescence (PL) spectra (A, C) and enlarged view of PL spectrums at 370—400 nm (B, D) of pure ZnO and ZnO-rGO nanorods grown from seed layers fabricated by electrospinning using different spinning voltages and amounts of rGO, respectively. *Reprinted with permission from P.G. Ramos, C. Luyo, L.A. Sánchez, E.D. Gomez, J.M. Rodriguez, The spinning voltage influence on the growth of ZnO-rGO nanorods for photocatalytic degradation of methyl orange dye, Catalysts 10 (2020) 660. Copyright 2020, MDPI and with permission from P.G. Ramos, E. Flores, C. Luyo, L.A. Sánchez, J. Rodriguez, Fabrication of ZnO-RGO nanorods by electrospinning assisted hydrothermal method with enhanced photocatalytic activity, Mater. Today Commun. 19 (2019) 407—412. Copyright 2019, Elsevier.*

observed for ZnO-rGO NRs compared to ZnO NRs, due to the presence of rGO in the fabricated ZnO-rGO NRs [48]. The emission peak in the visible region at ∼600 nm arises from single ionized oxygen vacancy in ZnO nanorods and due to the recombination of electron—hole pairs from the large number of localized sp^2-carbon clusters embedded within the sp^3 matrix [50]. While, the emission peak in the near-infrared region at ∼760 nm was assigned to the second-order diffraction of the NBE emission band [42]. Moreover, the emission peaks for ZnO–rGO nanorods decrease in intensities compared to ZnO emission peaks, which is associated with the presence of rGO in the nanostructures [51,52] and the decreasing in the diameters of NRs (see Fig. 4.11) due to the increase of spinning voltages.

Fig. 4.12C shows the band diagram demonstrating interfacial charge transfer that takes place from the trapping states of ZnO to rGO. The enhanced UV and the suppressed visible emission in ZnO–rGO hybrid nanostructures are attributed to the Zn^{2+} ions on the surface of the nanorods bonded with the negatively charged rGO sheets through the electrostatic interactions leading to the local creation of $Zn-O-C$ bonding. The interfacial charge transfer from ZnO to rGO takes place through the defect levels due to the favorable matching of energy levels. In the energy level diagram of ZnO, the CB and VB of ZnO lie about 4.05 and 7.25 eV (vs vacuum), respectively, defect states such as Zn_i and V_o^{++} lies 0.22 and 2 eV below the CB, and the work function of rGO is 4.42 eV [48]. The energy level of Zn_i in ZnO nanostructures is comparable with the Fermi energy of the rGO facilitates the electron transfer from the Zn_i trap states of ZnO to rGO, thus the synergetic effect from both the components enhances the effective radiative recombination of electrons results in improved UV and blue emissions. The quenching of visible emission in ZnO–rGO hybrid nanostructures refers to the charge transfer from the trap states of ZnO to rGO occurs at the interface, which can enhance the photocatalytic activity of the obtained nanorods [51,52].

4.5 Photocatalytic zinc oxide-based nanostructures

4.5.1 Zinc oxide nanofibers

The photocatalytic activities of the ZnO nanofibers fabricated by electrospinning using spinning solutions with different zinc acetate/PVA ratios were shown in Fig. 4.14A. The results indicate that the ZnO nanofibers fabricated with a zinc acetate/PVA mass ratio of 2:3 show the highest photocatalytic activity compared with other ZnO nanostructures fabricated with zinc acetate/PVA mass ratios of 1:2, 1:1, and 3:2. The degradation efficiency of the photocatalyst fabricated with a zinc acetate/PVA mass ratio of 2:3 shows a maximum degradation of ~83% at 6 h, whereas the degradation efficiency of ZnO nanofibers fabricated with zinc acetate/PVA mass ratios of 1:2, 1:1, and 3:2 is ~66%, ~79%, and ~43% at 6 h of irradiation time, respectively. The highest photocatalytic activity for the ZnO nanostructures fabricated from a mass ratio between zinc acetate and PVA of 2:3 is due to its larger specific surface area, which in turn relies on its small particle size (~47 nm), compared with the specific surface area and particle size of the ZnO nanostructures fabricated with zinc acetate/PVA mass ratios of 3:2 and 1:1 (see Fig. 4.2).

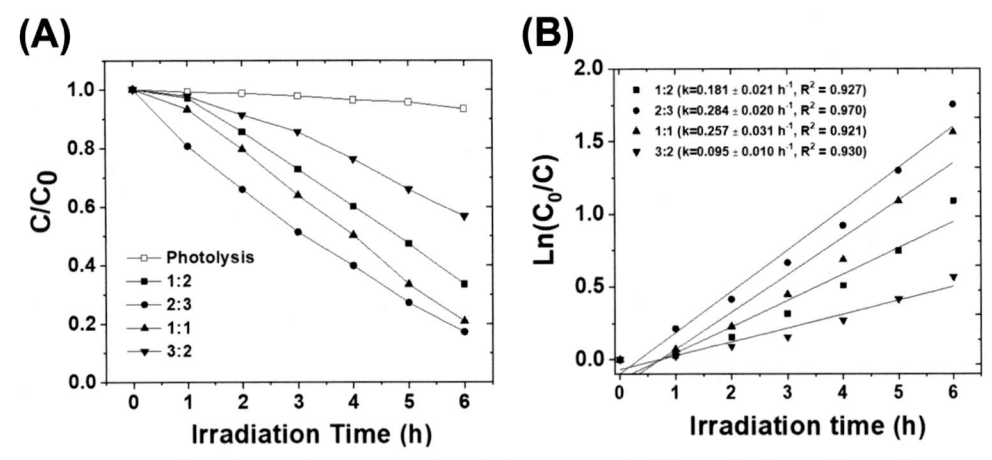

Figure 4.14 (A) Photodegradation curves of methyl orange with different ZnO nanofibers photocatalysts. (B) Kinetics plots calculated from (A) for ZnO nanofibers photocatalysts. *Reprinted with permission from P.G. Ramos, E. Flores, L.A. Sánchez, R.J. Candal, M. Hojamberdiev, et al., Enhanced photoelectrochemical performance and photocatalytic activity of ZnO/TiO₂ nanostructures fabricated by an electrostatically modified electrospinning, Appl. Surf. Sci. 426 (2017) 844–851. Copyright 2017, Elsevier.*

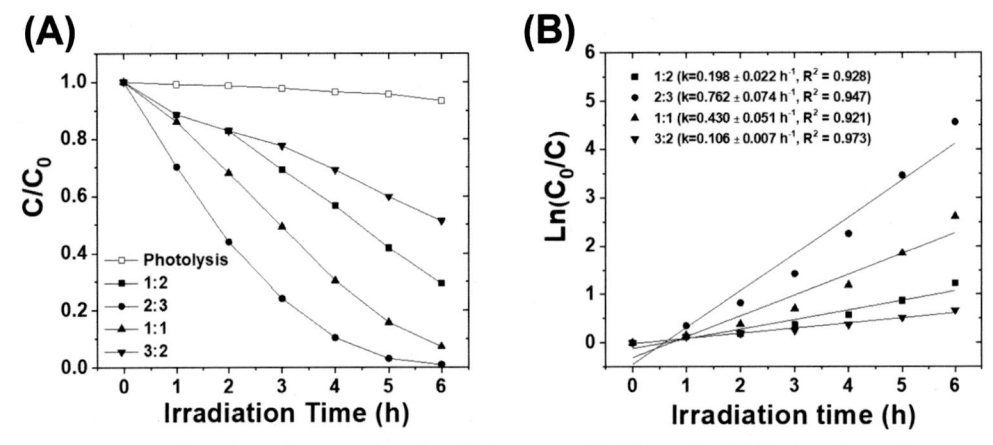

Figure 4.15 (A) Photodegradation curves of methyl orange with different ZnO/TiO$_2$ nanofibers photocatalysts. (B) Kinetics plots calculated from (A) for ZnO/TiO$_2$ nanofibers photocatalysts. *Reprinted with permission from P.G. Ramos, E. Flores, L.A. Sánchez, R.J. Candal, M. Hojamberdiev, et al., Enhanced photoelectrochemical performance and photocatalytic activity of ZnO/TiO$_2$ nanostructures fabricated by an electrostatically modified electrospinning, Appl. Surf. Sci. 426 (2017) 844–851. Copyright 2017, Elsevier.*

4.5.2 Zinc oxide—titanium dioxide nanofibers

The photocatalytic activities of ZnO/TiO$_2$ films fabricated by electrospinning using spinning solutions with different zinc acetate/PVA ratios were evaluated and the results are plotted in Fig. 4.15A. The results indicate an improvement in the photocatalytic activity for the ZnO/TiO$_2$ nanofibers compared with the pure ZnO nanofibers. The enhancement of photocatalytic efficiency is due to effective charge separation caused by the coupling of ZnO nanofibers with TiO$_2$-P25 nanoparticles. The ZnO/TiO$_2$ nanofibers fabricated with a zinc acetate/PVA mass ratio of 2:3 show the highest photocatalytic activity compared with the other ZnO/TiO$_2$ nanofibers. The degradation efficiency of the ZnO/TiO$_2$ photocatalyst fabricated with a zinc acetate/PVA mass ratio of 2:3 shows a maximum degradation of ∼99% at 6 h, whereas the degradation efficiency of ZnO/TiO$_2$ nanofibers fabricated with zinc acetate/PVA mass ratios of 1:2, 1:1 and 3:2 is ∼71%, ∼93%, and ∼49% at 6 h of irradiation time, respectively. The highest photocatalytic activity for the ZnO/TiO$_2$ nanostructures fabricated from a mass ratio between zinc acetate and PVA of 2:3 is due to the best inhibition of photogenerated charge carriers recombination and by its larger specific surface area [53], which in turn relies on its small particle size (∼36 nm), compared with the ZnO/TiO$_2$ nanostructures fabricated with zinc acetate/PVA mass ratios of 3:2 and 1:1.

4.5.3 Zinc oxide-reduced graphene oxide nanorods

The photocatalytic dye degradation performances of the ZnO and ZnO-rGO nanorods grown from seed layers fabricated by electrospinning using different spinning voltages and amounts of rGO are showed in Fig. 4.16A and C, respectively. The degradation profile is plotted as C/C_0 versus irradiation time, where C is the concentration of MO at the irradiation time (t in h) and C_0 represents the original concentration of MO. The ZG40 photocatalyst shows the highest photocatalytic activity compared with Z20, ZG20, and ZG30 photocatalysts. The degradation efficiency of the photocatalyst based on ZG40 nanorods shows a maximum degradation of ∼99% at 6 h, whereas the degradation efficiency of Z20, ZG20, and ZG30 is ∼77%, ∼95%, and ∼97% at 7 h of irradiation time, respectively. Meanwhile, the ZnO-rGO nanorods fabricated with 0.2 wt.% rGO show the highest photocatalytic activity compared with the other ZnO-rGO nanorods fabricated with 0.1 and 0.3 wt.% rGO. The enhancement of photocatalytic performance of ZnO-rGO NRs as compared to pristine ZnO NRs is mainly attributed to the fact that the photogenerated electrons from ZnO excited by a light source are trapped by the rGO, avoiding recombination of electron−hole pairs [54], and also to the increase in specific surface area that in turn relies on the decrease in particle size (see in Fig. 4.10), caused by the increase of voltage during electrospinning [55]. However, it should be noted that there is no linear dependence between the rGO amount added and the photocatalytic activity, a high content of rGO (0.3 wt.%) led to a decrement in the photocatalytic efficiency. Indeed, as reported by Fang et al. [56], an excessive rGO amount, can promote the recombination of pairs in rGO, reducing the photocatalytic activity of the photocatalysts. The results indicate that an optimal rGO amount is beneficial to achieve greater dye degradation efficiency [56].

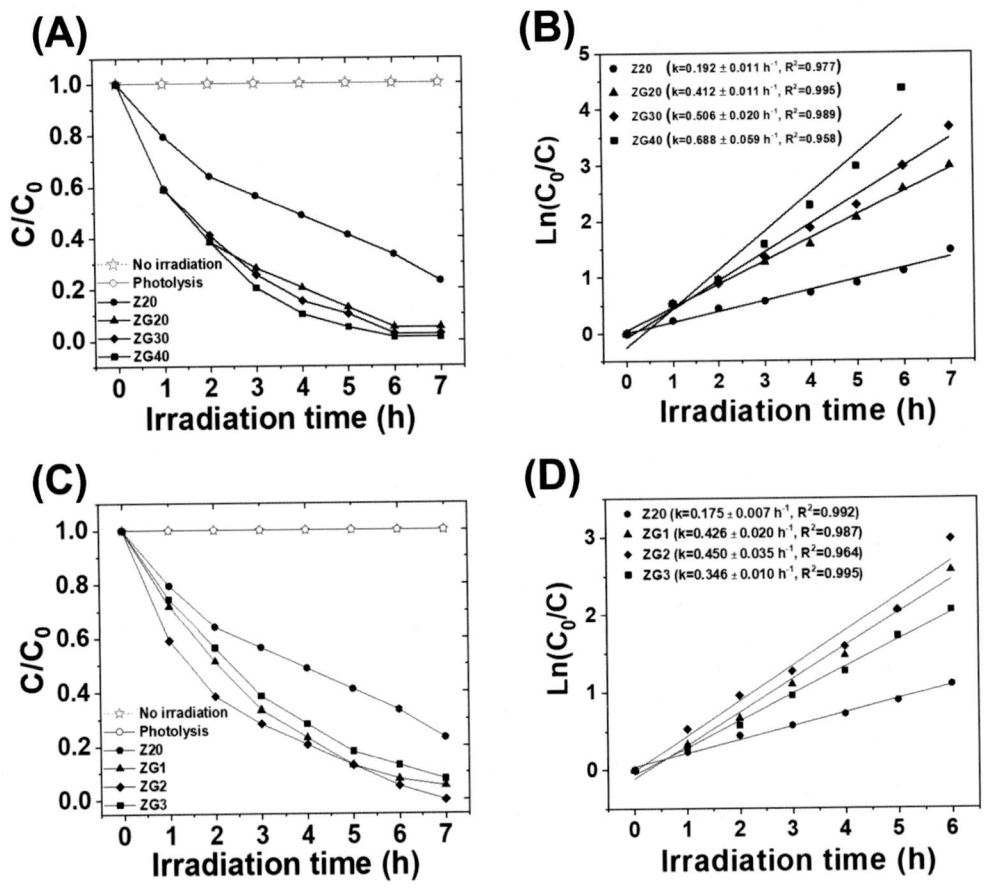

Figure 4.16 Photodegradation curves of methyl orange (A, C) and rate constants of photodegradation of MO (B, D) over ZnO-rGO nanorods photocatalysts grown from seed layers fabricated by electrospinning using different spinning voltages and amounts of rGO. *Reprinted with permission from P.G. Ramos, C. Luyo, L.A. Sánchez, E.D. Gomez, J.M. Rodriguez, The spinning voltage influence on the growth of ZnO-rGO nanorods for photocatalytic degradation of methyl orange dye, Catalysts 10 (2020) 660. Copyright 2020, MDPI and with permission from P.G. Ramos, E. Flores, C. Luyo, L.A. Sánchez, J. Rodriguez, Fabrication of ZnO-RGO nanorods by electrospinning assisted hydrothermal method with enhanced photocatalytic activity, Mater. Today Commun. 19 (2019) 407–412. Copyright 2019, Elsevier.*

4.6 Photocatalytic kinetic rate

The photodegradation reactions can be classified as first-order kinetic reactions [57], when the data are plotted as logarithms of the time-dependent pollutant concentration, linear plots are obtained their equation is given by $\ln(C_0/C_t) = kt$ where C_0 is the initial absorbance, C_t is the absorbance after a time t, and k, is the first-order rate constant. A plot between $\ln(C_0/C)$ and reaction time for ZnO nanofibers, ZnO/TiO$_2$ nanofibers are shown in Figs. 4.14B and 4.15B, respectively. While for ZnO-rGO nanorods grown from seed layers fabricated by electrospinning using different spinning voltages and amounts of rGO, are shown in Fig. 4.16B and D, respectively. Fig. 4.14B shows that depending on zinc acetate/PVA mass ratio, the estimated MO degradation rate constants for the ZnO nanostructures increases in the following order: $0.095 \pm 0.010 \text{ h}^{-1} < 0.181 \pm 0.021 \text{ h}^{-1} < 0.257 \pm 0.031 \text{ h}^{-1} < 0.284 \pm 0.020 \text{ h}^{-1}$ for zinc acetate/PVA mass ratios of 3:2, 1:2, 1:1 and 2:3, respectively. Furthermore, the calculated values of the correlation coefficient (R^2) for ZnO nanostructures, fabricated from spinning solutions with different zinc acetate/PVA ratios of 1:2, 2:3, 1:1, and 3:2, were 0.927, 0.970, 0.921, and 0.930, respectively. The estimated MO degradation rate constants for the ZnO/TiO$_2$ nanostructures are show in Fig. 4.15B and increases in the following order: $0.106 \pm 0.007 \text{ h}^{-1} < 0.198 \pm 0.022 \text{ h}^{-1} < 0.430 \pm 0.051 \text{ h}^{-1} < 0.762 \pm 0.074 \text{ h}^{-1}$ for zinc acetate/PVA mass ratios of 3:2, 1:2, 1:1, and 2:3, respectively. In addition, the calculated R^2 for ZnO/TiO$_2$ nanostructures, fabricated from spinning solutions with different zinc acetate/PVA ratios of 1:2, 2:3, 1:1 and 3:2, were 0.928, 0.947, 0.921, and 0.973, respectively. The estimated degradation rate constants for the Z20, ZG20, ZG30, and ZG40 nanorods exhibit a linear dependence with the spinning voltage and the values obtained were found to be $0.192 \pm 0.011 \text{ h}^{-1}$, $0.412 \pm 0.011 \text{ h}^{-1}$, $0.506 \pm 0.020 \text{ h}^{-1}$, and $0.688 \pm 0.059 \text{ h}^{-1}$, respectively (Fig. 4.16B). The calculated values of R^2 for Z20, ZG20, ZG30, and ZG40 nanorods were 0.977, 0.995, 0.989, and 0.958, respectively, whereas for the ZnO-rGO nanorods grown from seed layers fabricated by electrospinning using different amounts of rGO, the estimated degradation rate constants for the ZnO-rGO NRs were found to be 0.175 ± 0.007, 0.426 ± 0.020, 0.450 ± 0.035, and $0.346 \pm 0.010 \text{ h}^{-1}$ for 0, 0.5, 1.0, and 1.5 wt.% rGO, respectively (Fig. 4.16D). The calculated values of R^2 for ZnO-rGO NRs grown on ZnO-rGO seed layers obtained by electrospinning using spinning solutions with 0, 0.5, 1.0, and 1.5 wt.% rGO were 0.992, 0.987, 0.964, and 0.995, respectively. The R^2 values close to 1 for ZnO, ZnO/TiO$_2$ nanofibers and for all the ZnO-rGO nanorods obtained demonstrates that the degradation process of MO through these nanostructures follows first-order reaction kinetic.

4.7 Photocatalytic mechanism for nanostructures

4.7.1 ZnO/TiO$_2$ nanofibers photocatalytic mechanism

The mechanism of charge separation and photocatalytic reaction for ZnO/TiO$_2$ nanostructures was proposed and represented in Fig. 4.17. As illustrated in the scheme, when the nanostructure is irradiated with UV-light irradiation with a photon energy higher or equal to bandgap energy of ZnO and TiO$_2$, the electrons (e$^-$) are excited from the VB to the CB with generation simultaneous of the same number of holes (h$^+$) in the VB (Eq. 4.2). The electron transfer occurs from the CB of ZnO to the CB of TiO$_2$, and conversely, the hole transfer takes place from the VB of TiO$_2$ to the VB of ZnO

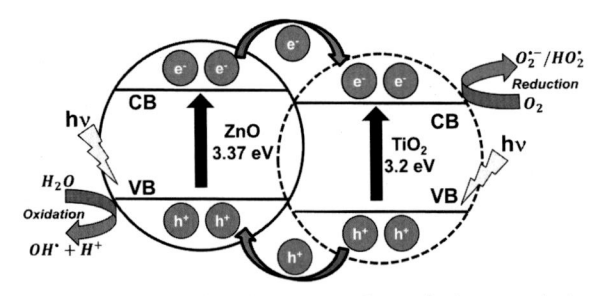

Figure 4.17 A schematic diagram for the charge-transfer and photocatalytic process of ZnO/TiO$_2$ nanofibers for methyl orange degradation. *Reprinted with permission from P.G. Ramos, E. Flores, L.A. Sánchez, R.J. Candal, M. Hojamberdiev, et al., Enhanced photoelectrochemical performance and photocatalytic activity of ZnO/TiO$_2$ nanostructures fabricated by an electrostatically modified electrospinning, Appl. Surf. Sci. 426 (2017) 844–851. Copyright 2017, Elsevier.*

(Eq. 4.3) [15]. The photogenerated electrons may be transferred from the CB to the dissolved molecular oxygen to produce superoxide anion radicals $O_2^{\bullet-}$, while the photogenerated holes at the VB of the photocatalyst oxidized adsorbed water (H$_2$O) molecules to yield •OH radicals. The hydroxyl radical (•OH) is a highly reactive, strong oxidizing chemical species for the partial or complete mineralization of organic dyes. Moreover, due to the synergistic effects between ZnO and TiO$_2$, the recombination rate of the photogenerated electrons and holes was suppressed [58] and the lifetime of the photogenerated charge carriers was increased in the composite nanostructures. Therefore increases the photocatalytic activity of ZnO/TiO$_2$ nanostructures obtained in this research. The overall reactions under the UV–light and photocatalytic reactions are shown below:

$$ZnO/TiO_2 + h\nu(UV) \rightarrow ZnO/TiO_2(e_{CB}^- + h_{VB}^+), \tag{4.1}$$

$$ZnO(e_{CB}^-) \rightarrow TiO_2(e_{CB}^-), \tag{4.2}$$

$$TiO_2(h_{VB}^+) \rightarrow ZnO(h_{VB}^+), \tag{4.3}$$

$$TiO_2(e_{CB}^-) + O_{2(ads)} \rightarrow O_2^{\bullet-}, \tag{4.4}$$

$$O_2^{\bullet-} + H_2O \rightarrow \bullet OH + HO_2^{\bullet} + H_2O_2, \tag{4.5}$$

$$ZnO(h_{VB}^+) + H_2O/OH^- \rightarrow H^+ + OH^{\bullet}, \tag{4.6}$$

$$\bullet OH \ or \ h_{VB}^+ + pollutant \ (MO) \rightarrow degradation \ products. \tag{4.7}$$

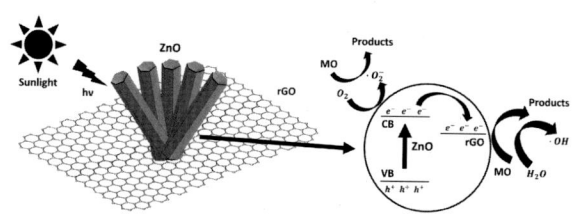

Figure 4.18 A schematic diagram illustrating the principle of charge separation and photocatalytic degradation of MO over ZnO-rGO nanorods. *Reprinted with permission from P.G. Ramos, C. Luyo, L.A. Sánchez, E.D. Gomez, J.M. Rodriguez, The spinning voltage influence on the growth of ZnO-rGO nanorods for photocatalytic degradation of methyl orange dye, Catalysts 10 (2020) 660. Copyright 2020, MDPI.*

4.7.2 ZnO-rGO nanorods photocatalytic mechanism

The possible photocatalytic degradation mechanism of MO in the presence of ZnO-rGO NRs was proposed, as illustrated in Fig. 4.18. When the ZnO nanorods are irradiated from the UV-light source, the excited electrons (e$^-$) migrate from the VB of ZnO to its CB with the simultaneous generation of the same number of holes (h$^+$) in the VB (Eq. 4.9). The rGO sheets receive these photoexcited electrons from the CB of ZnO nanorods and act as an electron transporter phase hindering the electron–hole recombination [51,59]. This is mainly due to the difference in energy levels between rGO and ZnO. The value obtained for the ZnO CB is -4.05 eV, which is higher than the value of the work function of rGO (-4.42 eV) [59]. The photogenerated electron and holes react with oxygen molecules and with the surface absorbed water molecules, respectively, as described in Eqs. (4.8)–(4.14). These chemical reactions produce the strongly oxidizing hydroxyl radical ($\bullet$OH) and oxygen radical anion ($O_2^{\bullet-}$), which react with the MO dye molecules in solution and degrade them [58,59]. Based on these results we conclude that the variation of spinning voltages and the presence of rGO enhance the photocatalytic activity performance of pure ZnO nanorods, suppressing photoinduced charge recombination effectively. Finally, the overall chemical and photocatalytic reactions are shown below:

$$ZnO/rGO + h\nu \rightarrow ZnO/rGO(e_{CB}^- + h_{VB}^+), \tag{4.8}$$

$$ZnO + h\nu \rightarrow ZnO\left(e_{CB}^- + h_{VB}^+\right), \tag{4.9}$$

$$ZnO(e_{CB}^-) \rightarrow rGO(e_{tr}^-), \tag{4.10}$$

$$ZnO/rGO(h_{VB}^+) + OH^- \rightarrow \frac{ZnO}{rGO} + \bullet OH, \tag{4.11}$$

$$rGO\left(e_{tr}^{-}\right) + O_{2(ads)} \rightarrow O_2^{\bullet-}, \tag{4.12}$$

$$O_2^{\bullet-} + \text{pollutent (MO)} \rightarrow \text{decomposed products}, \tag{4.13}$$

$$\bullet OH + \text{pollutent (MO)} \rightarrow \text{decomposed products}. \tag{4.14}$$

References

[1] S. Natarajan, H.C. Bajaj, R.J. Tayade, Recent advances based on the synergetic effect of adsorption for removal of dyes from waste water using photocatalytic process, J. Environ. Sci. 65 (2018) 201–222.

[2] E.C. Pastrana, V. Zamora, D. Wang, H. Alarcón, Fabrication and characterization of α-Fe_2O_3/CuO heterostructure thin films via dip-coating technique for improved photoelectrochemical performance, Adv. Nat. Sci. Nanosci. Nanotechnol. 10 (2019) 035012.

[3] L. Wang, J. Zhao, H. Liu, J. Huang, Design, modification and application of semiconductor photocatalysts, J. Taiwan Inst. Chem. Eng. 93 (2018) 590–602.

[4] I.J. Ani, U.G. Akpan, M.A. Olutove, B.H. Hameed, Photocatalytic degradation of pollutants in petroleum refinery wastewater by TiO_2- and ZnO-based photocatalysts: Recent development, J. Clean. Prod. 205 (2018) 930–954.

[5] A.C. Gandhi, W.-S. Yeoh, M.-A. Wu, C.-H. Liao, D.-Y. Chiu, W.-L. Yeh, et al., New insights into the role of weak electron–phonon coupling in nanostructured ZnO thin films, Nanomaterials 8 (2018) 632.

[6] N. Geetha, S. Sivaranjani, A. Ayeshamariam, J.S. Kissinger, M. Valan Arasu, M. Jayachandran, ZnO doped oxide materials: mini review, Fluid Mech. Open Access 3 (2016) 1000141.

[7] J. Aliaga, N. Cifuentes, G. González, C. Sotomayor-Torres, E. Benavente, Enhancement photocatalytic activity of the heterojunction of two-dimensional hybrid semiconductors ZnO/V_2O_5, Catalysts 8 (2018) 374.

[8] X. Dong, P. Yang, Y. Liu, C. Jia, D. Wang, J. Wang, et al., Morphology evolution of one-dimensional ZnO nanostructures towards enhanced photocatalysis performance, Ceram. Int. 42 (2016) 518–526.

[9] S. Goel, B. Kumar, A review on piezo-/ferro-electric properties of morphologically diverse ZnO nanostructures, J. Alloys Compd. 816 (2020) 152491.

[10] S. Danwittayakul, S. Songngam, S. Sukkasi, Enhanced solar water disinfection using ZnO supported photocatalysts, Environ. Technol. 41 (2020) 349–356.

[11] P. Dash, A. Manna, N.C. Mishra, S. Varma, Synthesis and characterization of aligned ZnO nanorods for visible light photocatalysis, Physica E Low Dimens. Syst. Nanostruct 107 (2019) 38–46. Available from: https://doi.org/10.1016/j.physe.2018.11.007.

[12] J. Rodríguez, G. Feuillet, F. Donatini, D. Onna, L. Sanchez, R. Candal, et al., Influence of the spray pyrolysis seeding and growth parameters on the structure and optical properties of ZnO nanorod arrays, Mater. Chem. Phys. 151 (2015) 378–384.

[13] J. Rodríguez, D. Onna, L. Sánchez, M.C. Marchi, R. Candal, S. Ponce, et al., The role of seeding in the morphology and wettability of ZnO nanorods films on different substrates, Appl. Surf. Sci. 279 (2013) 197–203.

[14] P.G. Ramos, N.J. Morales, R.J. Candal, M. Hojamberdiev, J. Rodriguez, Influence of zinc acetate content on the photoelectrochemical performance of zinc oxide nanostructures fabricated by electrospinning technique, Nanomater. Nanotechnol. 6 (2016) 1–7.

[15] P.G. Ramos, E. Flores, L.A. Sánchez, R.J. Candal, M. Hojamberdiev, et al., Enhanced photoelectrochemical performance and photocatalytic activity of ZnO/TiO_2 nanostructures fabricated by an electrostatically modified electrospinning, Appl. Surf. Sci. 426 (2017) 844–851.

[16] J. Walser, S.J. Ferguson, Oriented nanofibrous membranes for tissue engineering applications: electrospinning with secondary field control, J. Mech. Behav. Biomed. Mater 58 (2016) 188–198.

[17] P.G. Ramos, N.J. Morales, S. Goyanes, R.J. Candal, J. Rodríguez, Moisture-sensitive properties of multi-walled carbon nanotubes/polyvinyl alcohol nanofibers prepared by electrospinning electrostatically modified method, Mater. Lett. 185 (2016) 278–281.

[18] P.G. Ramos, L. Sánchez, J. Rodriguez, Photoactive hybrid ZnO/N-Ag-TiO$_2$ films for photocatalytic water purification: nanofibers vs nanorods, in: P. Banerjee, K. Gudmundsson, A. Lakhtakia, G. Subramanyam (Eds.), Proceedings of the Third International Workshop on Thin Films for Electronics, Electro-Optics, Energy, and Sensors, Reykjavik, June 24–June 26, 2019, Proc. SPIE 11371 (2019) 1137106.

[19] K. Siwińska-Stefańska, A. Kubiak, A. Piasecki, J. Goscianska, et al., TiO$_2$-ZnO binary oxide systems: comprehensive characterization and tests of photocatalytic activity, Materials. 11 (5) (2018) 841.

[20] J. Chen, W. Liao, Y. Jiang, D. Yu, M. Zou, H. Zhu, et al., Facile fabrication of ZnO/TiO$_2$ heterogeneous nanofibres and their photocatalytic behaviour and mechanism towards rhodamine B, Nanomater. Nanotechnol. 6 (2016) 1–8.

[21] D. Zang, S. Tarafdar, Y.Y. Tarasevich, M.D. Choudhury, T. Dutta, Evaporation of a droplet: from physics to applications, Phys. Rep. 804 (2019) 1–56.

[22] L. Sanchez, L. Guz, P. García, S. Ponce, S. Goyanes, M.C. Marchi, et al., Influence of pyrolytic seeds on ZnO nanorod growth onto rigid substrates for photocatalytic abatement of *Escherichia coli* in water, Water Supply 14 (2014) 1087–1094.

[23] S.A. Bilmes, P. Mandelbaum, F. Alvarez, N.M. Victoria, Surface and electronic structure of titanium dioxide photocatalysts, J. Phys. Chem. B 104 (2000) 9851–9858.

[24] R. Yatskiv, J. Grym, Luminescence properties of hydrothermally grown ZnO nanorods, Superlattices Microstruct 99 (2016) 214–220.

[25] K. Davis, R. Yarbrough, M. Froeschle, J. White, H. Rathnayake, Band gap engineered zinc oxide nanostructures via a sol–gel synthesis of solvent driven shape-controlled crystal growth, RSC Adv 9 (2019) 14638–14648.

[26] S. Erten-Ela, S. Cogal, S. Icli, Conventional and microwave-assisted synthesis of ZnO nanorods and effects of PEG400 as a surfactant on the morphology, Inorganica Chim. Acta 362 (2009) 1855–1858.

[27] A. Pieniążek, B.S. Witkowski, A. Reszka, M. Godlewski, B.J. Kowalski, Optical properties of ZnO microrods grown by a hydrothermal method – a cathodoluminescence study, Opt. Mater. Express 6 (2016) 3741–3750.

[28] Y. Jiao, H.J. Zhu, X.F. Wang, L. Shi, Y. Liu, L.M. Peng, et al., A simple route to controllable growth of ZnO nanorod arrays on conducting substrates, CrystEngComm 12 (2010) 940–946.

[29] E.H.H. Hasabeldaim, O.M. Ntwaeaborwa, R.E. Kroon, E. Coetsee, H.C. Swart, Cathodoluminescence degradation study of the green luminescence of ZnO nanorods, Appl. Surf. Sci. 484 (2019) 105–111.

[30] J. Wang, R. Chen, L. Xiang, S. Komarneni, Synthesis, properties and applications of ZnO nanomaterials with oxygen vacancies: a review, Ceram. Int. 44 (2018) 7357–7377.

[31] S. Kundu, S. Sain, B. Satpati, S.R. Bhattacharyya, S.K. Pradhan, Structural interpretation, growth mechanism and optical properties of ZnO nanorods synthesized by a simple wet chemical route, RSC Adv 5 (2015) 23101–23113.

[32] J.A. Maldonado-Arriola, R. Sánchez-Zeferino, M.E. Álvarez-Ramos, Photoluminescent properties of ZnO nanorods films used to detect methanol contamination in tequila, Sens. Actuator A Phys 312 (2020) 112142.

[33] B. Efafi, H. Mazandarani, M.H.M. Ara, B. Ghafary, Improvement in photoluminescence behavior of well-aligned ZnO nanorods by optimization of thermodynamic parameters, Physica B Condens. Matter 579 (2020) 411915.

[34] M. Fang, Z.W. Liu, Controllable size and photoluminescence of ZnO nanorod arrays on Si substrate prepared by microwave-assisted hydrothermal method, Ceram. Int. 43 (2017) 6955–6962.

[35] S.V. Kurudirek, H. Menkara, B.D.B. Klein, N.E. Hertel, C.J. Summers, Effect of annealing temperature on the photoluminescence and scintillation properties of ZnO nanorods, Nucl. Instrum. Methods Phys. Res. B 877 (2018) 80–86.

[36] H.-J. Li, N.-Q. Ou, X. Sun, B.-W. Sun, D.-J. Qian, M. Chen, et al., Exploitation of the synergistic effect between surface and bulk defects in ultra-small N-doped titanium suboxides for enhancing photocatalytic hydrogen evolution, Catal. Sci. Technol. 8 (2018) 5515−5525.

[37] P.K. Samanta, Band gap engineering, quantum confinement, defect mediated broadband visible photoluminescence and associated quantum States of size tuned zinc oxide nanostructures, Optik 221 (2020) 165337.

[38] N. Han, P. Hu, A. Zuo, D. Zhang, Y. Tian, Y. Chen, Photoluminescence investigation on the gas sensing property of ZnO nanorods prepared by plasma-enhanced CVD method, Sens. Actuators B Chem 145 (2010) 114−119.

[39] F. Kayaci, S. Vempati, I. Donmez, N. Biyikli, T. Uyar, Role of zinc interstitials and oxygen vacancies of ZnO in photocatalysis: a bottom-up approach to control defect density, Nanoscale 6 (2014) 10224−10234.

[40] Y. Lv, Z. Zhang, J. Yan, W. Zhao, C. Zhai, J. Liu, Growth mechanism and photoluminescence property of hydrothermal oriented ZnO nanostructures evolving from nanorods to nanoplates, J. Alloys Compd. 718 (2017) 161−169.

[41] P.G. Ramos, C. Luyo, L.A. Sánchez, E.D. Gomez, J.M. Rodriguez, The spinning voltage influence on the growth of ZnO-rGO nanorods for photocatalytic degradation of methyl orange dye, Catalysts 10 (2020) 660.

[42] P.G. Ramos, E. Flores, C. Luyo, L.A. Sánchez, J. Rodriguez, Fabrication of ZnO-RGO nanorods by electrospinning assisted hydrothermal method with enhanced photocatalytic activity, Mater. Today Commun. 19 (2019) 407−412.

[43] N. Theophile, H.K. Jeong, Electrochemical properties of poly(vinyl alcohol) and graphene oxide composite for supercapacitor applications, Chem. Phys. Lett. 669 (2017) 125−129.

[44] A. Haider, S. Haider, I.-K. Kang, A comprehensive review summarizing the effect of electrospinning parameters and potential applications of nanofibers in biomedical and biotechnology, Arab. J. Chem. 11 (2018) 1165−1188.

[45] J. Ding, M. Wang, J. Deng, W. Gao, Z. Yang, C. Ran, et al., A comparison study between ZnO nanorods coated with graphene oxide and reduced graphene oxide, J. Alloys Compd. 582 (2014) 29−32.

[46] X. Zhou, T. Shi, H. Zhou, Hydrothermal preparation of ZnO-reduced graphene oxide hybrid with high performance in photocatalytic degradation, Appl. Surf. Sci. 258 (2012) 6204−6211.

[47] W. Kang, X. Jimeng, W. Xitao, The effects of ZnO morphology on photocatalytic efficiency of ZnO/RGO nanocomposites, Appl. Surf. Sci. 360 (2016) 270−275.

[48] G. Jayalakshmi, K. Saravanan, J. Pradhan, P. Magudapathy, B.K. Panigrahi, Facile synthesis and enhanced luminescence behavior of ZnO:reduced graphene oxide(rGO) hybrid nanostructures, J. Lumin. 203 (2018) 1−6.

[49] G. Singh, A. Choudhary, D. Haranath, A.G. Joshi, N. Singh, S. Singh, et al., ZnO decorated luminescent graphene as a potential gas sensor at room temperature, Carbon 50 (2012) 385−394.

[50] G. Jayalakshmi, K. Saravanan, B.K. Panigrahi, B. Sundaravel, M. Gupta, Tunable electronic, electrical and optical properties of graphene oxide sheets by ion irradiation, Nanotechnology 29 (2018) 185701.

[51] Z.K. Bolaghi, S.M. Masoudpanah, M. Hasheminiasari, Photocatalytic activity of ZnO/RGO composite synthesized by one-pot solution combustion method, Mater. Res. Bull. 115 (2019) 191−195.

[52] A.A. Othman, M.A. Osman, A.G. Abd-Elrahim, The effect of milling time on structural, optical and photoluminescence properties of ZnO nanocrystals, Optik 156 (2018) 161−168.

[53] X. Zhang, G. Chen, W. Li, D. Wu, Preparation and photocathodic protection properties of ZnO/TiO$_2$ heterojunction film under simulated solar light, Materials 12 (2019) 3856.

[54] T. Yang, B. Sun, L. Ni, X. Wei, T. Guo, Z. Shi, et al., The mechanism of photocurrent enhancement of ZnO ultraviolet photodetector by reduced graphene oxide, Curr. Appl. Phys. 18 (2018) 859−863.

[55] K.N. Abbas, N. Bidin, Morphological driven photocatalytic activity of ZnO nanostructures, Appl. Surf. Sci. 394 (2017) 498−508.

[56] D. Fang, X. Li, H. Liu, W. Xu, M. Jiang, W. Li, et al., BiVO$_4$-rGO with a novel structure on steel fabric used as high-performance photocatalysts, Sci. Rep. 7 (2017) 7979.

[57] M. Arabnezhad, M.S. Afarani, A. Jafari, Co-precipitation synthesis of ZnO−TiO$_2$ nanostructure composites for arsenic photodegradation from industrial wastewater, Environ. Sci. Technol. 16 (2019) 463−468.

[58] X. Zheng, D. Li, X. Li, J. Chen, C. Cao, J. Fang, et al., Construction of ZnO/TiO$_2$ photonic crystal heterostructures for enhanced photocatalytic properties, Appl. Catal. B 168 (2015) 408−415.

[59] H.H. Mohamed, Sonochemical synthesis of ZnO hollow microstructure/reduced graphene oxide for enhanced sunlight photocatalytic degradation of organic pollutants, J. Photochem. A Chem 353 (2018) 401−408.

CHAPTER 5

Orthotropic friction at the edges and interior of graphene and graphene fluoride and frictional anisotropy of graphene at the nanoscale

Sergei F. Lyuksyutov[1], Liudmyla V. Barabanova[1,2], Alper Buldum[3] and Jeffrey A. McCausland[1,4]

[1]Physics Department, University of Akron, Akron, OH, United States
[2]Department of Chemistry, University of Akron, Akron, OH, United States
[3]Department of Mechanical Engineering, University of Akron, Akron, OH, United States
[4]Molecular Biology and Microbiology, School of Medicine, Case Western Reserve University, Cleveland, OH, United States

5.1 Introduction

Graphene is a two-dimensional material with novel electronic properties such as high thermal and electrical conductivity due to a zero band gap (the occupied valence band in contact with an unoccupied conduction band), which is useful in vacuum nanoelectronics [1] and practical sensors. Due to its optical transparency, graphene can be used in transparent electrodes.

Recently, increased attention gained unusual mechanical and tribological properties of graphene. The possibility to use successfully graphene as a new solid lubricant in both humid and dry environments [2] was confirmed by various studies. To investigate the lubricating properties of graphene, it is important to analyze the interatomic interaction at the nanoscale.

Frictional characteristics of graphene are found to be dependent on the structural defects [3,4], surface roughness [5], and material thickness, which indicates that friction decreases with increases in the number of graphene layers [6,7]. Atomic force microscopy (AFM) investigations show that the tribological properties of graphene appear to have a sliding directionality dependence that leads to friction anisotropy of the graphene layers [8].

The physics of scanning a probe sliding across the surface is very complicated. Recent computational study of friction on graphene suggests that the contact area between the probe and a surface is substantially smaller than it appears, and the quality of a contact area may influence how the friction force is transferred across the contact area [9].

Characterization of nanoscale friction based on lateral force microscopy, or friction force microscopy is a technique utilizing AFM as an indispensable tool. Tribological

Thin Film Nanophotonics
DOI: https://doi.org/10.1016/B978-0-12-822085-6.00010-8

properties of graphene require more investigation and also can be influenced by factors such as the adhesion during sliding between the AFM tip and sample surface and the difference in electronic structure.

For experimental considerations, in Part 3 we describe the orthotropic friction of graphene, and graphene fluoride at the edges of each material. We rely upon the Coulomb model of friction, $F = \mu N$, utilizing the sliding regime. This model holds well for one-dimensional motion. An extension of this model can be applied to a ball rolling on a surface. Extending to a ball moving through air, friction forces in three-spatial dimensions arise. The friction in these extended dimensions can be accounted for by treating the friction coefficient μ as a tensor.

We assume that the friction between a tip and surface can be accounted for through orthotropic friction. Orthotropic friction assumes that friction varies along the orthogonal principal axes. Theoretically, this model was described in Refs. [10,11]. The lateral (friction) force can be given by:

$$L = - N[C_0 + C_1\cos2\theta]v, \tag{5.1}$$

where L is the lateral force, N is the normal applied load, v is the velocity unit vector, and θ describes the tip's travel direction relative to the fast scan axis. The tensors C_0 and C_1 represent the two, second-rank friction tensors [10,11]. The signal utilized was the trace-minus-retrace (TMR) for both the height (TMR-H) and friction (TMR-F). A conversion for the TMR-F from volts to nm can be experimentally obtained by measuring the deflection sensitivity. A minimum of three data points is needed to independently define the tensor coefficients. The friction coefficient can be described as the function of the main diagonal tensor coefficients, and angle θ:

$$\mu = \sqrt{\left[C_0^{11}\cos\theta + C_1^{11}\cos2\theta\cos\theta\right]^2 + \left[C_0^{22}\sin\theta + C_1^{22}\cos2\theta\sin\theta\right]^2}. \tag{5.2}$$

This work explores the directional dependence of the coefficient of friction in both the interior, and at the edge of various graphene materials. Here, we investigate these materials experimentally attempting to elucidate peculiarities in the friction coefficients. Diagram of the interactions of tip—surface junction with a thin film of graphene/functionalized graphene is shown in Fig. 5.1.

For theoretical considerations, in Part 2 we report the anisotropic behavior of graphene at the nanoscale and demonstrate that graphene can change its tribological properties as the AFM silicon tip travels in the different sliding directions with respect to the C—C bonds. We constructed the theoretical model of AFM silicon tip in contact mode with a single layer of pristine graphene and performed atomistic molecular dynamics simulations (MDS) at 0.3 and 0.4 nm graphene-AFM tip separation. To investigate the anisotropic behavior, we analyzed different areas of graphene at the edges and on the interior.

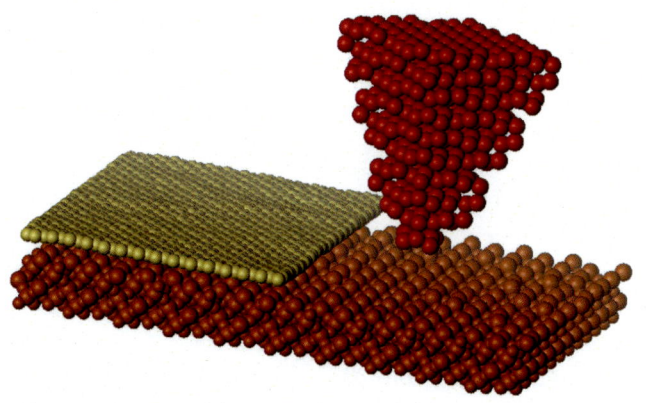

Figure 5.1 Schematic presentation of an AFM tip at the edge of graphene or graphene fluoride film [12]. *AFM*, Atomic force microscopy.

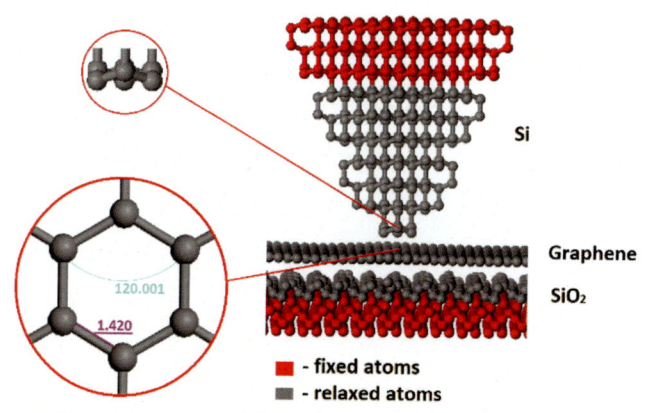

Figure 5.2 A constraint representation of the AFM sharp silicon asperity placed on the single layer of graphene sitting on the silicon oxide substrate (SiO_2) and graphene unit cell (left) with the bond length of 1.42 and an interbond angle of 120 degrees. *AFM*, Atomic force microscopy.

5.2 Frictional anisotropy of pristine graphene deposited on SiO_2 substrate at the nanoscale

5.2.1 Model description

The model of the AFM tip–graphene–SiO_2 substrate system was created and analyzed using the Material studio 4.3 Visualizer software package [13]. The computational study of the AFM friction measurements on the single-layer graphene at the nanoscale was performed using a sharp silicon tip with nine atoms at the apex (Fig. 5.2). A central atom of the tip was placed in the middle of the graphene hexagonal cell. The graphene structure is represented by a group of relaxed atoms. The AFM tip and SiO_2 substrate contain both fixed and relaxed atoms as shown in Fig. 5.2. The simulations parameters of the model were set as follows: for the Si atoms hybridization was tetrahedral with the bond length of ~ 2.2 Å; for the graphene structure the hybridization was trigonal with the bond length of 1.42 Å and the interbond angle of 120 degrees.

The simulations were performed to investigate frictional force in the interior of graphene monolayer at 0.3−0.4 nm and at the edge of graphene at 0.3 nm height. The z-direction represents the height coordinate of the tip placed between the tip and G-SiO$_2$ layer. For the graphene interior case, the apex of the silicon tip was placed in the middle of G-SiO$_2$ surface by using hexagonal geometry of graphene and positioning the sharpest point of the tip in the middle of graphene unit cell (Fig. 5.3A).

During the edge investigation, the height between the tip and G-SiO$_2$ layer was set to 0.3 nm. We modeled graphene nanoribbons assuming that the graphene sheet was cut in half. The tip apex was placed 1 nm away from the graphene nanoribbon

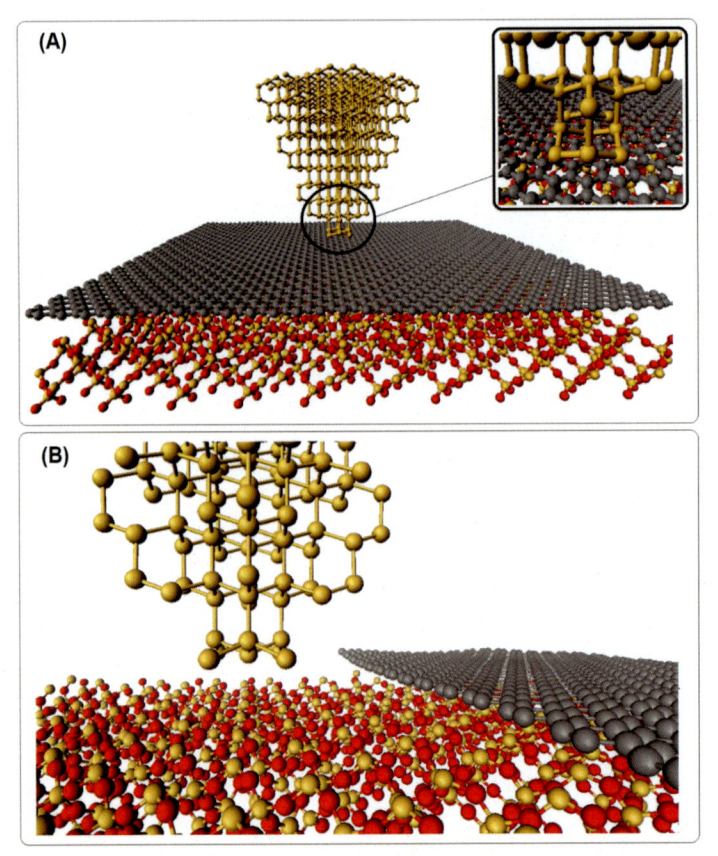

Figure 5.3 (A) An AFM tip−surface junction representation. A sharp (radius ~ 20 nm) Si tip is placed on a single layer of graphene (~ 4 Å height) above a SiO$_2$ substrate; (B) A side view of the Si tip placed on SiO$_2$ substrate 1 nm away from the edge of G-SiO$_2$ layer with 0.3 nm height. *AFM,* Atomic force microscopy.

edge suspended at a 0.3-nm height above the graphene sheet as shown in Fig. 5.3B. The height separation between the Si tip and the SiO_2 substrate was set to 1.3 nm.

For both cases, the sliding direction of the tip was along L_x so that it was crossing the middle of carbon—carbon bond (C—C bond) of the graphene layer. The distance L_x between the C—C bonds is equal to 0.246 nm (Fig. 5.4).

5.2.2 Methods of analysis

To investigate specific tribological properties of the single-layer graphene and its behavior in the interior and at the edges the MDS were performed. First, to minimize the potential energy and to determine the best molecular configuration, the optimization of the structure was performed using Material studio software package. It was carried out using the Universal forcefield [14], Ewald van der Waals summation method [15] with Convergence tolerance energy cut-off of 10^{-5} eV and force cut-off 0.2 eV nm^{-1} for maximum of 500 iterations. During the energy minimization the total enthalpy of the system at the optimization step 274 becomes constant, the energy reached the convergence criteria, and the gradient norm became smaller than 0.2 eV nm^{-1} as required. After fulfiling the convergence requirements the energy was minimized and the best structure was obtained.

The next step, equilibration, was established to stabilize the kinetic and potential energy distribution during the MDS. The canonical NVT ensemble with the temperature set to 298K and the Control method [16] were used as the initial parameters during this step. The simulations were performed for the 0.3—0.4 nm heights between the silicon AFM tip and the graphene surface. The total potential energy of the tip displacement in horizontal direction was calculated.

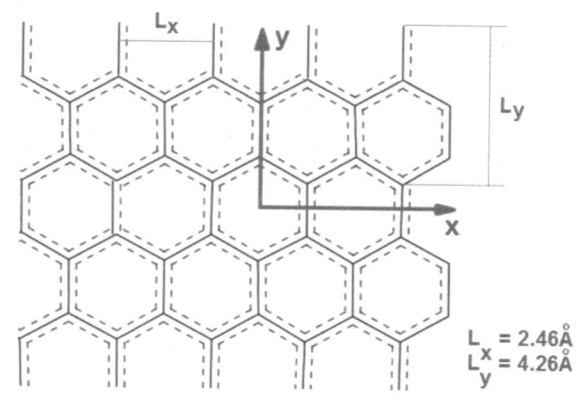

Figure 5.4 The armchair and zig-zag chirality, where L_x and L_y define the dimensions of the graphene unit cell used in computations; the direction along L_x was selected as the sliding direction during simulations.

During the simulation on graphene interior, the tip moved horizontally in the "positive" direction at the vertical tip—surface separation of 0.3 nm. Then the simulation was repeated for the separation of 0.4 nm.

For the simulations performed at the graphene edge, the tip was positioned on SiO_2 substrate (Fig. 5.3B). The distance between the tip and the graphene edge was 1 nm. To simulate the AFM tip "trace—retrace" mode, the tip moved horizontally towards the graphene edge until it reached it. Then, the tip stopped and returned to its original position.

5.2.3 Results and discussion

Anisotropic behavior of the graphene friction was observed during the MDS.

First, we calculated the total potential energy of the tip displacement on the graphene interior at 0.3—0.4 nm of vertical height. Second, we calculated the total potential energy at the graphene edge at 0.3 nm separation distance from the surface.

By comparing the energies for the two different separations on the graphene interior, we found that the potential energy of the system increases as the tip moves closer to the surface. This occurs due to the stronger atomic interactions between the tip and graphene-SiO_2 layer at 0.3 nm height. The normal load FN of the tip sliding on the graphene interior at 0.3 nm separation was estimated as 2.52 nN. Therefore, the friction coefficient (μ) in the interior of G-SiO_2 layer was found to be 2.3×10^{-3} (Fig. 5.5A).

Finally, we calculated the sliding frictional force on graphene interior at 0.3—0.4 nm range. The sliding frictional force was determined as a derivative $F = -\frac{\partial U}{\partial x}$, of the total potential energy with respect to the sliding direction x of the AFM tip at the graphene-SiO_2 surface. The average of the lateral force during the tip displacement on graphene interior is shown in Fig. 5.5A. The higher frictional force is

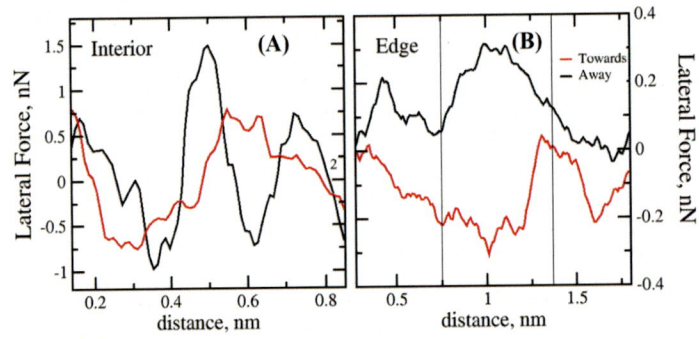

Figure 5.5 (A) Sliding frictional force on the graphene interior at 0.3 nm (red (dark in print version)) and 0.4 nm (black) separation between the tip and G-SiO_2; (B) Sliding frictional force on the graphene edge at 0.3 nm height: (red (dark in print version))—Si tip, placed on the SiO_2, moving toward the G-SiO_2 edge; (black)—Si tip, placed on G-SiO_2, sliding off the edge on SiO_2 substrate.

observed for the 0.3 nm vertical separation due to the stronger interaction between the Si tip and SiO_2 substrate.

The sliding frictional force was calculated at the graphene edges. In this procedure, the AFM tip "trace−retrace" motion was set as "towards−away" from the graphene edge as shown in Fig. 5.5B. The tip was initially positioned 1 nm away from the edge. Then, the tip moved toward the edge at a distance interval between 0 and 0.75 nm. The region when the tip approached the edge and stopped was determined at the distance of 0.75−1.25 nm. The exact position of the edge where the tip stopped was a 1 nm as presented in Fig. 5.5B. Lastly, the AFM tip retracted back to its original position away from the edge at a distance between 1.25 and 1.80 nm.

An average lateral force has increased as the tip approaches the edge of the surface, and the sliding lateral force equals to the sliding frictional force when the tip steps off the edge. The normal load F_N of the tip sliding on the graphene interior at 0.3 nm separation estimated as 2.52 nN. Therefore, the friction coefficient (μ) in the interior of G–SiO_2 layer was found to be 2.3×10^{-3}.

Similar to the graphene interior, the friction coefficient was calculated at the graphene edge. It was found to be 2.3×10^{-2} at the proximity of the edge as the tip reached 1nm distance from its initial position   as shown Fig. 5.5B.

A significant increase of the friction coefficient was observed at the graphene edge with respect to the friction in the graphene interior. This indicates that the higher potential energy is required for the tip to overcome the interatomic interaction to "step-on" or "step-off" at the graphene edge, while the potential energy to perform "stick-slip" motion in the graphene interior was observed significantly lower.

The impact of adhesion on graphene lubricity was taken into consideration. The unterminated π-bonds at the graphene edges may stimulate adhesion affecting the interatomic interactions between the tip and the graphene surface. The raised atoms of the graphene thus may interact chemically with the AFM tip promoting charge transfer. Strong interatomic interactions were observed if the tip approaches graphene surface due to the van der Waals interactions.

5.3 Orthotropic friction and the edges and interior of graphene and graphene fluoride

5.3.1 Atomic force microscopy measurements of orthotropic friction

In the experiments, a Digital Instruments Dimension 3100 AFM controlled by a Nanoscope IV was used in contact mode with the tip deflection set at 1.3 V. All measurements were performed under ambient conditions at temperature between 18°C and 20°C and the relative humidity between 45% and 60%. The tips used were the CSG-10/Au tips manufactured by the NT-MDT Spectrum Instruments with the spring constant ~ 0.5 nN nm^{-1}.

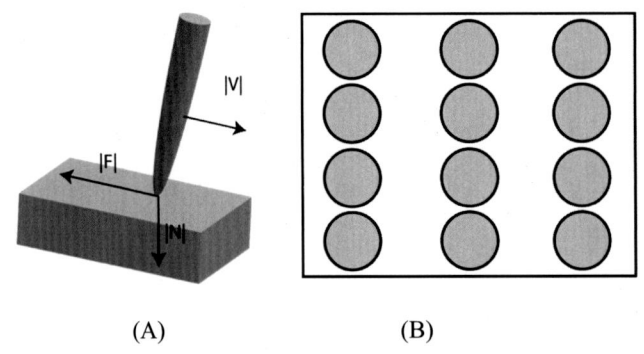

(A)   (B)

Figure 5.6 A Schematic of the tip—surface system including the magnitude of the vectors and direction (denoted via arrow orientation) of each vector. B Sample surface where the features of the surface (*circles*) are regular but differ in the two orthogonal $\hat{x}$- and $\hat{y}$-directions. In the case of graphene, orthotropic formalism is naturally introduced when traversing parallel or perpendicular to a C—C bond.

Tip—surface junction is presented in Fig. 5.6A. The sample surface consists of regular features (circles) that differ in the two orthogonal $\hat{x}$- and $\hat{y}$-directions shown in Fig. 5.6B. If a tip were to slide in the $\hat{x}$-direction, a different coefficient of friction (μ) than that in the $\hat{y}$-direction arises due to differences in the number of interactions per unit area. When this occurs, it is treated by using μ as a second-rank tensor. Orthotropic formalism is naturally introduced when traversing parallel or perpendicular to a C—C graphene bond as illustrated in Fig. 5.4 where L_x and L_y define the dimensions of the graphene unit cell used in computations above in Part 2.

The three sets of samples studied were highly ordered pyrolytic graphite (graphite), graphene fluoride, and chemical vapor deposited graphene (graphene). The graphene fluoride samples were created by taking chemical vaopr deposition (CVD) graphene which was subsequently fluorinated using XeF_2 [17]. CVD graphene used was created through a wet-transfer process. Graphene and graphene fluoride samples were measured on a silicon oxide substrate. To determine the friction on a surface, the deflection sensitivity of the tip was determined by measuring the slope of the force—distance curves with the Nanoscope software in the hard contact regime ten times and averaged for each data set. This value was used to convert the TMR-F signal into nm, the units of the TMR-H.

The criteria for an imaged area depended if the sampling was to be taken in the middle of a sheet or at an edge. If in the middle, the area needed to be free of tears or major defects in roughly $1 \times 1\,\mu m^2$ area. When measuring friction at the edge, an edge site is selected if it has an approximate $1\,\mu m$ length with no visible tears and no folding of the edge either over or under the monolayer. Folding was only an issue for the CVD graphene. In all cases, the scan size was set to 500 nm and the aspect ratio was adjusted to 256:1 minimizing the tip drift along the surface. Concurrently for each surface and angle for 100 sequential TMR-F and TMR-H data points were collected. For a particular surface, the data points for each individual angle were averaged to obtain three values for TMR-F and TMR-H signals corresponding to the particular angle scanned. The average values and corresponding angles were used to determine the tensor coefficients through the lateral force equation above and the angle-dependent value of μ using Eqs. (5.1) and (5.2).

It is assumed to a first order that the spring constant of the tip is the torsional spring constant of the tip. This allows for, with the conversion of the TMR-F signal into nm through the deflection sensitivity, a calculation of a lateral force. The results of the experiments are presented in Fig. 5.7. The direction of the tip motion versus graphene lattice vectors could not be determined due to working in ambient conditions. When measuring the friction coefficient at the edges, the coordinate system chosen is illustrated in Fig. 5.7A and B insets. Altering the scan angle was controlled through the Nanoscope software and it was set at 0, 60, and 90 degrees.

As expected, graphite has the lowest friction coefficient. The values obtained are within an order of magnitude or of the same order of magnitude of previously measured values [18]. The values obtained for graphene are higher relative to previously measured values [19,20]. Interestingly, it appears that graphene fluoride has a lower overall friction coefficient than graphene at both the edge and interior. This runs contrary to previously measured coefficients in Refs. [21,22].

5.3.2 Peculiarities in friction measurements

To explain the differences in friction between the different samples as well as sliding direction, one may immediately assume that the surface roughness may play a large role in explaining the differences between the various substrates. Utilizing Gwyddion SPM software, we calculated the surface roughness for each sample type. The surface roughness for each substrate was calculated to be 0.021, 0.037, and 0.033 nm for graphite, graphene, and graphene fluoride, respectively. The calculated surface roughness (Fig. 5.7) correlates well with the interior/edge friction coefficients.

A separate possibility is the difference in wettability of the substrates. When an AFM tip is brought close to a surface under ambient conditions a water meniscus forms. This can obviously induce the attractive capillary force. This capillary force could be the source of the difference in friction coefficient between the various substrates. It has been shown in that with increasing tip velocity the contribution from the capillary force and thus adhesive force decreases and approaches the van der Waals only potential with increasing tip velocity when the substrate and tip are made of the same material. The critical velocity can be expressed for the contribution of a capillary force as a function of scan frequency and maxim deflection of the tip due to surface roughness. This intuitively makes sense thermodynamically as there will be minimum

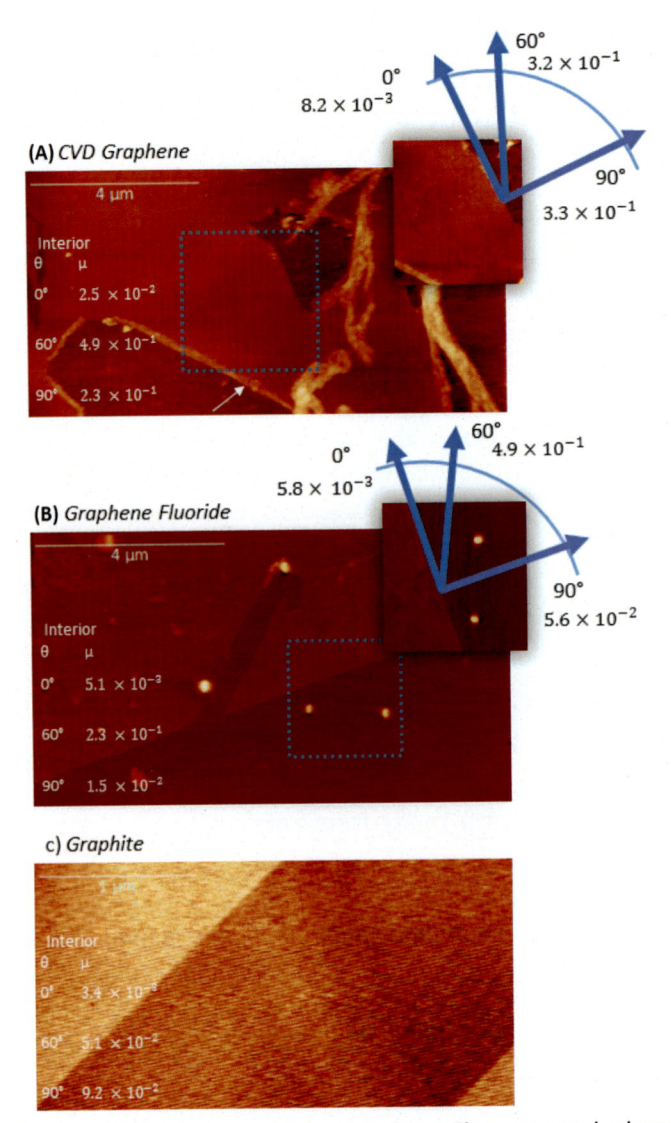

Figure 5.7 (A) Contact AFM mode image of CVD graphene. The arrow at the bottom highlights an edge section where the graphene has rolled onto itself. Edge measurements were performed at edges with no rolling or debris. Debris and rolled sections from the transfer process at the edge can be seen to the right. The granular structure of the CVD graphene is clearly seen in the inset. Interior measurements were made within a single grain. (B) AFM image of graphene fluoride. (C) Contact image of highly oriented pyrolytic graphite (HOPG). Interior measurements were made on facets without traversing an edge. *AFM*, Atomic force microscopy.

dwell time for the formation of a meniscus. If we consider the surface roughness parameters, our scan velocity far exceeds this critical velocity leading to the conclusion that a capillary force should not occur and cause a significant, if any, increase in the friction.

Normalizing the interior coefficients for each sample by the measured value of graphite at each particular angle for the interior, the coefficients are larger for graphene versus the graphene fluoride for each angle as indicated in Table 5.1. Normalization by graphite's values essentially normalizes to an "ideal" friction coefficient for graphene. Further normalization of the interior values by 90 degrees coefficients still shows an increase in the friction coefficient of graphene versus the graphene fluoride at any angle. Comparing the interior of a particular sample, there is also a maximum in the double normalized coefficients at 60 degrees. This corresponds on the graphene and graphene fluoride lattice with crossing an atom versus a bond (or along the armchair and zig-zag directions). We thus conclude, while not hypothesizing a source and under our experimental conditions, graphene, graphene fluoride, and, by extension, layered graphene structures such as graphite exhibit measurable orthotropic friction at the nanoscale.

Interpreting the results for the edge coefficients requires a slightly different analysis. Due to there being no "ideal" friction coefficient for an edge, two different measures are used. First, looking at the directly calculated friction coefficients for graphene, we see (Fig. 5.7A) that the lowest friction coefficient occurs when scanning along the edge for graphene. It then increases dramatically by two orders of magnitude for both 60 and 90 degrees but, without much difference between the two angles though the largest coefficient is for 90 degrees.

In Fig. 5.7B, the inset shows similarly the lowest friction coefficient for graphene fluoride is when the tip scans along the edge. The largest coefficient occurs for 60 degrees scan angle followed by the 90 degrees scan angle. If we consider the general properties of graphene and graphene fluoride, the friction coefficient at 60 degrees for graphene fluoride being higher than the 90 degrees coefficient is plausibly due to the average number of

Table 5.1 Angular dependence of the friction coefficient for three samples: graphite, graphene, and graphene fluoride inside and also at the edges of the samples.

Sample	Total friction coefficient results μ		
	0 degrees	60 degrees	90 degrees
Graphite (inside the sample)	3.4×10^{-3}	5.1×10^{-2}	9.2×10^{-2}
GF (inside the sample)	5.1×10^{-3}	2.3×10^{-1}	1.5×10^{-1}
G CVD (inside the sample)	2.5×10^{-2}	4.9×10^{-1}	2.3×10^{-1}
GF (at the edge of the sample)	5.8×10^{-3}	4.9×10^{-1}	5.6×10^{-2}
G CVD (at the edge of the sample)	8.2×10^{-3}	3.2×10^{-1}	3.3×10^{-1}

fluorine atoms interacting with the tip. If one imagines as single benzene ring, fluorine decorates every other carbon around the ring. While graphene is sp^2 hybridized, graphene fluoride is sp^3 hybridized with three fluorine atoms interacting with the tip for each ring in graphene. This could provide the opportunity for the tip to "catch" and distort the lattice more easily at the edge.

Comparing the edge coefficients of graphene and graphene fluoride, we see the largest coefficient peculiarity occurs when the tip scans perpendicular to the surface. Here, there is an order of magnitude difference between graphene fluoride and graphene. It was previously reported [21,22] that the friction force on graphene fluoride is enhanced when compared with graphene due to increased out of plane bending in the graphene fluoride when under UHV [21]. This out of plane bending extends beyond the contact area of the tip. This supports our supposition that the coefficient at 60 degrees for the edge on graphene fluoride may be due to lattice distortion effects. Also, an increased friction due to electrostatic interactions arises between the Si tip and fluorine when friction is measured under dry nitrogen. There are two large differences between our experimental setup and the previous work, namely the use of a gold-coated tip and working in ambient conditions. We hypothesize the lower friction values arise due hydrophobic interactions. Graphene fluoride is hydrophobic in exfoliated state [6]. If the tip is hydrophilic and the substrate is hydrophobic, we can assume a strong hydrophobic repulsion by the surface caused by water naturally condensing at the apex of the tip. This condensation then should occur in small nanopores at the apex of the tip. The hydrophilic nature of gold has been well established previously though most notably, it occurs readily on clean gold surfaces [22,23]. Clean surfaces are considered to be those free of oxide species as well as organic species. As the experiments were conducted in a clean room, it is a reasonable assumption this criterion has been met for the AFM tips used in our laboratory. We hypothesize then, though cannot conclusively show, that water is the main culprit in lowering the observed friction force on graphene fluoride. The scan angle peculiarities observed at the edges and overall lowered friction of graphene fluoride warrants future investigative work.

The frictional force measurements were taken under an ambient humidity at the room temperature. The graphene fluoride friction coefficients on the graphene interior at the 60 and 90 degrees angles were found to be in the range of 10^{-1} similar to the experimental results reported in the literature for fluorinated graphene deposited on SiO_2 substrate [24]. Significantly lower friction coefficient was observed at the 0 degrees angle of $\sim 10^{-3}$. The deviation of our results and those in the literature can be explained by experimental conditions such as an ambient humidity and the temperature.

5.4 Conclusion

Sliding friction of graphene and graphene fluoride was investigated using MDS. Friction anisotropy of a pristine graphene was studied by creating an atomistic model involving an

AFM tip in contact with a monolayer of graphene for various tip–graphene separation distances. The MDS were performed to obtain friction coefficient of graphene. Simulations were focused on the differences in sliding friction in the interior, and also at the edges of graphene at the constant temperature for a 0.3 and 0.4 nm tip–surface separations. As a result of the atomistic simulations, the friction coefficients of pristine graphene in the interior and at the edges were found to be in the range of 10^{-3}–10^{-1}. It was proven that the AFM tip–surface interactions have a profound influence on the friction. The directionality dependence of the friction between silicon AFM tip and a monolayer of pristine graphene was analyzed at the edges and on the interior. The orthotropic friction coefficients were calculated at 90 degrees angle for the various tip–surface separations at the edges, as well as on the interior of graphene. It was found that the friction coefficient at a 0.4-nm distance was higher than that at a 0.3 nm when the tip approaches the interior of graphene. It is assumed that the change in the friction coefficients when the tip was closer to the graphene surface is due to the increased van der Waals interaction between the tip and graphene monolayer. The calculated friction coefficients on the interior and at the graphene edges were compared. The higher friction coefficients were obtained at the edges than on the graphene interior. It is assumed that the deformation of graphene layer at the edges was influenced by the tip sliding direction. The MDS show the pattern of an average frictional force increase if the Si tip approached the edge of G-SiO_2.

The anisotropic behavior of graphene at the edges as well as on the interior of graphene must be taken into consideration to obtain the level of friction associated with the superlubricity of graphene. Our simulations also show that the bending deformation of the tip and downward deformation of graphene edge resulted in large friction forces ($\sim$0.6 nN) and energy dissipation ($\sim$1 eV difference) for an AFM tip moving from SiO_2 over the graphene edge.

Graphene and graphene fluoride are widely considered as the potential future solid lubricants. We experimentally demonstrated that graphene and graphene fluoride manifest different coefficients of sliding friction at the edges (graphene fluoride from 5.8×10^{-3} to 4.9×10^{-1}; graphene from 8.2×10^{-3} to 3.3×10^{-1}) of a sheet sample versus the interior(graphene fluoride from 5.1×10^{-3} to 1.5×10^{-1}; graphene from 2.5×10^{-2} to 2.3×10^{-1}) under ambient humidity conditions ($RH \sim 40\%$–60%). Experimental AFM analysis confirmed the friction coefficients show distinct directional dependence between graphene and graphene fluoride. Graphene fluoride was shown to have lower friction coefficients than graphene at almost all angles either inside or at the edges of the samples. At the edges, graphene and graphene fluoride were shown to have a dependence on the scan direction. It is unexpected with respect to the results reported as of today. It is assumed that differences in friction coefficients between those in literature and our results for graphene fluoride were due to ambient experimental conditions. It is assumed that differences in friction coefficients between those in literature and our results for graphene fluoride were due to the ambient experimental conditions.

References

[1] S.W. Lee, et al., A study on field emission characteristics of planar graphene layers obtained from a highly oriented pyrolyzed graphite block, Nanoscale Res. Lett. 4 (2009) 1218−1221.

[2] D. Berman, A. Erdemir, A.V. Sumant, Graphene: a new emerging lubricant, Mater. Today 17 (2014) 31−42.

[3] X.Y. Sun, R.N. Wu, R. Xia, X.H. Chu, Y.J. Xu, Effects of Stone−Wales and vacancy defects in atomic-scale friction on defective graphite, Appl. Phys. Lett. 104 (2014) 183109.

[4] W.L. Guo, W.Y. Zhong, Y.T. Dai, S.N. Li, Coupled defect-size effects on interlayer friction in multiwalled carbon nanotubes, Phys. Rev. B 72 (2005) 075409.

[5] Y.L. Dong, X.W. Wu, A. Martini, Atomic roughness enhanced friction on hydrogenated graphene, Nanotechnology 24 (2013) 375701.

[6] Q. Li, X.-Z. Liu, S.-P. Kim, V.B. Shenoy, P.E. Sheehan, J.T. Robinson, et al., Fluorination on graphene enhances friction due to increased corrugation, Nano Lett. 14 (2014) 5212−5217.

[7] K.S. Novoselov, A.K. Geim, S.V. Morozov, D. Jiang, Y. Zhang, S.V. Dubonos, et al., Electric field in atomically thin carbon films, Science 306 (2004) 666−669.

[8] J.S. Choi, et al., Friction anisotropy—driven domain imaging on exfoliated monolayer graphene, Science 333 (2011) 607−610.

[9] S. Li, Q. Li, R.W. Carpick, P. Gumbsch, X.-Z. Liu, X. Ding, et al., The evolving quality of frictional contact with graphene, Nature 539 (2016) 541−545.

[10] A. Zmitrowicz, Mathematical descriptions of anisotropic friction, Int. J. Solids Struct. 25 (1989) 837−862.

[11] M. Campione, M.S. Trabattoni, M. Moret, Nanoscale mapping of frictional anisotropy, Tribol. Lett. 45 (2012) 219−224.

[12] S.F. Lyuksyutov, L. Barabanova, J.A. McCausland, Why orthotropic friction is important on graphene and graphene fluoride thin films? Proc. SPIE 11371 (2019) 1137104.

[13] BIOVIA Accelrys Software Inc. Materials studio (version 4.3). <https://www.3ds.com/products-services/biovia/products/molecular-modeling-simulation/biovia-materials-studio/>, 2008.

[14] K. Rappé, C.J. Casewit, K. Colwell, W.A. Goddard, W.M. Skiff, Uff, a full periodic table force field for molecular mechanics and molecular dynamics simulations, J. Am. Chem. Soc. 114 (1992) 10024−10035.

[15] J. Hautman, M.L. Klein, An Ewald summation method for planar surfaces and interfaces, J. Mol. Phys. 75 (1992) 379−395.

[16] S. Nosé, A unified formulation of the constant temperature molecular-dynamics methods, J. Chem. Phys. 81 (1984) 511−519.

[17] J.T. Robinson, J.S. Burgess, C.E. Junkermeir, S.C. Badescu, T.L. Reinecke, F.K. Perkins, et al., Properties of fluorinated graphene films, Nano Lett. 10 (2010) 3001−3005.

[18] C.M. Mate, G.M. McClelland, R. Erlandsson, S. Chiang, Atomic-scale friction of a tungsten tip on a graphite surface, Phys. Rev. Lett. 59 (1987) 1942−1945.

[19] Y.J. Shin, R. Stromberg, R. Nay, A.T.S. Wee, Andrew, H. Yang, et al., Frictional characteristics of exfoliated and epitaxial graphene, Carbon 49 (2011) 4059−4073.

[20] T. Filleter, J.L. McChesney, A. Bostwick, E. Rotenberg, K.V. Emtsev, T. Seyller, et al., Friction and dissipation in epitaxial graphene films, Phys. Rev. Lett. 102 (2009) 086102.

[21] S. Kwon, J.-H. Ko, K.-J. Jeon, Y.-H. Kim, J.Y. Park, Enhanced nanoscale friction on fluorinated graphene, Nano Lett. 12 (2012) 6043−6048.

[22] R.L. Wells, T. Fort, Adsorption of water on clean gold by measurement of work function changes, Surf. Sci. 32 (1972) 554−560.

[23] T. Smith, The hydrophilic nature of a clean gold surface, J. Colloid Interface Sci. 75 (1980) 51−55.

[24] X. Zeng, Y. Peng, M. Yu, H. Lang, X. Cao, K. Zou, Dynamic sliding enhancement on the friction and adhesion of graphene, graphene oxide, and fluorinated graphene, ACS Appl. Mater. Interfaces 10 (2018) 8214−8224.

PART II

Applications

CHAPTER 6

Optical manipulation of nanoparticles with structured light

Guanghao Rui[1], Ying Li[1], Bing Gu[1], Yiping Cui[1] and Qiwen Zhan[2]
[1]Advanced Photonics Center, Southeast University, Nanjing, P.R. China
[2]School of Optical-Electrical and Computer Engineering, University of Shanghai for Science and Technology, Shanghai, P.R. China

6.1 Introduction

Since light carries linear and angular momentum, the interaction between light and matter leads to the radiation pressure and torque exerted on physical objects, enabling the optical manipulation techniques that utilize optical forces to realize optical cooling [1], trapping [2], binding [3–5], transporting, and sorting [6,7]. After more than 30 years of steady development as a noncontact and powerful tool in biological and medical science, Arthur Ashkin was awarded Nobel Prize for his seminal work on optical tweezers and their applications to biological systems. Back to 1970 Arthur Ashkin's pioneering work showed that the radiation pressure from lasers could trap and manipulate micron-sized particles such as accelerating, decelerating, and steering [2]. After that, he successfully developed single-beam gradient force optical tweezers and demonstrated its feasibility in trapping viruses, bacteria, and single cells [8,9]. Nowadays, optical trapping has been successfully implemented in two main size regimes: the subnanometer (e.g., cooling of atoms, ions, and molecules) and micrometer scale (such as cells). However, it has been difficult to apply these techniques to nanoscale (between ~ 1 and 100 nm). When the size of the particle is much smaller than the wavelength of the trapping light, both the polarizability of the particle and the optical force would be reduced, leading to the destruction of the optical trapping. Over the past decade, new techniques have been developed to stably trap nanostructures and successfully applied to a variety of objects, such as metallic nanoparticles [10,11], carbon nanotubes [12,13], quantum dots [14,15], and chiral particles [16,17]. For example, the trapping light is not limited to fundamental Gaussian mode but usually composed of complex spatial distributions, such as cylindrical vector (CV) beams, nondiffraction Bessel beams, and optical vortex beams generated by a spatial light modulator. These complex optical fields provide more degrees of freedom to modulate the characteristics of the optical force and are demonstrated to be helpful to realize optical force with special characteristics, such as optical tractor beams and nonconservative force [18—24], which can pull the objects toward the light source. The continuous development of optical tweezers has revolutionized the experimental study of small particles and become an important tool for research in biology, physical chemistry, and soft matter physics. In this chapter, we shall discuss the basic theory of optical tweezers and give some applications of optical tweezers using structured light.

6.2 Theory of optical tweezers

The physical mechanism of the optical force is very easy to understand. Consider light incident on an interface between two mediums with different refractive indices. The propagation path of the light would be deflected due to reflection or refraction. Consequently, optical force would be generated by the momentum change of light, which would be impacted by not only the size and material of the particle, but also the intensity, phase, polarization, and wavelength of the light.

Generally, the optical force can be categorized into gradient force and scattering force for particles with size much smaller than the light wavelength. The magnitude and direction of the gradient force are strongly determined by the magnitude of the field strength and the direction of the spatial gradient of the optical intensity. As for the scattering force, its origin is net forward linear momentum transferred from the forward moving photons to the particle, which usually push the particle along the light propagation direction. To achieve a stable trap in three-dimensional space, the gradient force needs to dominate over the scattering force; therefore an objective lens with high numerical aperture (NA) is appreciated to focus the light as tightly as possible. If the refractive index of the particle is larger than that of the surrounding medium, the gradient force points toward the position with the highest intensity, therefore a trapping light with Gaussian distribution would confine the particle at the center of the light beam. Any displacement of the particle would lead to the restoring force that pulls the particle back to the equilibrium position. Conversely, particle with refractive index lower than the surrounding medium would be trapped at the position having the lowest field intensity.

The magnitude of the optical force exerted on a particle usually is the range of piconewtons. Since the optical force can be strongly affected by the size and shape of the particle, the theoretical description of the optical tweezers can be categorized into two regions by considering the ratio between particle size and the wavelength of the trapping light. If the diameter of the particle is much smaller than the wavelength of light, Rayleigh theory is suitable to treat the particle as a dipole [25], and the interaction between light and the Rayleigh particle can be described by wave optics as a dipole interacting with inhomogeneous electric field. On the other hand, if the size of the particle is comparable to or larger than the wavelength of light, Mie theory and ray optics should be adopted to study the optical trapping phenomenon [26].

6.2.1 Rayleigh region (dipole approximation)

On assuming a dipole consists of two opposite charges, the total force acting on this system can be expressed as:

$$\vec{F} = (\vec{p} \cdot \nabla) \vec{E} + \dot{\vec{p}} \times \vec{B} + \vec{r} \times (\vec{p} \cdot \nabla) \vec{B}, \tag{6.1}$$

where the first and last terms arise from the interaction of the dipole with the inhomogeneous electric and magnetic field respectively, and the second term is due to the magnetic Lorentz force. It is worthy of noting that the dipole approximation method indicates that the existence of the dipole would not change the light field, therefore the field E and B in Eq. (6.1) are exactly the exciting fields.

To derive the expression of the optical force exerted on a particle, the second term in Eq. (6.1) can be rewritten as:

$$\dot{\vec{p}} \times \vec{B} = -\vec{p} \times \frac{d}{dt}\vec{B} + \frac{d}{dt}(\vec{p} \times \vec{B}) = \vec{p} \times (\nabla \times \vec{E}) + \frac{d}{dt}(\vec{p} \times \vec{B}). \tag{6.2}$$

By employing Eq. (6.2) into Eq. (6.1), the force equation becomes:

$$\vec{F} = \sum_{i=x,y,z} p_i \nabla E_i + \frac{d}{dt}(\vec{p} \times \vec{B}). \tag{6.3}$$

Note that the last term in Eq. (6.1) usually is negligible compared with other terms and can be omitted. The time-averaged force can be expressed as:

$$\left\langle \vec{F} \right\rangle = \sum_{i=x,y,z} \left\langle p_i(t)\nabla E_i(t) \right\rangle, \tag{6.4}$$

where the last term in Eq. (6.3) vanished during the time averaging operation. It is known that there is a linear relationship between induced dipole moment p, electric field E, and polarizability $a : \vec{p} = \alpha(\omega)\vec{E}(\vec{r_0})$, where r_0 is the position of the particle in the electric field. By representing the light field in its paraxial form as $\vec{E}(\vec{r}) = \vec{E}_0(\vec{r})e^{i\vec{k}\cdot\vec{r}}$, Eq. (6.4) can be reformatted as:

$$\left\langle \vec{F} \right\rangle = \frac{1}{4}Re\{\alpha(\omega)\}\nabla\left|\vec{E}_0\right|^2 + \frac{1}{2}\vec{k}\ Im\{\alpha(\omega)\}\nabla\left|\vec{E}_0\right|^2$$
$$- \frac{1}{2}Im\{\alpha(\omega)\}Im\{\vec{E}_0 \cdot \nabla\vec{E}_0^*\}. \tag{6.5}$$

The first term is the gradient force that is proportional to the gradient of the field intensity, while the second and the last terms are scattering forces due to the momentum transfer from the light to the particle. Note that the last term would be zero for nonabsorbing particles and plane waves, in which cases the imaginary part of α and E_0 is assumed to be zero, respectively. For a Rayleigh metallic particle, the polarizability can be approximated as [27]:

$$\alpha(\omega) = \frac{\alpha_0(\omega)}{1 - (2/3)ik^3\alpha_0(\omega)}, \tag{6.6}$$

where $\alpha_0 = a^3(\varepsilon - 1)/(\varepsilon + 2)$ is the Clausius−Mossotti relation, a is the radius of the particle, and $\epsilon = \epsilon_p/\epsilon_m$ is the ratio of relative permittivity of the particle and the surrounding medium. By substituting Eq. (6.6) into Eq. (6.5), the expression of the gradient force becomes:

$$\langle \vec{F}_{grad} \rangle = 4\pi\varepsilon_m a^3 \left(\frac{\varepsilon - 1}{\varepsilon + 2}\right) \frac{1}{2} \nabla \vec{E}_0^2 = 4\pi n_m^2 \varepsilon_0 a^3 \left(\frac{m^2 - 1}{m^2 + 2}\right) \frac{1}{2} \nabla \vec{E}_0^2$$

$$= 4\pi n_m^2 \varepsilon_0 a^3 \left(\frac{m^2 - 1}{m^2 + 2}\right) \frac{1}{2} \nabla I(\vec{r}), \tag{6.7}$$

where $m = n_p/n_m$ is the ratio of the refractive index of the particle and the surrounding medium, $I(r)$ is the time-averaged intensity. It can be clearly seen that this force depends on the gradient of the field intensity, and the direction of the gradient force would point toward the highest/lowest region of the light field for nanoparticle with refractive index larger/smaller than the medium.

Similarly, the scattering force can be expressed as [28,29]:

$$\vec{F}_{abs+scatt} = \frac{\left|\vec{E}_0\right|^2}{8\pi} (\sigma_{abs} + \sigma_{scatt}) \frac{\vec{k}}{k},$$

$$\sigma = \sigma_{abs} + \sigma_{scatt} = 4\pi ka^3 Im\left\{\frac{\varepsilon - 1}{\varepsilon + 2}\right\} + \frac{8\pi}{3} k^4 a^6 \left|\frac{\varepsilon - 1}{\varepsilon + 2}\right|^2, \tag{6.8}$$

where the first term in Eq. (6.8) is related to radiation pressure exerted on the particle caused by optical absorption, and the second term is contributed by photon scattering. If the absorption cross-section of the nanoparticle is negligible, the expression of the time-averaged scattering force can be rewritten as [30]:

$$\langle \vec{F}_{scatt}(\vec{r}) \rangle = \hat{z} \frac{n_m}{c} \frac{8\pi}{3} (ka)^4 a^2 \left(\frac{m^2 - 1}{m^2 + 2}\right)^2 I(\vec{r}). \tag{6.9}$$

Clearly, the scattering force would deteriorate a stable trap since it points toward the propagation direction of light. In addition, the size-dependent (a^6) characteristic determines that the magnitude of the scattering force would increase with increasing size of the particle.

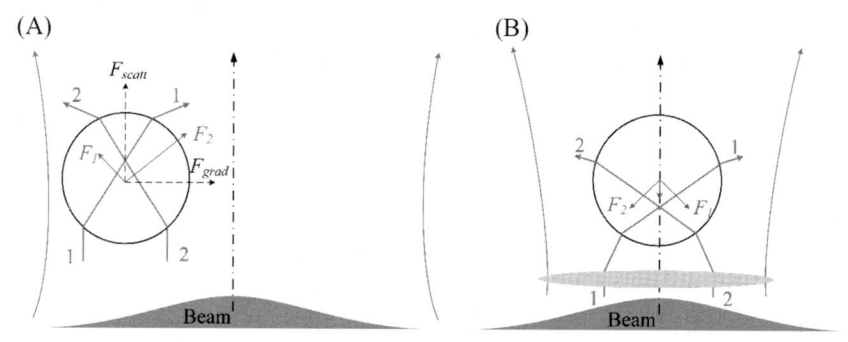

Figure 6.1 Schematic diagram of optical trapping in (A) transverse and (B) axial direction.

6.2.2 Geometrical regime (ray-optics approximation)

The concept of optical tweezers is much easier to explain using ray optics. As shown in Fig. 6.1A, a spherical particle displaced from the beam axis is stroked by two light rays from one side. Assuming the refractive index of the particle is larger than the ambient, it will work as a converging lens to refract light rays and produce force F_1 and F_2. The longitudinal and transversal components of F_1 and F_2 give rise to the scattering force and gradient force, respectively. Since the particle is displaced from the axis of the light beam, the transverse component of F_1 is smaller than that of F_2, lead to the gradient force pointing to the high-intensity region of the laser. If the particle moves along the axis of the light, as shown in Fig. 6.1B, the counterforce from the deflected photons will push the particle back to the focus of the laser.

With the Fresnel equations for reflection and transmission, the exact expression for the gradient and scattering forces can be derived accordingly [26]:

$$F_s = \frac{n_m n P}{C_0} \left[1 + R\cos2\theta - T^2 \frac{\cos(2\theta - 2\varphi) + R\cos2\theta}{1 + R^2 + 2R\cos2\varphi} \right], \tag{6.10}$$

$$F_g = \frac{n_m n P}{C_0} \left[1 + R\sin2\theta - T^2 \frac{\sin(2\theta - 2\varphi) + R\sin2\theta}{1 + R^2 + 2R\cos2\varphi} \right], \tag{6.11}$$

where θ and φ are the angles of the incidence and refraction, and R and T are the Fresnel reflection and refraction coefficients, respectively. Therefore one can easily calculate both the gradient and scattering forces with the coefficients R and T and the angle of incidence.

6.2.3 Maxwell-stress-tensor method

The Maxwell-stress-tensor (MST) method is a common approach to calculate the optical force exerted on a particle, regardless of the shape of the particle and the ratio between particle size and the wavelength of the light. By integrating the momentum flux over any closed surface surrounding the particle, the time-average force on a particle of any size can be derived as [31]:

$$\langle F \rangle = \oint \left\{ \frac{\varepsilon}{2} Re[(E \cdot n)E^*] - \frac{\varepsilon}{4}(E \cdot E^*)n + \frac{\mu}{2} Re[\mu(H \cdot n)H^*] - \frac{\mu}{4}(H \cdot H^*)n \right\} ds, \quad (6.12)$$

where ε and μ are the relative permittivity and relative permeability of the medium around the particle, and n is the unit normal perpendicular to the integral area ds. However, conventional MST method only gives the total electromagnetic force exerted on the particle. To understand the different mechanisms underlying the optical tweezers, the total electromagnetic force can be further separated into gradient and scattering forces as [32]:

$$\langle F_{grad} \rangle = \left\langle \int_s T_{grad} \cdot nds \right\rangle, \quad (6.13)$$

$$\langle F_{scat} \rangle = \left\langle \int_s T_{scat} \cdot nds \right\rangle, \quad (6.14)$$

which are given in

$$T_{grad} = \frac{\varepsilon}{2} EE = \frac{\varepsilon}{2} \begin{pmatrix} E_x E_x & E_x E_y & E_x E_z \\ E_y E_x & E_y E_y & E_y E_z \\ E_z E_x & E_z E_y & E_z E_z \end{pmatrix}, \quad (6.15)$$

$$T_{scat} = T - T_{grad}$$

$$= \begin{bmatrix} \frac{1}{2}\varepsilon E_x E_x - \frac{\varepsilon}{2}E^2 + \frac{\mu}{2}\left(H_x H_x - \frac{H^2}{2}\right) & \frac{1}{2}\varepsilon E_x E_y - \frac{\varepsilon}{2}E^2 + \frac{\mu}{2}\left(H_x H_y - \frac{H^2}{2}\right) & \frac{1}{2}\varepsilon E_x E_z - \frac{\varepsilon}{2}E^2 + \frac{\mu}{2}\left(H_x H_z - \frac{H^2}{2}\right) \\ \frac{1}{2}\varepsilon E_y E_x - \frac{\varepsilon}{2}E^2 + \frac{\mu}{2}\left(H_y H_x - \frac{H^2}{2}\right) & \frac{1}{2}\varepsilon E_y E_y - \frac{\varepsilon}{2}E^2 + \frac{\mu}{2}\left(H_y H_y - \frac{H^2}{2}\right) & \frac{1}{2}\varepsilon E_y E_z - \frac{\varepsilon}{2}E^2 + \frac{\mu}{2}\left(H_y H_z - \frac{H^2}{2}\right) \\ \frac{1}{2}\varepsilon E_z E_x - \frac{\varepsilon}{2}E^2 + \frac{\mu}{2}\left(H_z H_x - \frac{H^2}{2}\right) & \frac{1}{2}\varepsilon E_z E_y - \frac{\varepsilon}{2}E^2 + \frac{\mu}{2}\left(H_z H_y - \frac{H^2}{2}\right) & \frac{1}{2}\varepsilon E_z E_z - \frac{\varepsilon}{2}E^2 + \frac{\mu}{2}\left(H_z H_z - \frac{H^2}{2}\right) \end{bmatrix}. \quad (6.16)$$

6.3 Structured light

In the last decade, structured light with unconventional spatial distributions in terms of amplitude, polarization, and phase is rapidly becoming a current trend due to its potential applications in optical tweezers, microscopic and nanoscopic imaging, materials micromachining and processing [33−35], etc. Compared to conventional light with homogeneous distribution, structured light is demonstrated to improve the performance of current optical devices. As an important property of light, most past research dealt with optical beams with spatially homogeneous states of polarization (SOP), such as linear, elliptical, and circular polarizations. For these cases, the spatial dependence of SOP in the beam cross-section has been largely ignored. With the recent rapid advances in high power computing and micro and nanofabrication, optical beams with spatially variant SOPs are increasingly available and the theoretical and experimental research of their properties has become more accessible. One class of vector beams that have been intensively studied comprises optical beams with cylindrical symmetry in polarization, the so-called CV beams [36−38]. Particularly for radial and azimuthal polarization (shown in Fig. 6.2A and B), two special cases of CV beams have attracted lots of attention due to their remarkable focusing properties. Due to the orthogonality, these two modes form a complete basis for CV beams. Consequently, any generalized CV beam can be represented by the combination of radial and azimuthal modes with different weights (shown in Fig. 6.2C).

When a radially polarized beam is focused by high NA objective lens, the field distribution near the focus can be analyzed with the Richard Wolf vectorial diffraction method as [36]:

$$E_r(r, \phi, z) = 2A\cos\varphi_0 \int_0^{\theta_{max}} P(\theta)\sin\theta \cos\theta J_1(kr\sin\theta)e^{ikz\cos\theta} d\theta, \tag{6.17}$$

$$E_z(r, \phi, z) = i2A\cos\varphi_0 \int_0^{\theta_{max}} P(\theta)\sin^2\theta J_0(kr\sin\theta)e^{ikz\cos\theta} d\theta, \tag{6.18}$$

where θ_{max} is the maximal focusing angle determined by the NA of the objective lens, k is the wavenumber of the illumination, $J_n(r)$ is the nth order Bessel function of the

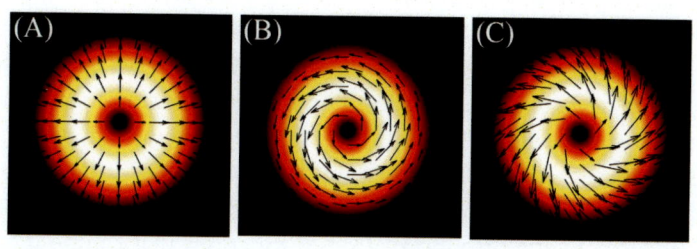

Figure 6.2 Intensity distribution superimposed with spatial polarization distribution of CV beams. (A) radial polarization; (B) azimuthal polarization; and (C) generalized CV beams. *CV, Cylindrical vector.*

first kind, $P(\theta)$ is the pupil apodization function of the objective lens. The constant A is given by $A = \pi f l_0 / \lambda$, where f is the focal length, λ is the wavelength of incident wave in the ambient environment, and l_0 is associated with the laser beam power. Similarly, the focal field of a tightly focused azimuthally polarized light can be expressed as [36]:

$$E_\phi(r, \phi, z) = 2A\sin\varphi_0 \int_0^{\theta_{max}} P(\theta)\sin\theta J_1(kr\sin\theta)e^{ikz\cos\theta}\,d\theta. \qquad (6.19)$$

For radially polarized incidence, the focal field consists of a longitudinal component and a radial component. The longitudinal component has its peak on the optical axis, while the radial component has a donut shape with dark center on the optical axis. Compared to conventional scalar light with homogeneous SOP, the focal field of the radially polarized light exhibits strong longitudinal electric field and smaller focal spot, which have been utilized in numerous applications including imaging, machining, particle trapping, data storage, and sensing. As for azimuthally polarized incident light, only a donut shape azimuthal component exists near the focal plane.

Besides the vectorial optical field, scalar vortex beam is another example of structured light that has been intensively studied, which is characterized by a helical wavefront along with an optical singularity in the center [39,40]. The optical vortex is a position in the beam cross-section around which the optical phase advances or retards by a multiple of 2π. Generally, a scalar vortex beam carries orbital angular momentum (OAM) and is characterized by its topological charge l (l can be any integer), according to the number of twists that the light does in one wavelength. The higher the number of the twist is, the faster the light is spinning around the axis. Due to the phase singularity at the center, a scalar vortex beam exhibits a donut-shaped distribution with a dark center when projecting onto a flat surface. Fig. 6.3 illustrates the helical phase patterns and corresponding intensity distribution at the beam waist for different Laguerre−Gauss mode (LG_{pl}), which is a well-known class of optical scalar vortex beam with azimuthal symmetric intensity pattern and a helical phase structure described by $e^{il\varphi}$. It can be clearly seen that there is a dark center in the intensity profiles, while the phase pattern would jump from 0 to $l \cdot 2\pi$ around the singularity point. Besides, the size of the dark hole increases with the topological charge. Scalar vortex beams in various forms have been developed by a number of groups for different applications ranging from optical spanners in optical tweezers, and faster data manipulation in quantum computing, to twisted light for telecommunications. Different from the spin angular momentum (SAM) that only has two distinct values, the number of topological charge of the vortex beam is infinite theoretically. The OAM modes with different l values form a large orthonormal set of functions that can be used to encode information. Consequently, OAM encoding has advantages over SAM with increased

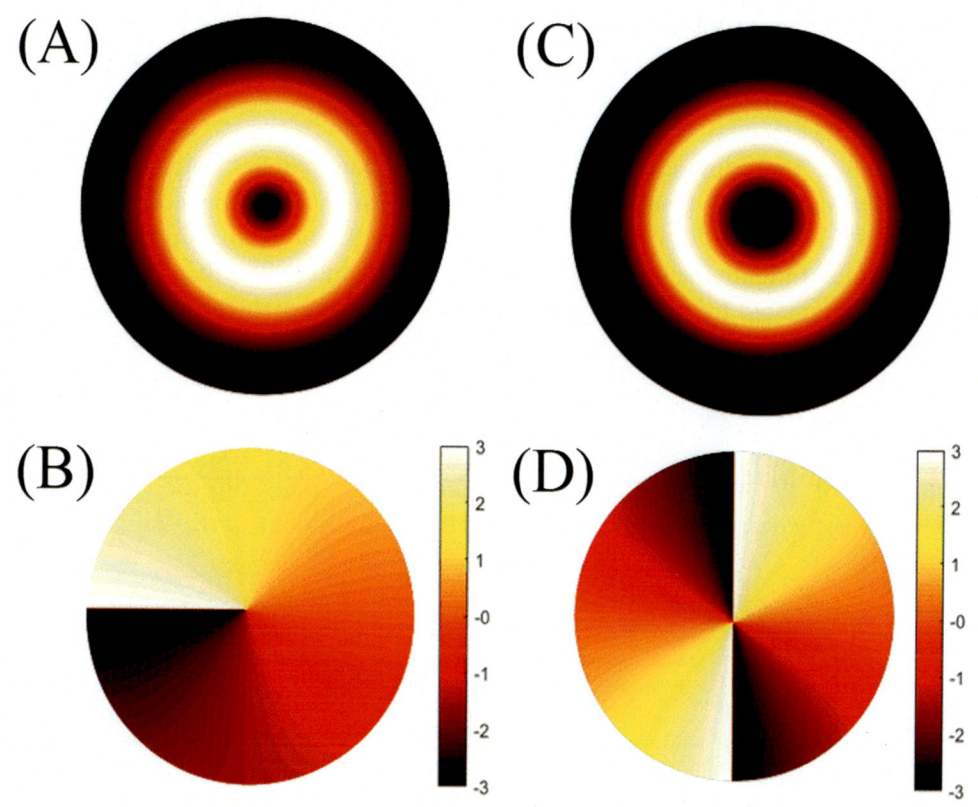

Figure 6.3 Intensity distributions (A, C) and corresponding helical phase patterns (B, D) at the focal plane of LG01 and LG02 beams.

channel capacity for communication systems. Besides, the light—matter interaction involving vortex beams is able to spin objects, create novel imaging systems, and behave within nonlinear material to give new insights into quantum optics [40].

6.4 Optical trapping of nonchiral particle

6.4.1 Trapping nonresonant metallic nanoparticle with cylindrical vector beam

Metallic nanoparticles are highly attractive in plenty of areas from biology to electronics due to their unique size-dependent properties. For example, the local field enhancement from metallic nanoparticles is adopted by surface-enhanced Raman spectroscopy (SERS) to enhance Raman signal and realize label-free detection of proteins, pollutants, and other molecules [41]. Consequently, the characteristics of optical tweezers such as noncontact and holding nature make it well suited to be combined with

SERS, enabling ultrasensitive molecular recognition in liquids. Different from dielectric particles that are relatively easier to trap due to the dominating gradient force, generally trapping metallic particles is considered difficult due to severe scattering and absorption.

To increase the trapping efficiency of metallic nanoparticles, radially polarized beam was proposed to replace the conventional linear polarization as the illumination [42]. Fig. 6.4 illustrates the line-scan of energy density distribution along the optical axis for a highly focused radial polarization. It can be seen that the focal field strength is dominated by the axial component, which has narrower full-width-half-maximum than that of linear polarization, leading to the stronger gradient force that pulls the particle toward the center of the focus. Note that the gradient force can be further increased by adopting an annulus pupil apodization function that only allows annular illumination focused by the lens. Moreover, since the axial component of the focal field is a pure imaginary value, this nonpropagating field does not contribute to energy flow along the optical axis. As shown in

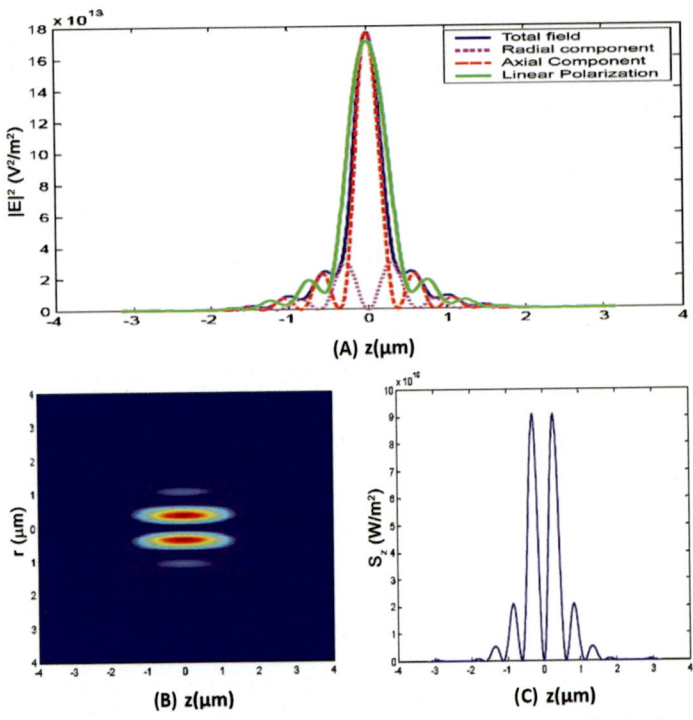

Figure 6.4 (A) Longitudinal line-scan of intensity distribution at the focal plane for a tightly focused radially and linearly polarized beam. (B) Spatial distribution of the time-averaged Poynting vector $\langle S_z \rangle$ in the $r-z$ plane for highly focused radial polarization. (C) Line scan of (B) at the focal plane.

Fig. 6.4B and C, it is clear that the axial component of the time-averaged Poynting vector along the optical axis is substantially zero, giving rise to negligible scattering force near the optical axis. As the wavelength of the trapping laser ($\lambda = 1.047$ μm) is assumed to be away from the resonance region of the gold particle, the optical force exerted on the particle is calculated using dipole approximation and its distributions are illustrated in Fig. 6.5. Compared to the case of linearly polarized light illumination, radial polarization provides higher gradient force as well as lower scattering forces. More importantly, the scattering/absorption force is substantially zero along the optical axis, eliminating the major potential cause of trap destabilization. Therefore the characteristics of highly focused radial polarization in terms of spatial force separation and dominating nonpropagating axial field allow one to trap metallic particles in three-dimensional space stably.

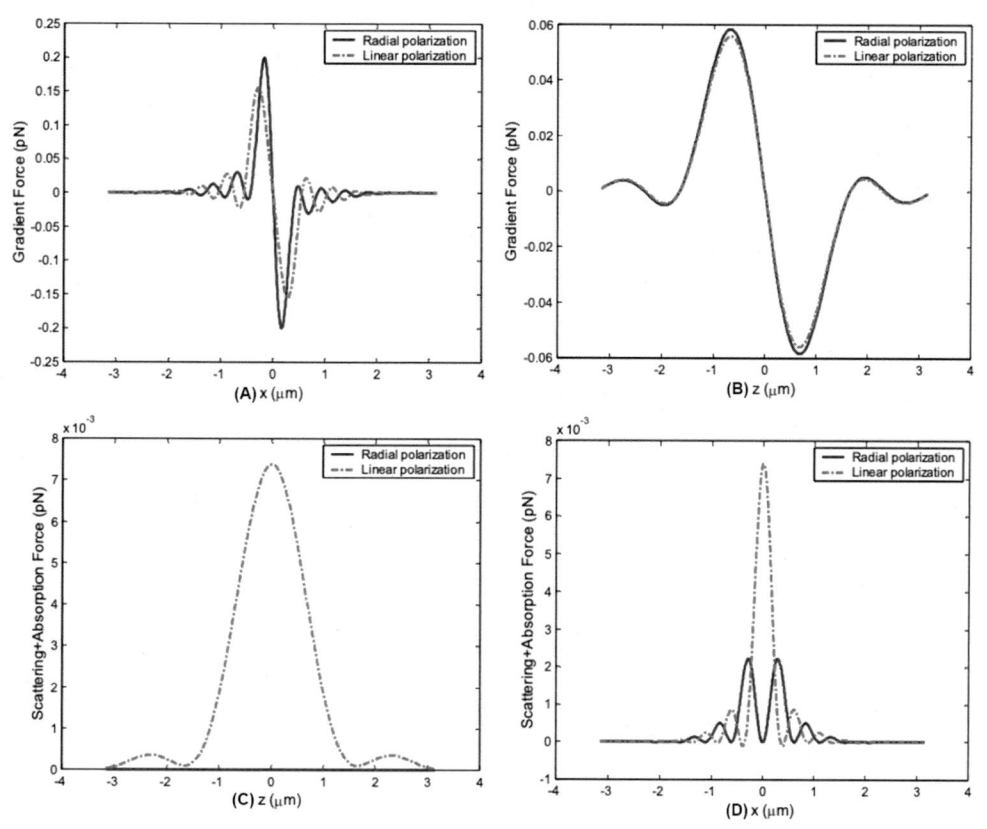

Figure 6.5 Optical forces exerted on a 38.2 nm (diameter) gold particle. (A) Transverse gradient force along x-axis; (B) axial gradient force along z-axis; (C) sum of axial scattering and absorption forces along z-axis; (D) sum of axial scattering and absorption forces on x-axis. For comparison, result for highly focused linearly polarized incident is also shown by the dotted lines.

The feasibility and advantage of using radial polarization to trap metallic nanoparticles have also been experimentally demonstrated [43]. As the optical trapping setup is shown in Fig. 6.6, a laser beam from a Nd-YVO4 laser ($\lambda = 1.064\,\mu m$) is expanded by lenses L1 and L2, and then introduced into a polarization converter, which converts the linearly polarized Gaussian beam into a radially or azimuthally polarized beam. To measure the trapping stiffness, a position sensitive detector (PSD) is adopted to detect the Brownian motion of a single trapped gold particle. In addition, the dark-field imaging method is used to observe and detect the gold particles by using an oil-immersion condenser (NA = 1.3) and the NA of the trapping objective is chosen to be 0.9.

Assuming a particle with mass m immersed in liquid with drag coefficient γ, its motion is determined by not only stochastic force but also viscous force, in which case a Langevin equation is suitable to describe the Brownian motion of particle:

$$m\frac{\partial^2 x}{\partial t^2} + \gamma\frac{\partial x}{\partial t} + \kappa x = F_f(t), \tag{6.20}$$

where the terms on the left side of the equation are inertial force, viscous damping force, and optical restoring force for a trap stiffness κ, respectively. On the right side of the equation, $F_f(t)$ is the fluctuating force induced by Brownian motion. Under the overdamping approximation, the inertial force can be ignored since the motion of the particle is dominated by viscous force. By performing a Fourier transformation to the Langevin equation, one could obtain a Lorentzian power spectrum for the motion trajectory of gold particles, which provides corner frequency f_c that determines the trap stiffness. The relationship between corner frequency and trap stiffness can be described via $\kappa = 2\pi\gamma f_c$, where $\gamma = 6\pi\eta r$, with η and r being the viscosity coefficient of water and the radius of the particle [44].

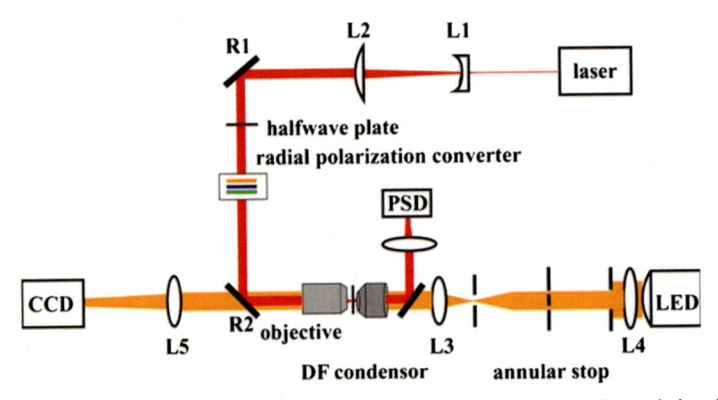

Figure 6.6 Optical tweezers setup. The trapping laser was introduced into the radial polarization converter after expanding and focused to form a trap. The trapped particles are imaged by *CCD*, and the Brownian motion is detected by PSD. *CCD*, Charge coupled Device; *PSD*, Position sensitive detector.

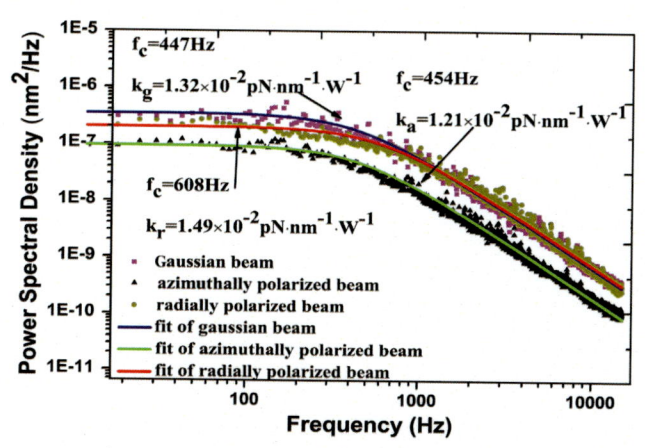

Figure 6.7 Power spectra of gold particles with a diameter of 90 nm trapped by radially polarized, azimuthally polarized, and Gaussian beams. The stiffness is normalized by laser power, which are 220, 200, and 180 mW for the radially polarized, the azimuthally polarized, and the Gaussian beam, respectively.

Fig. 6.7 shows the measured power spectrum of a trapped single gold nanoparticle with diameter of 90 nm. By fitting the power spectrum with a Lorentzian curve, the corner frequency and the trapping stiffness can be calculated accordingly. For comparison, the transverse trapping stiffness of linear, radially, and azimuthally polarized light illumination are considered, which are found to be $(1.34 \pm 0.13) \times 10^{-2}\,\mathrm{pN\,nm^{-1}\,W^{-1}}$, $(1.49 \pm 0.16) \times 10^{-2}\,\mathrm{pN\,nm^{-1}\,W^{-1}}$ and $(1.09 \pm 0.17) \times 10^{-2}\,\mathrm{pN\,nm^{-1}\,W^{-1}}$, respectively. The experimental results clearly show that the radially polarized light provides the highest trapping stiffness, while the azimuthally polarized beam gives the lowest.

6.4.2 Trapping resonant metallic nanoparticle with vectorial beam

Although radial polarization exhibits excellent performance in trapping metallic nanoparticle when the trapping wavelength is far from the plasmon resonance, trapping resonant metallic nanoparticle is still challenging due to two main reasons. First, both the induced polarization and the scattering force are enhanced at the resonant condition, which would strongly push particles away from the focal spot. Second, resonant illuminations of metallic nanoparticles give rise to severe heating effects due to absorption [45]. The high temperature of the trapped nanoparticles may destroy the trapping and release the nanoparticles via bubble formation at the particle surface, which remains the ultimate obstacle of realizing stable trapping of metallic nanoparticles under resonance conditions. With structured light for illumination, a novel strategy has been developed to form a stable three-dimensional trapping of metallic nanoparticles even under resonant conditions [46]. As shown in Fig. 6.8, the optical illumination is created by sculpting the amplitude and phase of a radially polarized optical field. In this

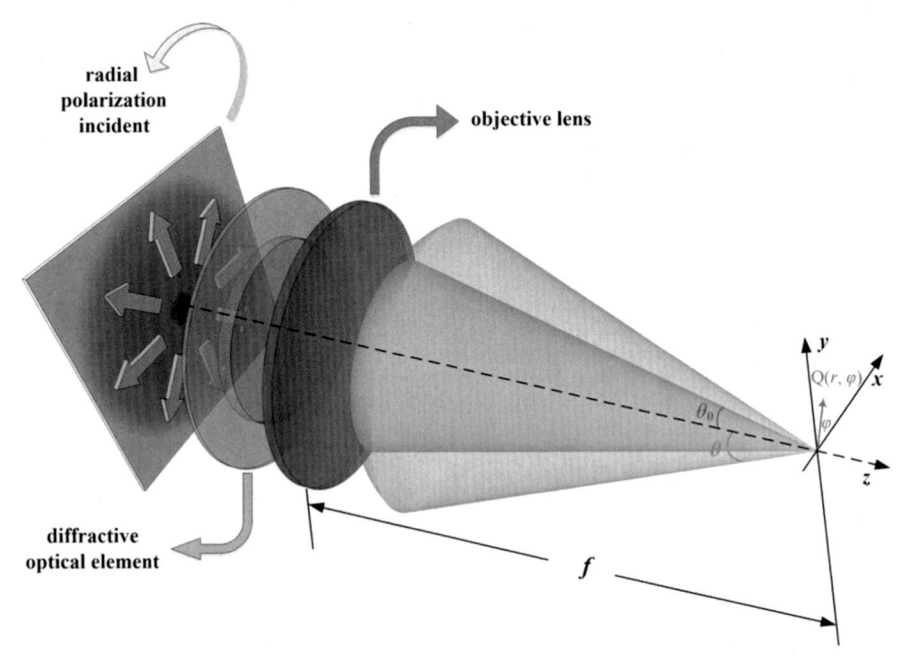

Figure 6.8 Diagram of the optical tweezers setup using spatially tailored radially polarized beam. A DOE is inserted at the entrance pupil plane of the objective lens. *DOE*, Diffractive optical element.

case, radial polarization is chosen because its dominating axial component under tightly focusing condition is capable of generating negative scattering force pointing against the optical power flow, which is similar to the Bessel beams [47]. The amplitude of the illumination gradually increases from the center to the edge in the beam cross-section, which can be described as $U(\theta) = 1/\cos^2 \theta$. This nonuniform amplitude distribution is helpful to increase the magnitude of negative scattering force, since which is mainly attributed to the optical field that comes from large converging angles. In addition, a two-zone diffractive optical element (DOE) is inserted at the entrance pupil plane of a high NA objective lens, which provides an opportunity to engineer the gradient and scattering forces separately. The modulation function $M(\theta)$ applied by the DOE is chosen to be:

$$M(\theta) = \begin{cases} 1, \theta > \theta_0 \\ i, \theta \leq \theta_0 \end{cases}, \tag{6.21}$$

where $\theta_0 = 1.47$ rad indicates the boundary between the inner and outer zones of the DOE.

Considering a gold nanoparticle with the radius of 50 nm is placed in water, and the wavelength of the illumination (532 nm) is near its resonant absorption peak, the optical forces exerted on the gold nanoparticle are numerically calculated and shown in Fig. 6.9. It can be seen that negative scattering force is generated along the optical axis, while the gradient force has a deep valley near the focus. The phase modulation leads to a dislocation between the centers of the distributions of the gradient and scattering forces, giving rise to an equilibrium point at $z \approx 0.1\lambda$. One can see that the key of this trapping technique is to create an optical focal field that can provide appropriate distribution of the optical forces, particularly the optical scattering force. To better understand these phenomena, the relationship between the characteristic of the optical focal field and the optical forces are further studied. Due to the interference between the optical fields transmitted from the inner and outer zones of the lens, complex optical vortices structures (shown in Fig. 6.10) are found within the focal volume that redirect the power flow from the axial direction to the radial direction locally, which leads to the creation of negative scattering forces.

Thermal mechanism in optical tweezers is another main reason that destabilizes the trap and a much more difficult factor to combat in conventional optical tweezers

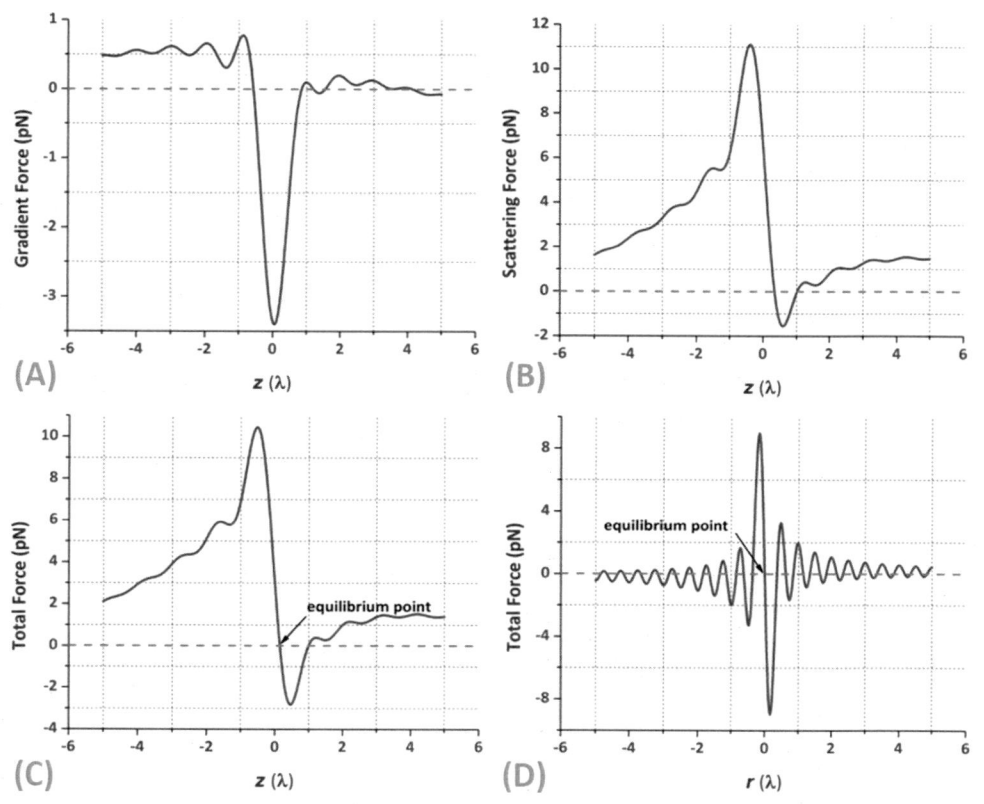

Figure 6.9 Optical forces on 50 nm (radius) resonant gold nanoparticle using the optical tweezers setup shown in Fig. 6.8 with phase difference of $\pi/2$ and $\theta_0 = 1.47$ rad° for the DOE. (A) axial gradient force; (B) axial scattering force; (C) sum of the axial gradient and scattering forces; (D) transverse plane distribution of total force at the axial equilibrium point. *DOE*, Diffractive optical element.

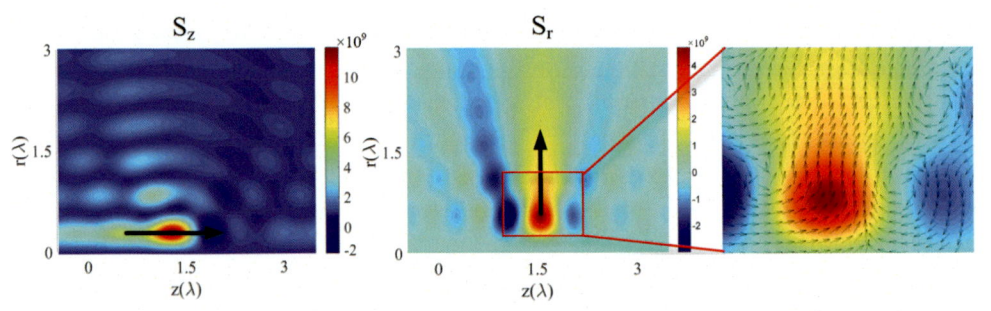

Figure 6.10 Redirection of the local power flow near the focus due to the interference between the optical fields from different zones of the illumination.

when the operating wavelength is close or at the resonant wavelength for the metallic nanoparticles, especially for the particles immersed in solution with low viscosity. For a gold nanoparticle immersed in water, the critical temperature of water is about 647K, above which the water at the surface of the nanoparticle would evaporate and form the nucleation of vapor bubble [48], leading to the escape of nanoparticle from the optical trap. Certainly, the temperature can be decreased by reducing the laser power; however, the potential depth would drop as well. To solve the dilemma between heating effect and trapping stability, the stability of the optical tweezers can be optimized by adjusting the phase difference and transition position of the DOE under the premise of avoiding vapor bubble formation. In this case, the highest allowed input power is calculated to be only 6.7 mW, and the potential depth is calculated to be $8.07 \times k_B T$ and $4.57 \times k_B T$ along axial and lateral axes, respectively. Consequently, optical overheating in this novel optical tweezers can be avoided while maintaining deep enough trapping potential, enabling stable trapping of metallic nanoparticle under the most challenging condition.

Although the resonant metallic nanoparticle is possible to be trapped by taking advantage of the subtle balance between gradient force and scattering force, it is still challenging to realize the required complex spatial distribution on the vectorial optical illumination. To solve the drawbacks arising from the use of multizone phase plate, 4Pi microscopy has been adopted to construct an optical tweezers that are also capable of trapping and manipulating metallic nanoparticle [49]. As shown in Fig. 6.11, two counter-propagating radially polarized beams with phase difference of π normally illuminate the focusing system consists of two high NA aplanatic lens. The electric field in the vicinity of the focus can be expressed as:

$$\vec{E}_f(r, \phi, z) = \vec{E}_L(r, \phi, z) + \vec{E}_R(r, \phi, -z), \tag{6.22}$$

where E_L and E_R are the focal field propagating from the left and right, respectively. Different from conventional focusing system with only one objective lens, 4Pi

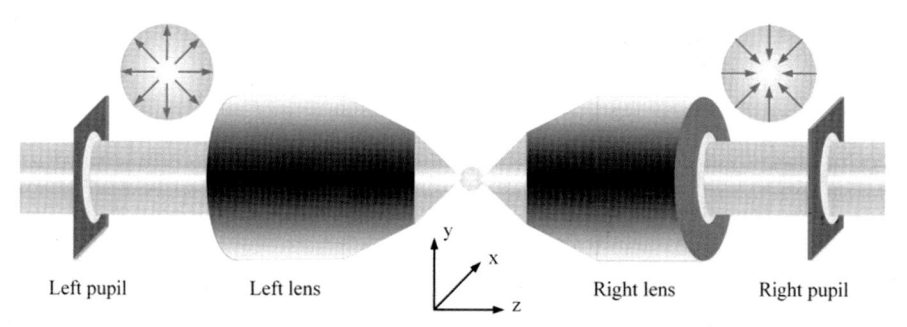

Figure 6.11 The diagram of the optical tweezers constructed around a 4Pi focusing system.

microscopy utilizes two counter-propagating beams to create complex interference field in the focal region, providing more degrees of freedom to tailor the optical force. As the intensity pattern and the corresponding line-scans shown in Fig. 6.12A and B, a spherical spot rather than an elliptical-shaped focal field is generated in the focal region. Considering a resonant gold nanoparticle, the exerted optical forces are illustrated in Fig. 6.12C and D. Due to the smaller focusing spot, both the axial and radial gradient forces are enhanced compared to the case of single objective lens. More importantly, the axial scattering force is canceled due to the counter-propagating light beam, leading to the equilibrium position along the optical axis at $z = 0$.

Besides metallic nanoparticle, three-dimensional trapping of dielectric nanoparticle with refractive index lower than ambient is also a challenging task, which is widely required in aqueous systems (such as air bubbles and hollow particles). As for a low-index nanoparticle interacting, the gradient force would point to the region of low intensity, consequently a beam with focusing feature of hollow intensity distribution is necessary [50]. However, light beams with donut-shaped intensity distributions such as high-order Bessel beams, optical vortex beams and CV beams can only trap the low-index particles at the focal plane [51−54], due to the absences of axial gradient force. To create an axial equilibrium position as well, three-dimensional optical chain and bottle beams are generated with complicated DOE and introduced to optical tweezers [55−58]. Similar to the technique mentioned in Fig. 6.11, a three-dimensional hollow beam can also be generated by illuminating the 4Pi focusing system by two radially polarized first-order Laguerre−Gaussian beams with wavelength of 808 nm and π phase difference [59]. As the focal field shown in Fig. 6.13A, the interference pattern has complete circular symmetry in the transverse plane with zero intensity at the center. Besides, this is accompanied by a nearly circularly symmetric intensity distribution along the optical axis (Fig. 6.13B), leading to a tightly focused hollow spherical spot, which can also be demonstrated by the corresponding line-scans of the axial and transversal intensity distributions shown in Fig. 6.13C and D, demonstrating that axial and transversal focal spots are nearly equal.

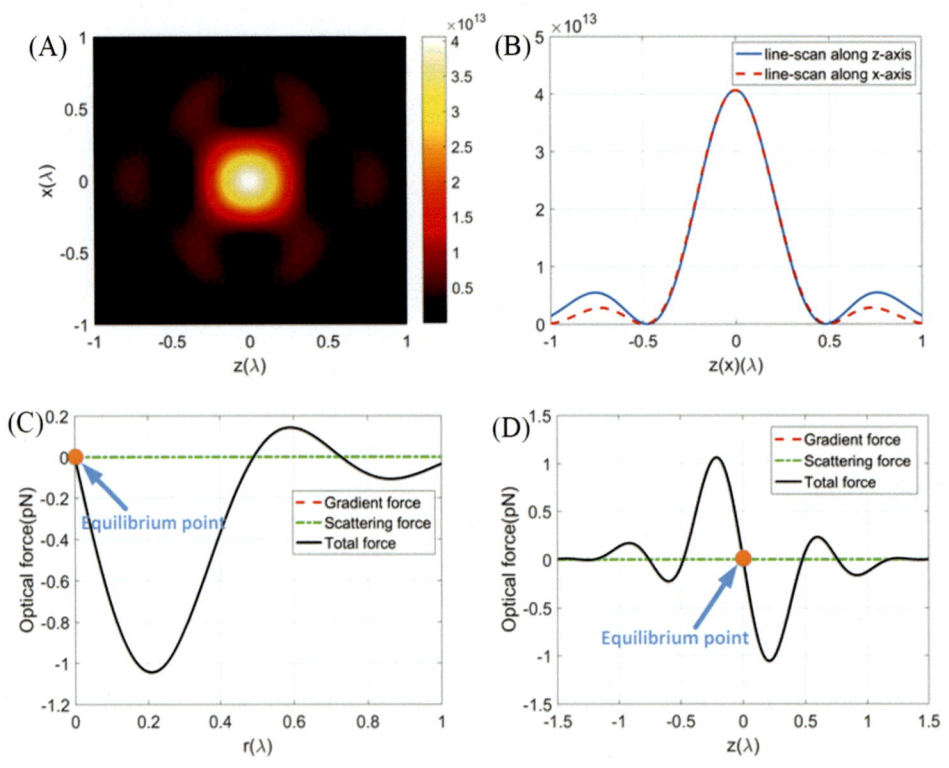

Figure 6.12 (A) Intensity distribution in the vicinity of focal point for radially polarized nonvortex beam focused by 4Pi focusing system. (B) Line-scans of corresponding axial and transversal intensity distributions. Optical forces exerted on 50 nm (radius) resonant gold nanoparticle along (C) radial and (D) longitudinal axes.

Considering a hollow spherical low-index nanoparticle with relative permittivity of 1 and radius of 50 nm immersed in water, the optical forces along the z- and x-axis are calculated and shown in Fig. 6.14A and B. It can be clearly seen that the axial scattering force is canceled by the counter-propagating focal fields. Besides, the transversal scattering force is also negligible for lossless particle. Consequently, the low-index nanoparticle would be confined at the dark center of the hollow spherical focal spot by the gradient force.

6.4.3 Manipulating the dynamic behavior of sphere/ellipsoidal nanoparticle

In addition to stably trapping particle at specific location in three-dimensional space, flexibly manipulating the dynamic behavior of the particles also is an important issue. When particle interacts with a vortex optical field, the OAM would be transferred from light to the particle, leading to the rotation of the particle around the optical axis. For example, high-order Bessel beam is one of the

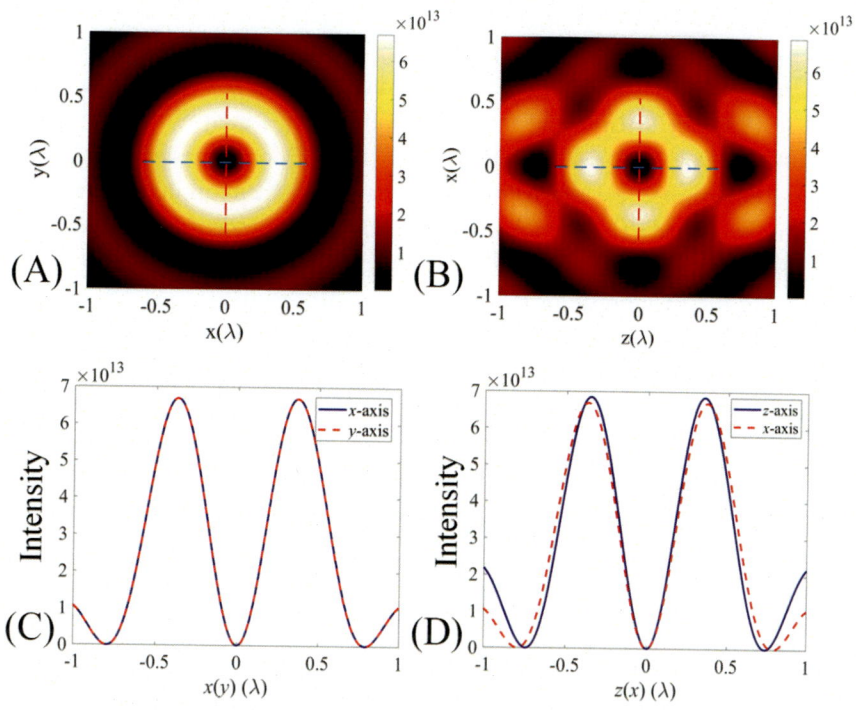

Figure 6.13 Intensity distribution of the focal field for radially polarized first-order Laguerre–Gaussian focused by 4Pi focusing system in the (A) x–y plane and (B) z–x plane. (C, D) Line-scans of intensity distributions indicated by the dashed line shown in (A) and (B).

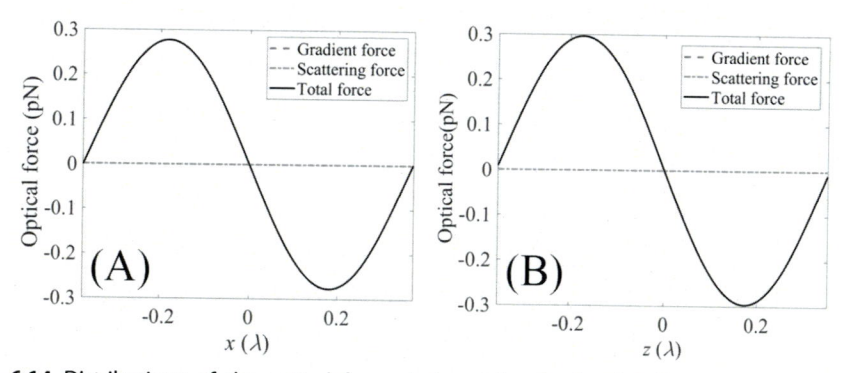

Figure 6.14 Distributions of the optical forces induced by the focal field in Fig. 6.13 exerted on 50 nm (radius) low-refractive index nanoparticle along (A) transverse and (B) longitudinal directions.

promising types of optical field, which has doughnut–shaped intensity profile and possess OAM associated with the spatial field distribution in vortex beam [60–62]. Conventional generation methods of Bessel beams involve the use of bulky elements (such as axicon), which are difficult to be integrated into a compact platform. It is known that the focal fields of tightly focused radial and azimuthal polarized light are dominated by J_0 and J_1 modes respectively, consequently zeroth and first-order Bessel beams can be generated by illuminating CV beam on a resonant nanostructure [63–66]. Moreover, since the dominating longitudinal electric field of tightly focused radially polarized vortex beam (RPVB) only possess Bessel function with order m determined by the topological charge of the illumination, as shown in Fig. 6.15, it is feasible to generate arbitrary high-order evanescent Bessel beam through coupling the focusing RPVB with a resonant photonics band gap (PBG) structure while depressing the magnitude of the transverse components [67].

Considering a gold nanoparticle with radius of 50 nm placed near the last interface of the PBG, Fig. 6.16 illustrates the distribution of the exerted optical force induced by evanescent second- and third-order vortex Bessel beam with wavelength of 808 nm. As the force along radial direction is shown in Fig. 6.16A and B, the particle is stably trapped by the gradient force at $r = 0.47\lambda$ and 0.65λ, respectively, corresponding to the locations that have the largest intensity. Due to the evanescent decay of the generated Bessel beam, the particles would experience a negative force along optical axis (Fig. 6.16C and D), which would drag the particle toward the PBG structure. Since the particle is trapped off the beam axis, the inclination of the helical phase fronts and the corresponding momentum result in a tangential force, leading to the particle rotating around the beam axis. As shown in Fig. 6.16E, the azimuthal force is constant thus the particle will orbit around the optical axis, and the direction of the orbital motion can be reversed by changing the handedness of the helical phase front. The torque of a transverse radiation pressure is written as

$$T_z(r, z = 0) = 2rF_{trans}^{scat}(r, \varphi, z = 0), \tag{6.26}$$

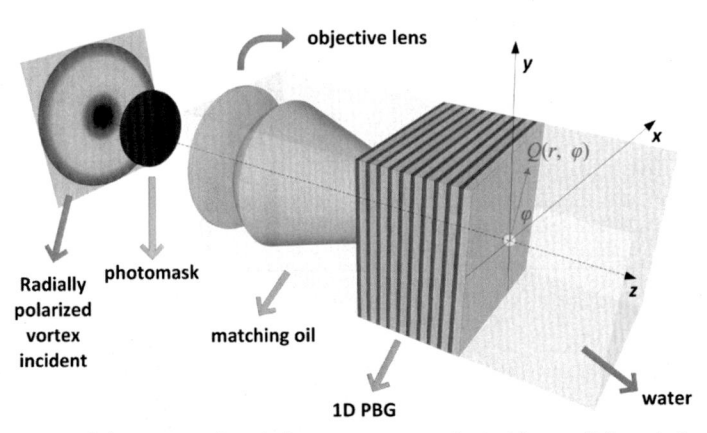

Figure 6.15 Diagram of the proposed optical tweezers setup. An incident radially polarized vortex beam is highly focused by an objective lens onto a one-dimensional PBG structure. *PBG*, Photonics band gap.

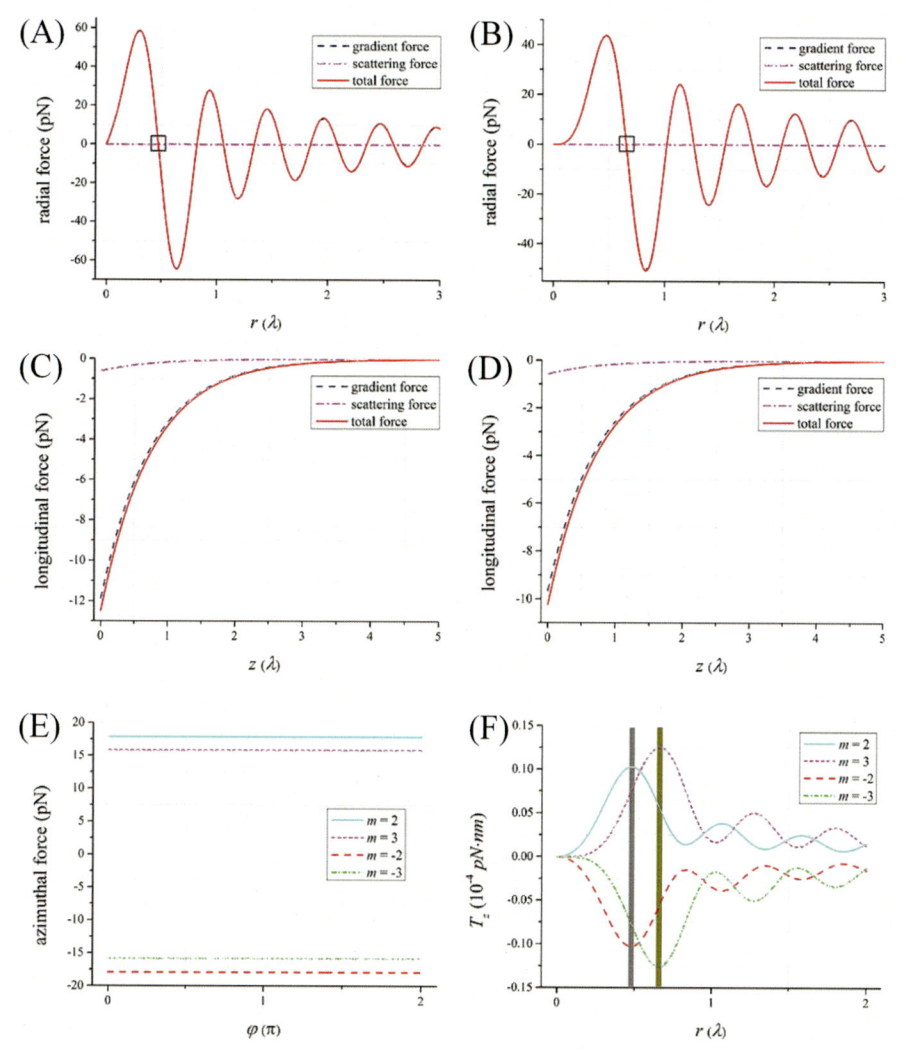

Figure 6.16 Calculated optical forces on 50 nm (radius) gold nanoparticle in evanescent vector Bessel beam with different topological charge. (A, C) Radial force along r-axis and (B, D) longitudinal force along z-axis for $m = 2$ and 3. Distribution of (E) Azimuthal force and (F) longitudinal component of torques for $m = \pm 2, \pm 3$. The gray and yellow bars denote the equilibrium position in the focal plane for the case of $|m| = 2$ and 3, respectively.

where F_{trans}^{scat} denotes the scattering force in the transverse plane. Fig. 6.16F illustrates the influence of the controllable factor m of the evanescent vortex Bessel beam on T_z. Note that the direction of T_z depends on the topological charge of the incident vector field. The magnitude of T_z is proportional to linear velocity of the trapped particle and its corresponding angular velocity is determined by both the torque and the orbital trajectory. Consequently, the orbital rotating radius and direction of the trapped particle can be tuned by the order and OAM of the Bessel beam, respectively.

Besides using vortex beam to control the revolution of the nanoparticle, the concepts of controlling the orientation and rotation of tiny objects have attracted increasing attentions. It is well-known that the circularly polarized light would induce a torque on metallic particle that make it rotate around the optical axis. Besides, SAM and OAM would be coupled under tight focusing condition, making it possible to control the rotation speed of the trapped particle [68]. It is demonstrated that the interactions between the tunable photonic spin and the nanoparticles lead to not only three-dimensional trapping, but also the precise control of the nanoparticle's movement in terms of stable orientation, rotational orientation and rotation frequency [69].

To achieve controllable spin axis orientation and ellipticity within a diffraction limited tightly focused beam, the required pupil field is analytically derived through reversing the radiation patterns from three dipoles located at the focal point of a high NA lens. Considering an ellipsoidal particle that is much smaller than the incident wavelength, the modified electric polarizability can be expressed as

$$\alpha = \frac{\alpha_0}{1 - i\alpha_0 k^3/(6\pi)}, \tag{6.27}$$

where α_0 is a matrix representing Claussius−Mossotti polarizability:

$$\alpha_0 = \begin{pmatrix} \alpha_{0,x'} & 0 & 0 \\ 0 & \alpha_{0,y'} & 0 \\ 0 & 0 & \alpha_{0,z'} \end{pmatrix}. \tag{6.28}$$

The coefficient $\alpha_{0,i}$ represent the components of the polarizability tensor:

$$\alpha_{0,i} = \frac{1}{3} abc \frac{\varepsilon_m(\omega)/\varepsilon - 1}{\left[1 + \varepsilon_m(\omega)/\varepsilon - 1\right] n_i}, \quad (i = a, b, c), \tag{6.29}$$

where n_i is the depolarizing factor [69]:

$$
\begin{aligned}
n_i &= \frac{1}{2} abc \int_0^\infty \left[\left(s+a^2\right)^2 \left(s+b^2\right)\left(s+c^2\right) \right]^{-1} ds, \quad (i = x') \\
&= \frac{1}{2} abc \int_0^\infty \left[\left(s+a^2\right)\left(s+b^2\right)^2\left(s+c^2\right) \right]^{-1} ds, \quad (i = y') \\
&= \frac{1}{2} abc \int_0^\infty \left[\left(s+a^2\right)\left(s+b^2\right)\left(s+c^2\right)^2 \right]^{-1} ds, \quad (i = z').
\end{aligned}
\tag{6.30}
$$

where k is the wave-vector, $\varepsilon_m(\omega)$ and ε are the relative permittivity of the particle from bulk material and the surrounding medium, respectively. Parameters a, b, and c are half the length of the principal axes, corresponding to the semimajor and semiminor axis of an ellipse.

As shown in Fig. 6.17, the orientation of the ellipsoid with respect to its center of mass can be described by two angular momenta (Θ_0, Φ_0), where Θ_0 and Φ_0 are the polar and azimuthal angles of the major axis, respectively. It is known that the polarizability tensor of the particle would be affected by its orientation in space as $\alpha_{0,ij} = R_{i\delta}\alpha_{0,\,\delta\gamma}R_{\gamma j}^{-1}$, where the rotation matrix R_{ij} is given by [70]:

$$R_{ij} = \begin{pmatrix} \cos\theta_0\cos\phi_0 & -\sin\phi_0 & \sin\theta_0\cos\phi_0 \\ \cos\theta_0\sin\phi_0 & \cos\phi_0 & \sin\theta_0\sin\phi_0 \\ -\sin\theta_0 & 0 & \cos\theta_0 \end{pmatrix}. \tag{6.31}$$

Assuming a particle immersed in the in the focal volume of a highly focused optical field, the particle would spin around its center of mass due to the transfer of the SAM from the light. This intrinsic torque $\mathbf{T}$ can be expressed as [29]:

$$\vec{T} = \frac{1}{2}|\alpha|^2 \mathfrak{R}\left\{ \frac{1}{\alpha_0^*}(\vec{E} \times \vec{E}^*) \right\}. \tag{6.32}$$

Without loss of generality, a dielectric spheroidal particle with size $(A, B, C) = (30\ \text{nm}, 30\ \text{nm}, 50\ \text{nm})$ and initial orientation $(\Theta_0 = 45\ \text{degrees}, \Phi_0 = 45\ \text{degrees})$ is placed near the focal field with photonic spin orientated at $(\alpha, \beta, \gamma) = (60\ \text{degrees}, 60\ \text{degrees}, 45\ \text{degrees})$, where α, β, and γ are the angles between the spin axis of the focal field and the positive direction of the Cartesian coordinate. The optical torque distribution and the corresponding three-dimensional view of the torque vector on the spheroid are also plotted in Fig. 6.18, where the position of the particle is marked by the asterisk. Note that (x', y', z') are the Cartesian coordinate in which z'-axis

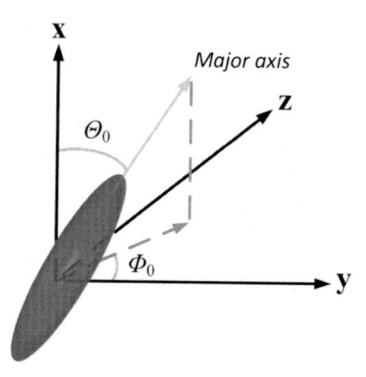

Figure 6.17 Spatial orientation of a spheroid particle.

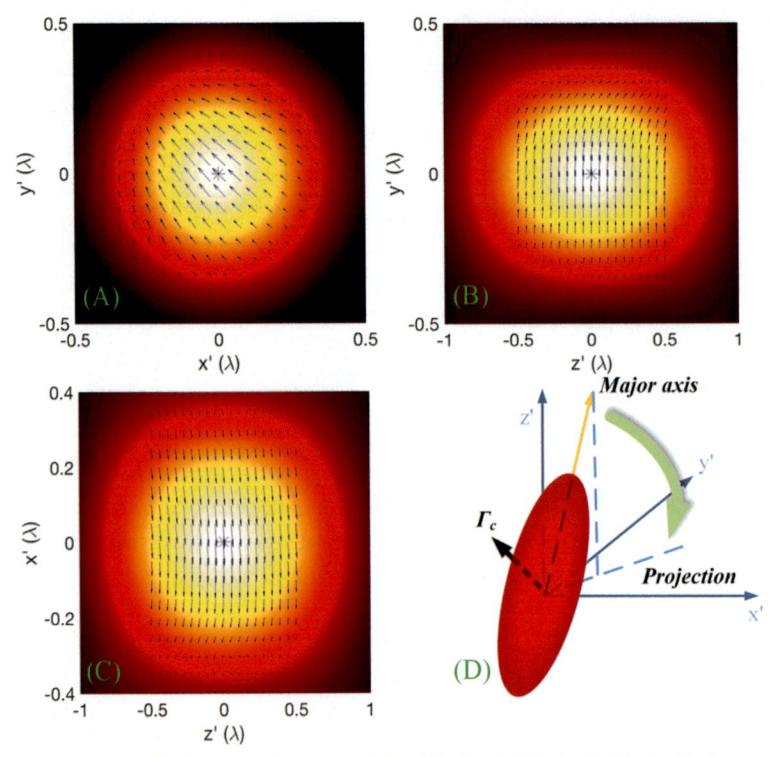

Figure 6.18 Distribution of the optical torque in the (A) $x'-y'$, (B) $y'-z'$, (C) $x'-z'$ plane, and (D) corresponding rotation schematic diagram of spheroid with orientation at ($\Theta_0 = 45$ degrees, $\Phi_0 = 45$ degrees). The equilibrium position is indicated by the asterisk.

coincides with the direction of the photonic spin. One can find that $\Gamma_{z'}$ is much smaller than the torque in the $x'-y'$ plane. The direction of the torque vector is calculated to be (135 degrees, 46 degrees, 84 degrees) with respect to the x'-, y'- and z'-axis, and the torque tends to rotate the major axis of the spheroid with increasing Θ_0. On the other hand, it can be seen from Eq. (6.33) that nonzero intrinsic torque is obtainable for spherical nanoparticle if the particle is absorptive and the optical field is not locally linear polarized.

Consequently, a gold spherical nanoparticle would experience a pure torque along z' direction at the equilibrium position, leading to the rotation of the particle along the spin axis (shown in Fig. 6.19). Therefore it is feasible to tune the rotation axis of the spherical particle by adjusting the orientation of the photonic spin. Furthermore, the relationship between the torque and the ellipticity of the focal field is studied and illustrated in Fig. 6.20A. It can be seen that the rotation direction of the particle is determined by the handedness of the ellipticity, and the decrease of the ellipticity could result in decreased intrinsic torque. Fig. 6.20B illustrates the relation between rotation

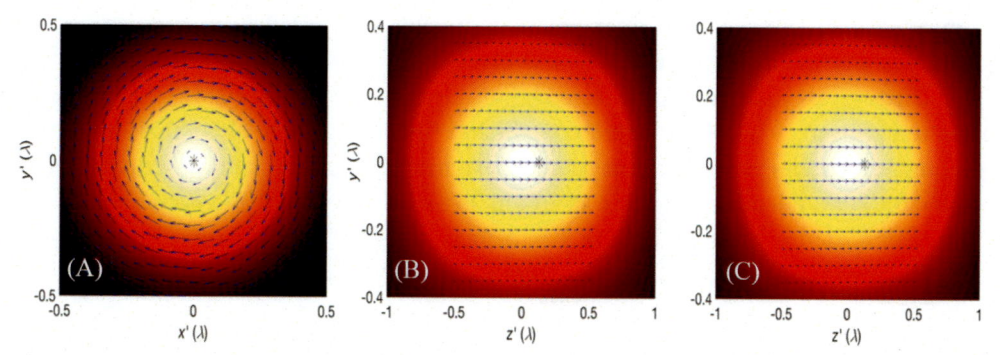

Figure 6.19 Distribution of the optical torque in the (A) $x'-y'$, (B) $y'-z'$, (C) $x'-z'$ plane for absorbing spherical nanoparticle. The equilibrium position is indicated by the asterisk.

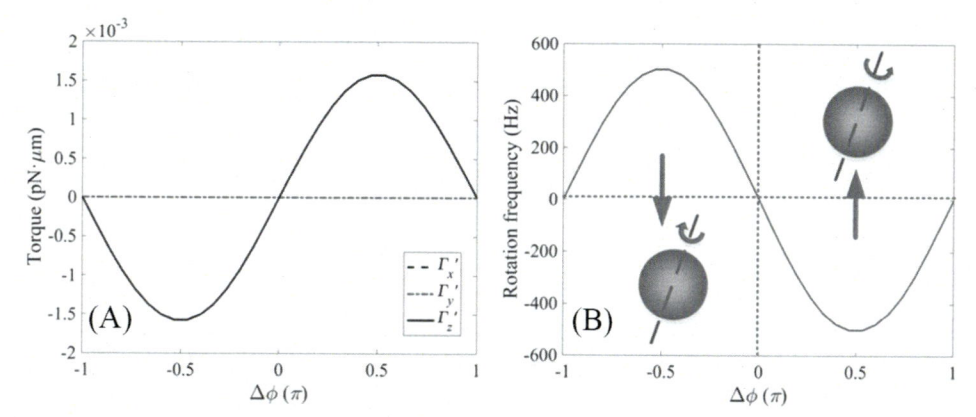

Figure 6.20 Particle rotation as a result of the torque from elliptically polarized light. (A) The torque and (B) the rotation frequency for absorbing nanoparticle is shown as a function of $\Delta\phi$, which is the phase difference determines the ellipticity of the focal field.

frequency and the ellipticity of the focal field. It confirms that the particle rotation is in fact a result of the torque from elliptically polarized light. Consequently, both the direction and the rate of the particle's rotation can be controlled by changing the orientation and the ellipticity of the photonic spin within the focal field.

6.5 Optical trapping of chiral particle

6.5.1 Modeling of optical force and torque on chiral particle

Chirality is a geometrical property where an object cannot be superposed onto its mirror image via either a translational or a rotational operation [71]. The mirror images of a chiral structure are enantiomers, and individual enantiomers are often designated as either right or left-handedness. In fact, this type of symmetry is much harder to be maintained than to be broken, consequently, chirality exists widely in various macroscopic and microscopic structures. For example, proteins and nucleic acids are built of chiral amino acids and chiral sugar. In addition, Deoxyribo Nucleic Acid (DNA) double helix, sugar, quartz, cholesteric liquid crystals, and biomolecules are also chiral structures. Although molecules with different handedness have the same chemical construction, usually they would possess distinct chemical behaviors. A chiral biomolecule can be inactive or toxic to cells if its original handedness is varied, which causes many diseases such as phocomelia, Parkinson's, Alzheimer's, type II diabetes and Huntington's [72]. Consequently, the sensitive detection and separation of substances by chirality are therefore of high demand in the fields of pharmacology, toxicology, and pharmacodynamics.

Different from conventional materials, the chirality of the electromagnetic material induces cross-polarization between electric and magnetic fields, leading to unique optical force effects, which derives from the momentum transfer associated with bending light when tightly focused laser beam interacts with the particles. Different from achiral metallic and dielectric materials, the optical response of chiral particle depends on not only the electric and magnetic polarizability, but also has close connection to the electromagnetic/magnetoelectric polarizability (also known as chiral polarizability), which provides additional degree of freedom to tailor the optical force and develop novel optical tweezers techniques.

The scattering behavior of the particle can be rigorously solved by the Mie theory, and the corresponding Mie scattering coefficients $(a_n^{(1)}, a_n^{(2)}, b_n^{(1)}, b_n^{(2)})$ govern the relationship between the incident and scattering fields [74,73–77]:

$$a_n^{(1)} = [A_n^{(2)} V_n^{(1)} + A_n^{(1)} V_n^{(2)}]Q_n, \quad a_n^{(2)} = [A_n^{(1)} W_n^{(2)} - A_n^{(2)} W_n^{(1)}]Q_n,$$
$$b_n^{(1)} = [B_n^{(1)} W_n^{(2)} + B_n^{(2)} W_n^{(1)}]Q_n, \quad b_n^{(2)} = a_n^{(2)}, \tag{6.33}$$

with

$$A_n^{(j)} = Z_S D_n^{(1)}(x_j) - D_n^{(1)}(x_0), \quad B_n^{(j)} = D_n^{(1)}(x_j) - Z_S D_n^{(1)}(x_0),$$
$$W_n^{(j)} = Z_S D_n^{(1)}(x_j) - D_n^{(3)}(x_0), \quad V_n^{(j)} = D_n^{(1)}(x_j) - Z_S D_n^{(3)}(x_0),$$
$$Q_n = \frac{\psi_n(x_0)/\xi_n(x_0)}{V_n^{(1)} W_n^{(2)} + V_n^{(2)} W_n^{(1)}}, \tag{6.34}$$

where $x_0 = kr_s$, $x_1 = k_1 r_s$, and $x_2 = k_2 r_s$ with r_s represents the radius of particle and wave number $k_1 = k(\sqrt{\varepsilon_r \mu_r} - \kappa)/\sqrt{\varepsilon_m}, k_2 = k(\sqrt{\varepsilon_r \mu_r} + \kappa)/\sqrt{\varepsilon_m}$, with ε_r and μ_r represents the relative permittivity and permeability of the chiral medium, and ε_m represents the relative permittivity of the immersion medium. $Z_S = \sqrt{\mu_r \varepsilon_m / \varepsilon_r}$ is the wave impedance of the particle, while $\psi_n(x)$ and $\xi_n(x)$ are the Riccati–Bessel functions of the first and third kinds and $D_n^{(1)}(x) = \psi_n'(x)/\psi_n(x)$, $D_n^{(3)}(x) = \xi_n'(x)/\xi_n(x)$ are the corresponding derivatives.

Considering a spherical chiral particle that is much smaller than the incident wavelength, it can be treated in quasistatic limit and represented by point polarizability. Within the dipole approximation, the induced electric dipole moment $\boldsymbol{p}$ and magnetic dipole moment $\boldsymbol{m}$ can be expressed as $\boldsymbol{p} = \alpha_{ee}\mathbf{E} + \alpha_{em}\mathbf{B}$ and $\boldsymbol{m} = -\alpha_{em}\mathbf{E} + \alpha_{mm}\mathbf{B}$, where $\mathbf{E}$ and $\mathbf{B}$ are the electric field and magnetic induction of the incident optical fields, α_{ee}, α_{mm}, and α_{em} describe the electric, magnetic and electromagnetic polarizability of the sample. Note that α_{em} is related to the chirality parameter κ of the material that the particle is made of, and the imaginary part of α_{em} will change its sign if the handedness of the chiral sample changes. The polarizability elements of the dipolar chiral particle relate to the scattering coefficients [73]:

$$\alpha_{ee} = \frac{i6\pi\varepsilon_0\varepsilon_m}{k^3}a_1^{(1)}, \alpha_{mm} = \frac{i6\pi}{\mu_0 k^3}b_1^{(1)}, \alpha_{em} = -\frac{6\pi}{Z_0 k^3}a_1^{(2)}, \tag{6.35}$$

where $Z_0 = \sqrt{\mu_0/\varepsilon_0\varepsilon_m}$ is the wave impedance in vacuum and ε_0 and μ_0 are the vacuum permittivity and permeability.

The optical force exerted on a chiral nanoparticle in the transverse optical needle field with transverse spin (TONFTS) can be derived in terms of the electric and magnetic dipoles [73]:

$$\langle\mathbf{F}\rangle = \frac{1}{2}Re\left[(\nabla\mathbf{E}^*)\cdot\mathbf{p} + (\nabla\mathbf{B}^*)\cdot\mathbf{m} - \frac{Z_0 k^4}{6\pi}(\mathbf{p}\times\mathbf{m}^*)\right]. \tag{6.36}$$

The force expression can be written as follows:

$$\langle\mathbf{F}\rangle = -\nabla\langle U\rangle + \frac{n_m}{c}(C_{ext} + C_{recoil})\langle\mathbf{S}\rangle + \mu_0\nabla\times Re[\alpha_{em}]\langle\mathbf{S}\rangle + \nabla$$

$$\times\left\{C_p\frac{c}{n_m}\langle\mathbf{L_p}\rangle + C_m\frac{c}{n_m}\langle\mathbf{L_m}\rangle\right\}$$

$$+\left\{2\omega^2\mu_0 Re[\alpha_{em}] - \frac{k^5}{3\pi\varepsilon_0^2\varepsilon_m^2}Im[\alpha_{ee}\alpha_{em}^*]\right\}\langle\mathbf{L_p}\rangle \tag{6.37}$$

$$+\left\{2\omega^2\mu_0 Re[\alpha_{em}] - \frac{k^5\mu_0}{3\pi\varepsilon_0\varepsilon_m}Im[\alpha_{mm}\alpha_{em}^*]\right\}\langle\mathbf{L_m}\rangle.$$

$$+\frac{ck^4\mu_0^2}{12\pi n_m}Im[\alpha_{ee}\alpha_{mm}^*]Im[\mathbf{E}\times\mathbf{H}^*],$$

where c is the speed of light in vacuum, $\langle S\rangle = Re[\mathbf{E}\times\mathbf{H}^*]/2$ denotes the time-averaged Poynting vector, $\langle U\rangle = -1/4(Re[\alpha_{ee}]|\mathbf{E}|^2 + Re[\alpha_{mm}]|\mathbf{B}|^2 - 2Im[\alpha_{em}]Im[\mathbf{B}\cdot\mathbf{E}^*])$ is an energy term due to the interaction of the dipolar chiral particle with light, $\langle L_p\rangle = Re[\varepsilon_0\varepsilon_m/(4i\omega)\mathbf{E}\times\mathbf{E}^*]$ and $\langle L_m\rangle = Re[\mu_0/(4i\omega)\mathbf{H}\times\mathbf{H}^*]$ represent the electric and magnetic part of the time-averaged SAM densities respectively, ω is the angular frequency. $C_{ext} = C_p + C_m = k(Im[\alpha_{ee}]/(\varepsilon_0\varepsilon_m) + \mu_0 Im[\alpha_{mm}])$ is a sum of contribution from the electric and magnetic dipole channel. $C_{recoil} = -k^4\mu_0/(6\pi\varepsilon_0\varepsilon_m) \, Re[a_{ee}a_{mm}^*] - k^4\mu_0/(6\pi\varepsilon_0\varepsilon_m)|\alpha_{em}|^2$ describes the recoil force and is related to the asymmetry parameter. The first and second terms in Eq. (6.37) correspond to the gradient force and the radiation pressure, respectively. The third term is a vortex force determined by the energy flow vortex around the particle and the optical activity, while the fourth and fifth term represents the scattering force associated with the curl of SAM densities. The sixth term is scattering forces solely from the particle chirality, which are known as curl-spin force associated with the curl of the spin densities. The particle chirality makes no explicit contribution to the last term, which is due to the alternating flow of the stored energy. Consequently, the movement of a chiral particle immersed in the optical focal field is subject to the induced time-averaged optical force, arising from the transfer of the linear momentum between the light and the material.

Moreover, due to the SAM transfer from light, the chiral particle would experience an intrinsic torque Γ and spin around its center of mass [78]:

$$
\begin{aligned}
\langle\boldsymbol{\Gamma}\rangle = & \left[-2\mu_0 Re(\alpha_{em}) + \frac{\mu_0 k^3}{3\pi\varepsilon_0\varepsilon_m} Im(\alpha_{ee}\alpha_{em}^*) + \frac{\mu_0^2 k^3}{3\pi} Im(\alpha_{mm}\alpha_{em}^*) \right]\langle\mathbf{S}\rangle \\
& + \left[\frac{\mu_0 k^3}{6\pi\varepsilon_0\varepsilon_m} Re(\alpha_{ee}\alpha_{em}^*) - \frac{\mu_0^2 k^3}{6\pi} Re(\alpha_{mm}\alpha_{em}^*) \right] Im(\mathbf{E}\times\mathbf{H}^*) \\
& + \left[\frac{2\omega}{\varepsilon_0\varepsilon_m} Im(\alpha_{ee}) - \frac{\omega k^3}{3\pi\varepsilon_0^2\varepsilon_m^2}\alpha_{ee}\alpha_{ee}^* - \frac{\omega\mu_0 k^3}{3\pi\varepsilon_0\varepsilon_m}\alpha_{em}\alpha_{em}^* \right]\langle\mathbf{L_p}\rangle \\
& + \left[2\omega\mu_0 Im(\alpha_{mm}) - \frac{\omega\mu_0^2 k^3}{3\pi}\alpha_{mm}\alpha_{mm}^* - \frac{\omega\mu_0 k^3}{3\pi\varepsilon_0\varepsilon_m}\alpha_{em}\alpha_{em}^* \right]\langle\mathbf{L_m}\rangle,
\end{aligned}
\tag{6.38}
$$

where the first term is the radiation torque proportional to the time-averaged Poynting vector, the second term is owing to an alternating flow of the so-called "stored energy," the third and fourth terms represent the spin torque from the time-averaged SAM densities viz the electric contribution L_p and the magnetic contribution L_m, respectively.

6.5.2 Sorting chiral nanoparticles using the structured interference field

It is demonstrated that special lateral optical forces (LOF) can be induced on chiral nanoparticles immersed in the structured field. The LOF can push chiral particles sideways, which can play a very wide role in chiral sorting and micromanipulation [79]. Analytical theory reveals that the simplest ways to generate LOF are to immerse chiral particles in the interference field of two plane waves. As illustrated in Fig. 6.21A, a LOF appears on the chiral particle immersed in the interference field of two plane waves with the same

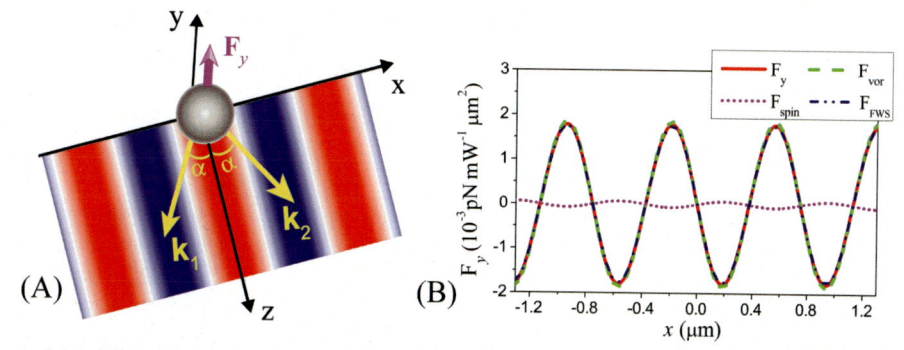

Figure 6.21 (A) Diagram of the setup for a LOF on a chiral particle immersed in the interference field of two plane waves. (B) The LOF F_y on a nanoparticle versus the particle displacement along the x-axis. *LOF*, Lateral optical forces.

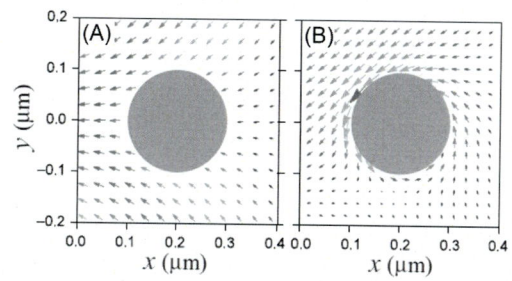

Figure 6.22 Time-averaged Poynting vectors for a nanoparticle with (A) nonchiral and (B) chiral particles immersed in the two interfering plane waves.

incident wavelength and incident angle of $\alpha = 45$ degrees. It is unambiguously shown that the LOF comes mostly from the vortex force as is shown in Fig. 6.21B, whereas the spin term F_{spin} has much smaller contribution and other terms without contributions of the y component. To further trace the origin of the LOF, the time-averaged Poynting vectors were stimulated as plotted in Fig. 6.22. For the case of the nonchiral particle, the time-averaged Poynting vectors always maintain the symmetry with respect to y-axis, suggesting that photon momenta above and below the particle balance each other. Differently, the time-averaged Poynting vectors for the chiral particle show the total momentum flux scattered to y-axis direction is not balanced. As a consequence, a net LOF can now exist and hence pushes the particle along the y direction. Therefore the LOF on a small chiral nanoparticle immersed in the interference field originates from the direct coupling of the optical vorticity with the particle chirality.

   In addition, an optical field that consists of two counter-propagating circularly polarized collimated beams with opposite handedness but equal power and waist has been demonstrated useful to separate chiral particles [17]. In a fluidic environment, chiral microparticles that differ only by opposite handedness passes perpendicularly

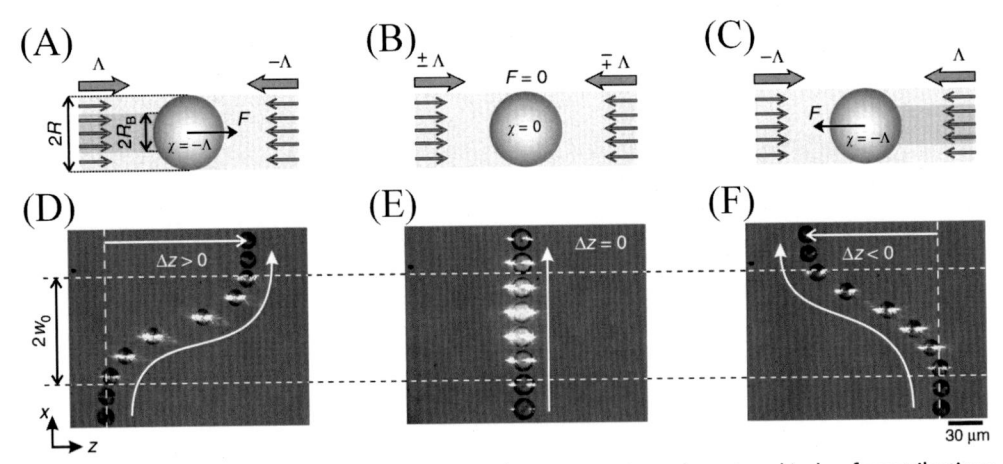

Figure 6.23 (A-C) Different light—matter interaction geometries, where two kinds of contributions to the net optical force F are identified for Bragg chiral droplets: red arrows refer to the 'Bragg optical rays' that are considered to be totally reflected due to circular Bragg reflection over a circular cross-section of radius, whereas blue arrows refer to the 'Fresnel optical rays' that are considered to be refracted/reflected on the droplet as is the case for an isotropic non-chiral dielectric sphere. (D—F) Demonstration of passive chiral optical sorting concept with chiral (D,F) and non-chiral (E) droplets that pass perpendicularly through the beams at a constant velocity V0 along the x axis.

through the two-beam configuration at constant velocity. For a nonchiral material, in the absence of additional external forces exerted on the particle, its trajectory is thus straight because of virtue of cancelation of the two individual beam contributions. In contrast, if the material is chiral, the force balance does not hold anymore and a nonzero net force, which originates either from circular birefringence or circular dichroism, is now exerted on the particle. The experimental demonstration of optical chiral sorting is illustrated in Fig. 6.23, where panels A—C depict the light-matter interaction geometry, whereas panels D—F compile snapshots of the droplet dynamics.

In practice, this optical chiral sorting scheme can process any random sequence of left-handed and right-handed chiral microparticles by using a Y-shaped bifurcating microfluidic output channel as is shown in Fig. 6.24. Moreover, chiral microparticles may also be used as "chiral conveyers" for molecular scale entities. When mixed with small particles to be sorted, chiral conveyers with appropriate functionalization may selectively bind with particles in the fluidic environment, then be sorted by light and finally sorted to two different output microchannels. This scheme was the first experimental demonstration of passive optical sorting of material chirality.

6.5.3 Sorting and manipulating chiral nanoparticles using tightly focused structured field

With the development of optical field regulation technology, a new approach to separating chiral particles was introduced by using not only the LOF but also a radial

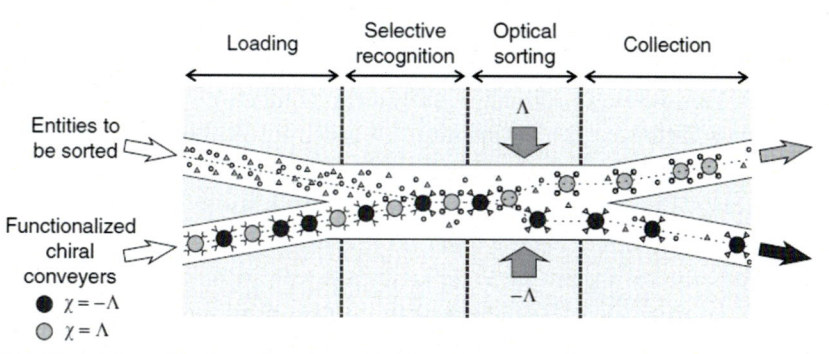

Figure 6.24 Illustration of an integrated optofluidic sorter.

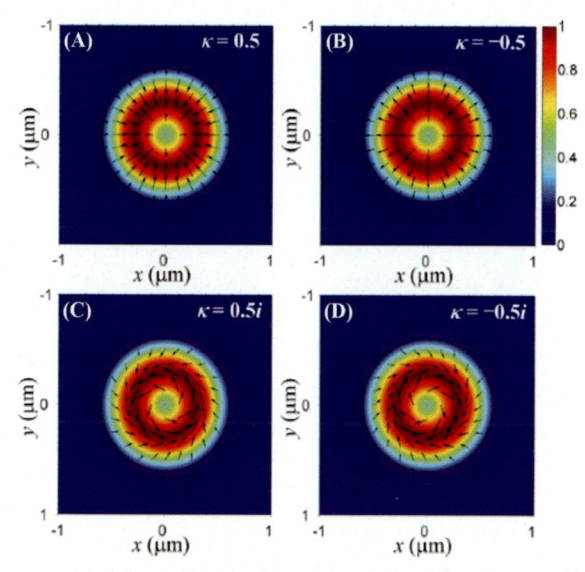

Figure 6.25 Transverse optical force distributions experienced by the particle with chirality parameters $\kappa = 0.5$ (A), -0.5 (B), $0.5i$ (C), and $-0.5i$ (D), respectively, in the focal plane illuminated by the focused vector beam.

optical force (ROF) induced by a tightly focused vector beam, which is composed of radial and azimuthal components with a $\pi/2$ phase difference [77]. Fig. 6.25A–D shows the transverse optical force distributions exerted on the chiral particle in the focal plane for the vector beam illumination, where the arrows denote the direction and magnitude of the transverse optical force and the background presents the intensity distribution of the focused field. When the chirality parameter $\kappa = 0.5$ or -0.5, the particle experiences only the ROF. The ROF vanishes on a circle somewhat smaller than that of the intensity maximum for the case of $\kappa = 0.5$ as shown in Fig. 6.25A, while the ROF on particle with $\kappa = -0.5$ is always positive for arbitrary positions, implying one kind of chiral particles ware trapped at the off-focus equilibrium position and the other was pushed away. Further, the case of $\kappa = 0.5i$ or $-0.5i$ was simulated as visualized in Fig. 6.25C or D. The particle experiences a strong azimuthal optical force (AOF) in addition to the ROF near the intensity maxima, indicating that the particle will be trapped off focus on a circle and undergo an orbital motion around the optical axis. It should be noted that there is neither intensity gradient nor wave propagation in the azimuthal direction according to the focused field.

To trace the physical origins of these transverse optical forces on the chiral particles, the contribution of each term of the optical forces was plotted in Fig. 6.26, where the chirality parameter $\kappa = 0.5 + 0.5i$. It can be seen that the radial force F_ρ totally comes from the gradient force F_{grad}, which contains the interaction of particle polarizabilities with the electric and magnetic energy densities as well as chirality density. However, the azimuthal force F_φ is dominated largely by the spin-density force F_{spin} and the vortex force F_{vor} which is directly related to the particle chirality, also associated with the radiation pressure F_{rad} and the curl-spin force F_{curl}, but almost not related to the gradient force and the alternating flow owing to the terms without the azimuthal component. As a result, it becomes clear that, in addition to the usual optical gradient force and radiation pressure as well as curl-spin force, the chiral particle indeed can be affected by new forces that directly depend on the particle chirality.

In addition, a family of vector beams with radially varied SOPs was demonstrated to separate enantiomers and manipulating paired enantiomers with the same incident beam [80]. The focused field exhibits bifocal spot intensity distribution and the two focal spots carry opposite optical chirality. They can achieve a three-dimensional stable optical trapping of a single chiral nanoparticle in both enantiomeric forms. When considering trapping chiral nanoparticles, such focused fields can stably trap the S enantiomer in one focal spot and the R enantiomer in the other one, accomplishing an effective separation of the chiral entities. Besides, the difference in the chirality can be recognized through the trapping potential of the particles. As shown in Fig. 6.27,

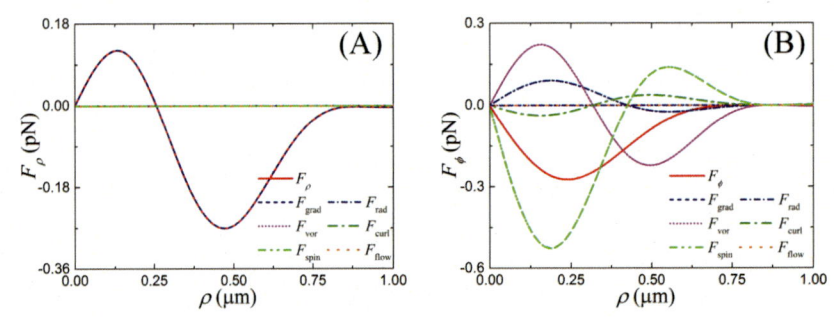

Figure 6.26 The contributions of decomposed radial (A) and azimuthal (B) optical forces defined in Eq. (6.7) acting on the chiral particle with $\kappa = 0.5 + 0.5i$ under the vector beam illumination.

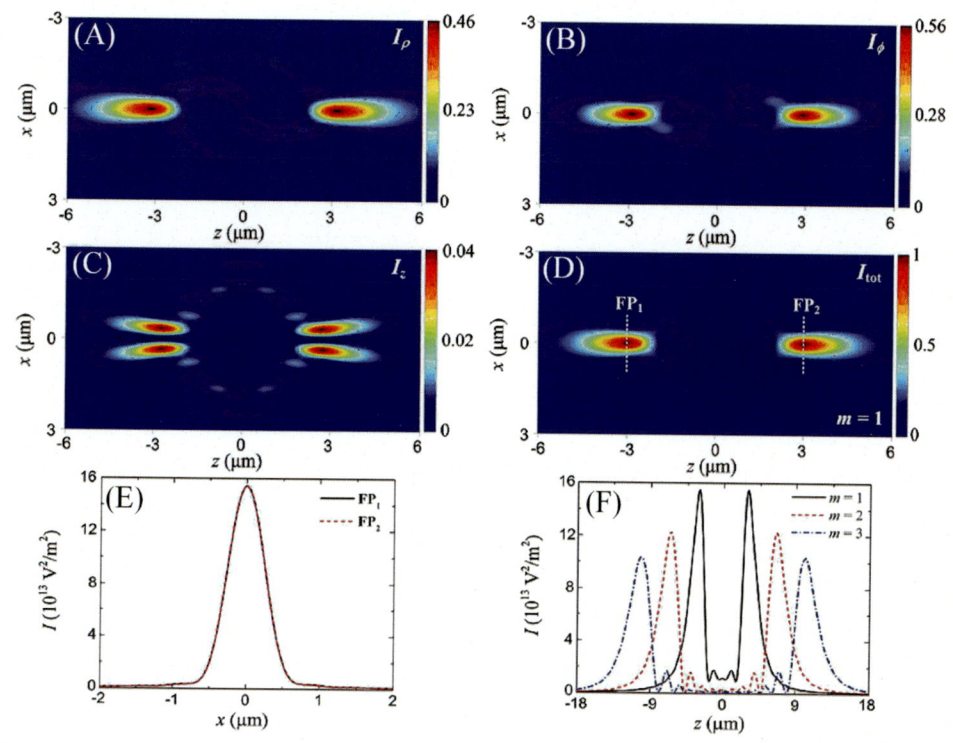

Figure 6.27 (A)–(D) Component intensity distributions in *xz* plane of tightly focused vector beam with PTC *m* = 1. (E) Lines scanning the total intensity along the *x*-axis in the FP1 and FP2 planes with PTC *m* = 1. (F) Lines scanning the total intensity along the *z*-axis with PTC *m* = 1, 2, and 3. *PTC*, Polarization topological charge.

the focusing properties of the incident vector beams show all these intensity components exhibit symmetrical patterns arranged on both sides of the geometric focal plane. Moreover, the lines scanning the total intensity along the *x*-axis in the FP1 and FP2 planes plotted in Fig. 6.27E show two curves are completely coincident which verify the symmetry of focal spots. By changing the value of polarization topological charge (PTC), the axial distance between the two focal spots is variable and growing, and the intensity maxima decrease as plotted in Fig. 6.27F. Consequently, the trapping distance of the enantiomeric pairs can be increased by increasing the magnitude of the PTC, and trapping position of the enantiomeric pairs can be exchanged by varying the sign of the PTC. Such a bifocal spot intensity profile indicates a possible application for trapping particles along different axial planes simultaneously.

   To trace the physical origins of the optical forces on the chiral particles, the contribution of each term of the optical forces was shown in Fig. 6.28. It is clearly seen that both the transverse F_x and longitudinal F_z optical forces overwhelmingly come from

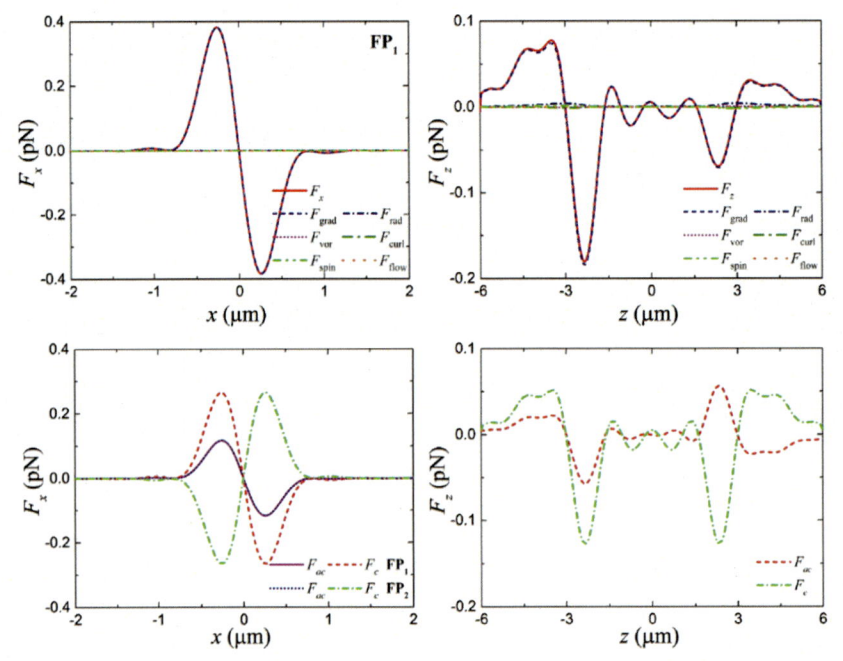

Figure 6.28 Contributions of decomposed transverse and longitudinal optical forces defined in (A and B) Eq. (6.10) and (C and D) Eq. (6.11) acting on the S enantiomer illuminated by the tightly focused vector beam with PTC $m = 1$. *PTC*, Polarization topological charge.

the gradient force F_{grad}. Clearly, both the sign and magnitude of the achiral force F_{ac} keeps the same as the focal plane changed from FP1 to FP2. For the chiral force F_c, however, the magnitude remains unchanged against the two focal planes, while the sign experiences an inversion, showing the two focal spots carry opposite chirality densities. In addition, the magnitude of the chiral force F_c is larger than that of the achiral force F_{ac}, resulting in the total optical force enhancing and pointing to the center of left focal spot, while weakening and pointing to the periphery for the right spot as shown in Fig. 6.28A and B. This indicates that this focused field may achieve a three-dimensional trapping of the S enantiomer in the left focal spot.

LOF phenomenon also occurs when a chiral particle is illuminated by an optical beam with a zero-order Bessel beam with transverse magnetic polarization [73], in which case LOF contains AOF and ROF. The former enables an orbital motion of particle, while the latter serves to confine the particle near the beam axis. It has been numerically demonstrated that even the azimuthally symmetric beam can exert an AOF on the particle, which enables an orbital motion around the optical beam axis and manifests a mechanical OAM around the beam axis in the system.

It can be observed from Fig. 6.29A that the ROF F_r shows no difference for achiral particle and chiral particles, indicating that the ROF F_r is nearly independent of the

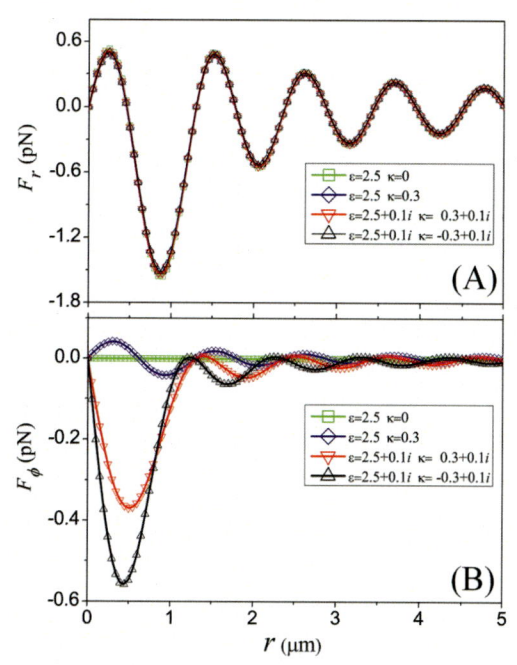

Figure 6.29 (A) The ROF and (B) the AOF versus the displacement r from the beam axis on a chiral particle illuminated by a zero-order Bessel beam with transverse magnetic (TM) polarization. *AOF,* Azimuthal optical force; *ROF,* radial optical force.

particle chirality and loss when $|\kappa|$ is small. However, the distribution is dramatically different for the AOF F_φ, as is shown in Fig. 6.29B. By substituting the expression of optical field into the formula of optical force, it shows that the concise expression of TOF contains one term corresponds to the ROF and the other two terms correspond to the AOF. The AOF comes from the coupling between the particle chirality and vorticity (the curl of energy flow) and the coupling of the particle chirality to the electric SAM density which implies a conversion of optical SAM to mechanical OAM of the chiral particle.

An intuitive understanding of the AOF exerting on a chiral particle immersed in the azimuthally symmetric vectorial beam can be developed from Fig. 6.30, where the time-averaged Poynting vector is shown for two cases with a particle having $(\varepsilon, \kappa) = (2.5, 0)$ and $(2.5, 0.3)$, respectively. The lengths of the arrows, together with their colors, denote the magnitudes of the energy flux. For the case of a conventional particle without chirality, the scattered field preserves the symmetry with respect to $-y$ and $+y$, so does the time-averaged Poynting vector. The photon momenta above and below the particle balance each other; there exists no AOF acting on the particle. Differently, an asymmetric distribution of photon momentum emerges near the chiral particle. As a result, an AOF comes into existence since the photon momenta scattered to the top and bottom of the particle do not balance. In addition, a vortex of the time-averaged Poynting vector develops near

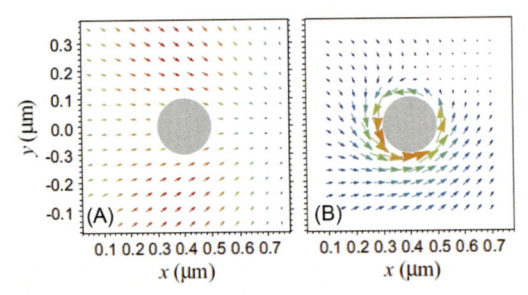

Figure 6.30 Time-averaged Poynting vectors for a particle with $\kappa = 0$ (A) and $\kappa = 0.3$ (B) immersed in a zero-order Bessel beam with TM polarization propagating along z.

the chiral particle, which is a manifestation of particle chirality. It is noted that the Poynting vector always shows an asymmetry to the left and right of the particle (even for a nonchiral particle), which is due to the radial inhomogeneity of the incident light field.

6.6 Conclusion

In this chapter, recent progresses on optical tweezers with the use of structured light as illumination are reviewed. The interaction between structured light and nanoparticle results in plenty of benefits, rendering the tailored optical complex field as a promising tool to explore novel applications in the field of optical manipulation, which will continue to rapidly progress in the future. The focus of optical tweezers techniques will be on the manipulation, sorting and separation of nanoscale objects in complex medium (such as chiral, bi-isotropic environment), in which cases the optical force calculations stems from the momentum exchange between light and matter would become sophisticated, and the object would become challenging to manipulation through minuscule gradient force. With the recent advances in optical engineering, we anticipate that more structure beams will emerge as the suitable illuminations for manipulating various kinds of objects, and the optical force effects can be used to monitor various physical parameters (such as electromagnetic field, temperature, and chiral parameter). Meanwhile, sub-10 nm of trapping scale is possible to be realized when miniaturized optical tweezers are implemented into integrated platform. Accurately trapping and spatially controlling objects at the nanometer scale in inhomogeneous background is a key issue to developing nanodevices with novel functionalities and significantly improved performance, which have abundant promising applications in biosciences, chemistry and engineering.

References

[1] A. Ashkin, J.M. Dziedzic, J.E. Bjorkholm, S. Chu, Observation of a single-beam gradient force optical trap for dielectric particles, Opt. Lett. 11 (1986) 288−290.

[2] A. Ashkin, Acceleration and trapping of particles by radiation pressure, Phys. Rev. Lett. 24 (1970) 156−159.

[3] K. Dholakia, P. Zemánek, Gripped by light: optical binding, Rev. Mod. Phys. 82 (2010) 1767−1791.

[4] M.M. Burns, J.-M. Fournier, J.A. Golovchenko, Optical binding, Phys. Rev. Lett. 63 (1989) 1233−1236.

[5] P.C. Chaumet, M. Nieto−Vesperinas, Optical binding of particles with or without the presence of a flat dielectric surface, Phys. Rev. B 64 (2001) 035422.

[6] E. Almaas, I. Brevik, Possible sorting mechanism for microparticles in an evanescent field, Phys. Rev. A 87 (2013) 063826.

[7] M.M. Wang, E. Tu, D.E. Raymond, J.M. Yang, H. Zhang, H. Norbert, et al., Microfluidic sorting of mammalian cells by optical force switching, Nat. Biotechnol. 23 (2005) 83−87.

[8] A. Ashkin, J.M. Dziedzic, Optical trapping and manipulation of viruses and bacteria, Science 235 (1987) 1517−1520.

[9] A. Ashkin, J.M. Dziedzic, T. Yamane, Optical trapping and manipulation of single cells using infra-red laser beams, Nature 330 (1987) 769−771.

[10] K. Svoboda, S.M. Block, Optical trapping of metallic Rayleigh particles, Opt. Lett. 19 (1994) 930−932.

[11] K.C. Toussaint, M. Liu, M. Pelton, J. Pesic, M.J. Guffey, P. Guyot−Sionnest, et al., Plasmon reso-nance−based optical trapping of single and multiple Au nanoparticles, Opt. Express 15 (2007) 12017−12029.

[12] S. Tan, H.A. Lopez, C.W. Cai, Y. Zhang, Optical trapping of single−walled carbon nanotubes, Nano Lett. 4 (2004) 1415−1419.

[13] T. Rodgers, S. Shoji, Z. Sekkat, S. Kawata, Selective aggregation of single−walled carbon nano-tubes using the large optical field gradient of a focused laser beam, Phys. Rev. Lett. 101 (2008) 127402.

[14] L. Jauffred, A.C. Richardson, L.B. Oddershede, Three−dimensional optical control of individual quantum dots, Nano Lett. 8 (2008) 3376−3380.

[15] Y.F. Chen, X. Serey, R. Sarkar, P. Chen, D. Erickson, Controlled photonic manipulation of pro-teins and other nanomaterials, Nano Lett. 12 (2012) 1633−1637.

[16] S.B. Wang, C.T. Chan, Lateral optical force on chiral particles near a surface, Nat. Commun. 5 (2014) 3307.

[17] G. Tkachenko, E. Brasselet, Optofluidic sorting of material chirality by chiral light, Nat. Commun. 5 (2014) 3577.

[18] A. Novitsky, C.-W. Qiu, A. Lavrinenko, Material-independent and size-independent tractor beams for dipole objects, Phys. Rev. Lett. 109 (2012) 023902.

[19] S. Sukhov, A. Dogariu, Negative nonconservative forces: optical 'tractor beams' for arbitrary objects, Phys. Rev. Lett. 107 (2011) 203602.

[20] S. Sukhov, A. Dogariu, On the concept of 'tractor beams', Opt. Lett. 35 (2010) 3847−3849.

[21] D.B. Ruffner, D.G. Grier, Optical conveyors: a class of active tractor beams, Phys. Rev. Lett. 109 (2012) 163903.

[22] A. Novitsky, C.-W. Qiu, H.F. Wang, Single gradientless light beam drags particles as tractor beams, Phys. Rev. Lett. 107 (2011) 203601.

[23] V. Kajorndejnukul, W.Q. Ding, S. Sukhov, C.-W. Qiu, A. Dogariu, Linear momentum increase and negative optical forces at dielectric interface, Nat. Photonics 7 (2013) 787−790.

[24] O. Brzobohaty, V. Karásek, M. Šiler, L. Chvátal, T. Čižmár, P. Zemánek, Experimental demonstration of optical transport, sorting and self−arrangement using a 'tractor beam', Nat. Photonics 7 (2013) 123−127.

[25] A. Ashkin, Forces of a single−beam gradient laser trap on a dielectric sphere in the ray optics regime, Methods Cell Biol. 55 (1998) 1−27.

[26] A. Ashkin, Forces of a single−beam gradient laser trap on a dielectric sphere in the ray optics regime, Biophys. J. 61 (1992) 569−582.

[27] B.T. Draine, The discrete−dipole approximation and its application to interstellar graphite grains, Astrophys. J. 333 (1988) 848.

[28] J.R. Arias−González, M. Nieto−Vesperinas, Optical forces on small particles: attractive and repul-sive nature and plasmon−resonance conditions, J. Opt. Soc. Am. A 20 (2003) 1201−1209.

[29] P.C. Chaumet, M. Nieto−Vesperinas, Time−averaged total force on a dipolar sphere in an electro-magnetic field, Opt. Lett. 25 (2000) 1065−1067.

[30] Y. Harada, T. Asakura, Radiation forces on a dielectric sphere in the Rayleigh scattering regime, Opt. Commun. 124 (1996) 529−541.

[31] D.J. Griffiths, Introduction to Electrodynamics, Prentice Hall, 1998, pp. 351−356.

[32] C. Min, Z. Shen, J. Shen, Y. Zhang, H. Fang, G. Yuan, et al., Focused plasmonic trapping of metallic particles, Nat. Commun. 4 (2013) 2891.

[33] C.J.R. Sheppard, A. Choudhury, Annular pupils, radial polarization, and superresolution, Appl. Opt. 43 (2004) 4322−4327.

[34] M. Meier, V. Romano, T. Feurer, Material processing with pulsed radially and azimuthally polar-ized laser radiation, Appl. Phys. A 86 (2007) 329−334.

[35] Q. Zhan, J.R. Leger, Microellipsometer with radial symmetry, Appl. Opt. 41 (2002) 4630−4637.

[36] Q. Zhan, Cylindrical vector beams: from mathematical concepts to applications, Adv. Opt. Photonics 1 (2009) 1−57.

[37] E. Hasman, G. Biener, A. Niv, V. Kleiner, Space−variant polarization manipulation, Prog. Opt. (2005) 215−289.

[38] T.G. Brown, Unconventional polarization states: beam propagation, focusing and imaging, Prog. Opt. (2011) 81−129.

[39] A.M. Yao, M.J. Padgett, Orbital angular momentum: origins, behavior and applications, Adv. Opt. Photonics 3 (2011) 161−204.

[40] M.R. Dennis, K. O'Holleran, M.J. Padgett, Singular optics: optical vortices and polarization singu-larities, Prog. Opt. 53 (2009) 293−363.

[41] K. Kneipp, M. Moskovits, H. Kneipp, Surface−Enhanced Raman Scattering, Springer, Berlin, 2006.

[42] Q. Zhan, Trapping metallic Rayleigh particles with radial polarization, Opt. Express 12 (2004) 3377−3382.

[43] L. Huang, H. Guo, J. Li, L. Ling, B. Feng, et al., Optical trapping of gold nanoparticles by cylindri-cal vector beam, Opt. Lett. 37 (2012) 1694−1696.

[44] M. Dienerowitz, M. Mazilu, K. Dholakia, Optical manipulation of nanoparticles: a review, J. Nanophotonics 2 (2008) 021875.

[45] A. Ohlinger, S. Nedev, A.A. Lutich, J. Feldmann, Optothermal escape of plasmonically coupled sil-ver nanoparticles from a three−dimensional optical trap, Nano Lett. 11 (2011) 1770−1774.

[46] G. Rui, Q. Zhan, Trapping of resonant metallic nanoparticles with engineered vectorial optical field, Nanophotonics 3 (2014) 351−361.

[47] J. Chen, J. Ng, Z. Lin, C.T. Chan, Optical pulling force, Nat. Photonics 5 (2011) 531−534.

[48] V. Kotaidis, C. Dahmen, G. von Plessen, F. Springer, A. Plech, Excitation of nanoscale vapor bub-bles at the surface of gold nanoparticles in water, J. Chem. Phys. 124 (2006) 184702.

[49] X. Wang, G. Rui, L. Gong, B. Gu, Y. Cui, Manipulation of resonant metallic nanoparticle using 4Pi focusing system, Opt. Express 24 (2016) 24143−24152.

[50] M. Miyazaki, Y. Hayasaki, Motion control of low−index microspheres in liquid based on optical repulsive force of a focused beam array, Opt. Lett. 34 (2009) 821−823.

[51] K.T. Gahagan, G.A. Swartzlander, Simultaneous trapping of low−index and high−index micropar-ticles observed with an optical−vortex trap, J. Opt. Soc. Am. B 16 (1999) 533−537.

[52] F. Peng, B. Yao, S. Yan, W. Zhao, M. Lei, Trapping of low−refractive−index particles with azi-muthally polarized beam, J. Opt. Soc. Am. B 26 (2009) 2242−2247.

[53] K.T. Gahagan, G.A. Swartzlander, Trapping of low−index microparticles in an optical vortex, J. Opt. Soc. Am. B 15 (1998) 524−534.

[54] Q. Zhan, Radiation forces on a dielectric sphere produced by highly focused cylindrical vector beams, J. Opt. A Pure Appl. Opt 5 (2003) 229−232.

[55] D. McGloin, G.C. Spalding, H. Melville, W. Sibbett, K. Dholakia, Three−dimensional arrays of optical bottle beams, Opt. Commun. 225 (2003) 215−222.

[56] B.P.S. Ahluwalia, W.C. Cheong, X.-C. Yuan, L.-S. Zhang, S.-H. Tao, J. Bu, et al., Design and fabrication of a double−axicon for generation of tailorable self−imaged three−dimensional intensity voids, Opt. Lett. 31 (2006) 987−989.

[57] L. Isenhower, W. Williams, A. Dally, M. Saffman, Atom trapping in an interferometrically generated bottle beam trap, Opt. Lett. 34 (2009) 1159–1161.

[58] P. Zhang, Z. Zhang, J. Prakash, S. Huang, D. Hernandez, M. Salazar, et al., Trapping and transporting aerosols with a single optical bottle beam generated by moiré techniques, Opt. Lett. 36 (2011) 1491–1493.

[59] G. Rui, Y. Wang, X. Wang, B. Gu, Y. Cui, Trapping of low–refractive–index nanoparticles in a hollow dark spherical spot, J. Phys. Commun. 2 (2018) 065015.

[60] D.S. Bradshaw, D.L. Andrews, Interactions between spherical nanoparticles optically trapped in Laguerre–Gaussian modes, Opt. Lett. 30 (2005) 3039–3304.

[61] V. Garcés–Chávez, K. Volke–Sepulveda, S. Chávez–Cerda, W. Sibbett, K. Dholakia, Transfer of orbital angular momentum to an optically trapped low–index particle, Phys. Rev. A 66 (2002) 063402.

[62] K. Volke–Sepúlveda, S. Chávez–Cerda, V. Garcés–Chávez, K. Dholakia, Three–dimensional optical forces and transfer of orbital angular momentum from multiringed light beams to spherical microparticles, J. Opt. Soc. Am. B 21 (2004) 1749–1757.

[63] G. Rui, Y. Lu, P. Wang, H. Ming, Q. Zhan, Generation of enhanced evanescent Bessel beam using band–edge resonance, J. Phys. Commun. 108 (2010) 074304.

[64] G. Rui, Y. Lu, P. Wang, H. Ming, Q. Zhan, Evanescent Bessel beam generation through filtering highly focused cylindrical vector beams with a defect mode one–dimensional photonic crystal, Opt. Commun. 283 (2010) 2272–2276.

[65] Q. Zhan, Evanescent Bessel beam generation via surface plasmon resonance excitation by a radially polarized beam, Opt. Lett. 31 (2006) 1726–1728.

[66] W. Chen, Q. Zhan, Realization of an evanescent Bessel beam via surface plasmon interference excited by a radially polarized beam, Opt. Lett. 34 (2009) 722–724.

[67] G. Rui, X. Wang, Y. Cui, Manipulation of metallic nanoparticle with evanescent vortex Bessel beam, Opt. Express 23 (2015) 25707–25716.

[68] G. Rui, Y. Li, S. Zhou, Y. Wang, B. Gu, Y. Cui, et al., Optically induced rotation of Rayleigh particles by arbitrary photonic spin, Photon. Res. 7 (2019) 69–79.

[69] L.D. Landau, J. Bell, M. Kearsley, L. Pitaevskii, E. Lifshitz, J. Sykes, Electrodynamics of Continuous Media, Second Edition, Pergamon Press, 1984.

[70] A. Hinojosa–Alvarado, J.C. Gutiérrez–Vega, Geometrical optics calculation of forces and torques produced by a ringed beam on a prolate spheroid, J. Opt. Soc. Am. B 27 (2010) 1651–1658.

[71] A. Hayat, J.P.B. Mueller, F. Capasso, Lateral chirality–sorting optical forces, Proc. Natl. Acad. Sci. U.S.A 112 (2015) 13190–13194.

[72] J. Lekner, Optical properties of isotropic chiral media, Pure Appl. Opt. 5 (1996) 417–443.

[73] H. Chen, N. Wang, W. Lu, S. Liu, Z. Lin, Tailoring azimuthal optical force on lossy chiral particles in Bessel beams, Phys. Rev. A 90 (2014) 043850.

[74] A. Lakhtakia, V.K. Varadan, V.V. Varadan, Time–Harmonic Electromagnetic Fields in Chiral Media, Springer–Verlag, Berlin, 1989.

[75] C.F. Bohren, D.R. Huffman, Absorption and Scattering of Light by Small Particles, John Wiley & Sons, New York, 1983.

[76] Ø. Farsund, B.U. Felderhof, Force, torque, and absorbed energy for a body of arbitrary shape and constitution in an electromagnetic radiation field, Physica A 227 (1996) 108–130.

[77] M. Li, S. Yan, Y. Zhang, Y. Liang, P. Zhang, B. Yao, Optical sorting of small chiral particles by tightly focused vector beams, Phys. Rev. A 99 (2019) 033825.

[78] H. Chen, W. Lu, X. Yu, C. Xue, S. Liu, Z. Lin, Optical torque on small chiral particles in generic optical fields, Opt. Express 25 (2017) 32867–32878.

[79] H. Chen, C. Liang, S. Liu, Z. Lin, Chirality sorting using two-wave-interference-induced lateral optical force, Phys. Rev. A 93 (2016) 053833.

[80] M. Li, S. Yan, Y. Zhang, P. Zhang, B. Yao, Enantioselective optical trapping of chiral nanoparticles by tightly focused vector beams, J. Opt. Soc. Am. B 36 (2019) 2099–2105.

CHAPTER 7

Design and nanophotonic thin film devices using phase change materials

Andrew M. Sarangan
Department of Electro-Optics and Photonics, University of Dayton, Dayton, OH, United States

7.1 Phase change materials

Phase change materials (PCMs) belong to a class of materials that exhibit changes in electrical or optical properties in response to an external stimulus. For example, vanadium dioxide (VO_2) abruptly switches from a semiconducting state to a metallic state at a temperature of $68°C$. Germanium—antimony—telluride (GST), which is a chalcogenide glass, switches from an amorphous state to an fcc crystal structure, and subsequently to a hexagonal crystal structure. All of these changes are easily reversible and are associated with very large changes in their electrical conductivity and optical constants. Compared to nonlinear refraction and electro-optic effects, PCMs produce changes that are several orders of magnitude larger. This has made it possible to realize novel electrical and optical devices that are not possible with ordinary dielectrics or semiconductor materials. In this chapter, we will review the basic properties of PCMs and their synthesis methods as applicable to the design of optical thin film devices. We will also review the basic principles of optical thin film design methods, focusing specifically on those that incorporate PCMs to accomplish unique tunable functions.

7.1.1 Vanadium dioxide (VO_2)

Vanadium dioxide (VO_2) in thin film form is a partially transparent tan colored film. It exhibits an abrupt change in crystal structure at the phase transition temperature. Below this transition temperature, the film will have a monoclinic crystal structure and will behave as a semiconductor. Above the transition temperature, the atoms spontaneously reorganize into a rutile structure and the film will exhibit a metallic behavior. This transition temperature is approximately $68°C$, but due to a built-in hysteresis the transition during the heating cycle is slightly higher than $70°C$, and the transition during the cooling cycle is slightly lower than $65°C$. The width of this hysteresis depends on process conditions and has been attributed to film strain and polycrystalline grain size [1,2]. The rutile phase of VO_2 is retained only as long as the temperature is held above the transition temperature. As a consequence, VO_2 transition is referred to as a

Thin Film Nanophotonics
DOI: https://doi.org/10.1016/B978-0-12-822085-6.00004-2

volatile transition. This is useful in applications where only a momentary switching is desired, such as for example, in optical limiting or optical bistability.

The most noticeable aspect of the semiconductor-to-metal transition is the change in resistivity. Below the transition temperature, the electrical resistivity is fairly high. It exhibits a slow decline in resistivity with increasing temperature that is characteristic of a semiconductor. Above the transition temperature, the resistivity drops precipitously by 2−5 orders of magnitude and then shows a relatively constant value that is characteristic of metallic behavior. This is shown in Fig. 7.1.

The refractive index of VO_2 has been measured by various research groups. Although there are slight discrepancies between the reported values (which are mostly attributable to the different synthesis methods), all of them consistently show the semiconductor-to-metal transition [3−5]. A representative optical dispersion plot for VO_2 is shown in Fig. 7.2, where the blue curves represent data taken at 25°C and the red curves represent data taken at 80°C. Additionally, we can see that the imaginary part κ of the refractive index increases with increasing wavelength in the hot state, whereas it decreases with increasing wavelength in the cold state, which is consistent with their metallic and semiconducting characteristics, respectively. This is the basic property that is exploited in the design of optical switches and bistable devices.

An important material property that captures optical absorption loss is the loss tangent, which is defined as [6]

$$\tan\delta = \frac{2n\kappa}{|n^2 - \kappa^2|},\qquad(7.1)$$

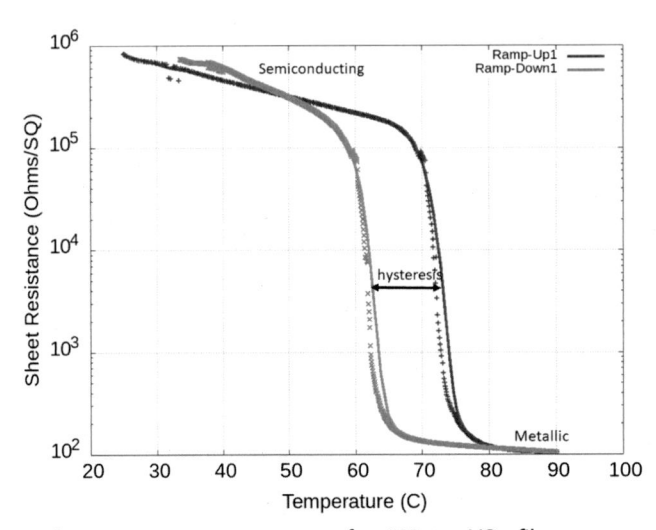

Figure 7.1 Sheet resistance versus temperature of a 100 nm VO_2 film measured in the author's laboratory.

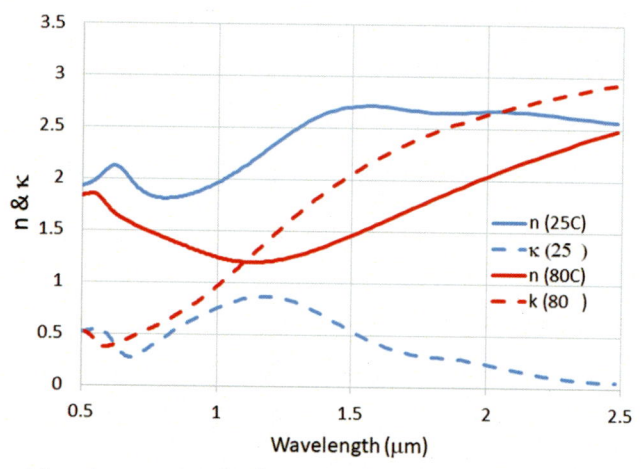

Figure 7.2 Real and imaginary parts of refractive index (n and κ) data of VO_2 measured at 25°C and 80°C (based on data originally reported in Ref. [3]).

where n is the real part of the refractive index and κ is the imaginary part. This is the ratio between the absorbed power and the transmitted power in the material. In a thin film structure, loss tangent and interference play a key role in the total absorption loss. A higher loss tangent indicates irreversible optical losses in the material that cannot be corrected by interference coatings. It is important to note that κ alone does not determine the optical losses. There are cases, such as in some metals, where a large κ actually results in lower optical losses.

The loss tangent of VO_2 is shown in Fig. 7.3. As evident from this plot, the hot state of VO_2 has a much higher loss tangent than the cold state, and the loss tangent asymptotically rises to a high value near the wavelength of 1.1 μm. This point corresponds to the real part of the permittivity ϵ crossing the zero value and becoming negative. An absorption modulator, for example, could be designed to operate near this wavelength to exploit this switching effect. However, the cold state also has a peak near this wavelength, which makes the contrast between the two loss tangents more prominent at longer wavelengths than at this point. As a result, absorption or reflective modulators using VO_2 are most effective at longer wavelengths.

The switching magnitude and transition temperature of VO_2 can be modified to a small extent by doping the VO_2 with various elements, such as Ti and W. In general, doping results in a reduction in both the switching magnitude and the switching temperature [7]. It has also been shown that the switching stimulus for VO_2 is not limited to the temperature. Electric field or light illumination can also induce switching without the associated thermal effect [8]. This opens up many interesting possibilities, such as field-effect devices and photodetection. Many of these areas are still in their infancy. The most exploited property of VO_2 is its temperature–induced phase change.

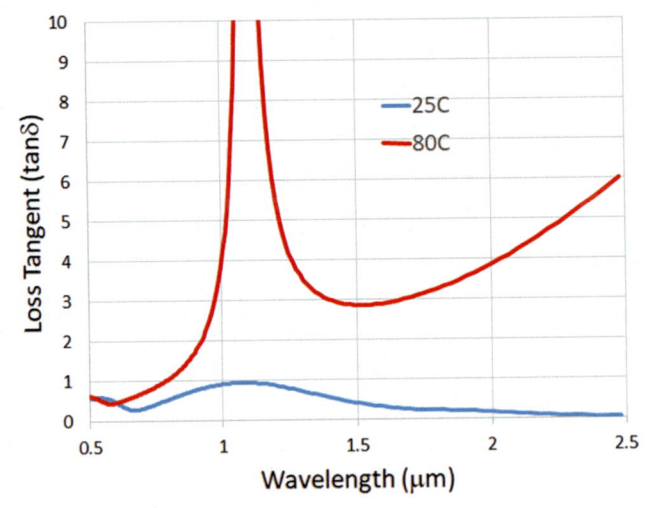

Figure 7.3 Calculated loss tangent of VO$_2$ in the cold and hot states.

7.1.2 Germanium–antimony–telluride

GST is a chalcogenide glass that is a pseudo-binary alloy between Ge$_a$Te$_b$ and Sb$_c$Te$_d$. Most of these exhibit some type of reversible amorphous to crystal transition. However, unlike VO$_2$, the switched state will be retained indefinitely until it is actively switched back to its original state. Hence, GST exhibits a nonvolatile phase change. This nonvolatile aspect is useful in optical memory, spatial light modulators, tunable color filters, and low-power or static displays.

Although the alloy Ge$_a$Te$_b$ + Sb$_c$Te$_d$ spans a very large composition space, the GeTe and Sb$_2$Te$_3$ tie-line is the most studied composition. This can be represented as

$$(GeTe)_x + (Sb_2Te_3)_{1-x} \rightarrow Ge_xSb_{2(1-x)}Te_{3-2x}.$$

In this range, the crystallization temperature has been found to increase with increasing x, from around 120°C for values of x near 0.1 to about 170°C for values of x approaching 0.9 [9]. The composition corresponding to $x = 0.67$ is by far the subject of the majority of studies. This is Ge$_{0.67}$Sb$_{0.67}$Te$_{1.67}$, or equivalently Ge$_2$Sb$_2$Te$_5$. This is also commonly abbreviated as GST-225. It's amorphous-to-crystal transition temperature is around 150°C.

Generally, the as-deposited film using physical vapor deposition methods (described later in this chapter) starts off as an amorphous film. When raised to a temperature above 150°C, it crystallizes into an fcc structure. This transition is associated with a very large increase in refractive index as well as its conductivity. Further increase in temperature beyond 200°C results in it reorganizing into a hexagonal crystal structure. The first two states (amorphous and fcc) are the ones of most interest because they are easily reversed. The hexagonal state is much more difficult to reverse. The amorphous-to-fcc crystallization is relatively straightforward to implement and to understand. The elevated temperature causes the atoms in the film to reorganize into a polycrystalline structure. The reversal requires a fast melt-quench technique. The film has to be raised to a high enough temperature to melt the material (which is typically above 600°C) and rapidly cooled on the order of a nanosecond without allowing the atoms to reorganize into a crystal structure. This is illustrated in Fig. 7.4. While the amorphous-to-fcc transition can be accomplished with virtually any type of heating, such as on a hot plate or using ohmic heating through an electrode, the fcc-to-amorphous transition requires a more careful design of the structure taking into account the thermal conductivities of the surrounding media and temporal shape of the heating pulse. As a result, the vast majority of switching devices demonstrated in the literature have been one-way only (amorphous-to-fcc), although in theory they could be reset back to the amorphous state with appropriate considerations to the temperature pulse.

The refractive index of GST-225 is shown in Fig. 7.5. As evident from this figure, the amorphous state exhibits a refractive index of around 4.0 and then increases to about 5.5 when switched to the fcc crystalline state (which is induced by annealing at 155°C). Upon further annealing at 250°C, the refractive index flattens at a value around 6 due to the hexagonal state. The imaginary part κ is relatively small in the amorphous state (about 0.1) and rises to about 1.3 in the fcc state and then to a very large value of about 4.2 in the hexagonal state.

Similar to VO_2, it is insightful to look at the intrinsic optical losses by plotting the loss tangent values of GST. This is shown in Fig. 7.6. We can see that the loss tangent is lowest in the amorphous state and increases dramatically in the fcc and hexagonal states. However, in comparison with the hot state of VO_2, the loss tangent of fcc-GST is significantly smaller. This allows the possibility of designing tunable transmissive devices using GST, whereas transmissive devices are significantly more difficult with VO_2.

Fig. 7.7 shows the electrical resistivity of GST-225 measured after annealing at various temperatures. We can see that it shows a similar behavior to the VO_2 transition, except in this case, the transition is nonvolatile. We also see two transition steps, one

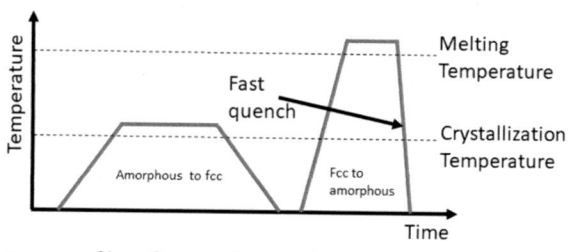

Figure 7.4 Temperature profiles for setting and resetting GST phases. *GST*, Germanium—antimony—telluride.

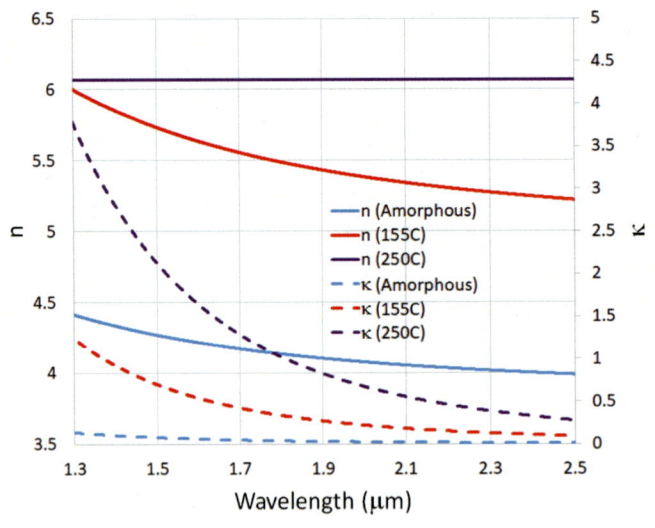

Figure 7.5 Refractive index (*n* and κ) data of GST-225 measured in the amorphous state, after annealing at 155°C and after annealing at 250°C (based on data originally reported in Ref. [10]). *GST*, Germanium−antimony−telluride.

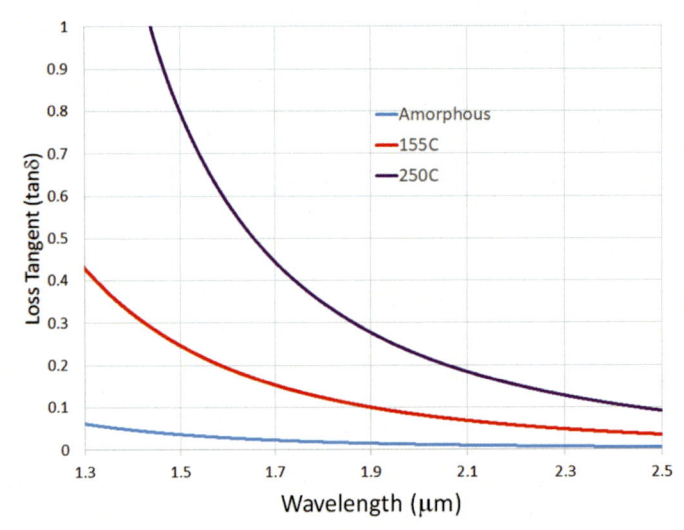

Figure 7.6 Calculated loss tangent of GST-225 in the amorphous state, and after annealing at 155°C and 250°C temperatures. *GST*, Germanium−antimony−telluride.

due to amorphous-to-fcc and the second one due to fcc–to-hexagonal. The plot also shows the impact of dopants (in this case, nickel) on the resistivity performance of GST-225. The objective of this study was to reduce the resistivity contrast while maintaining the optical contrast. The details of the study can be found in Ref. [11].

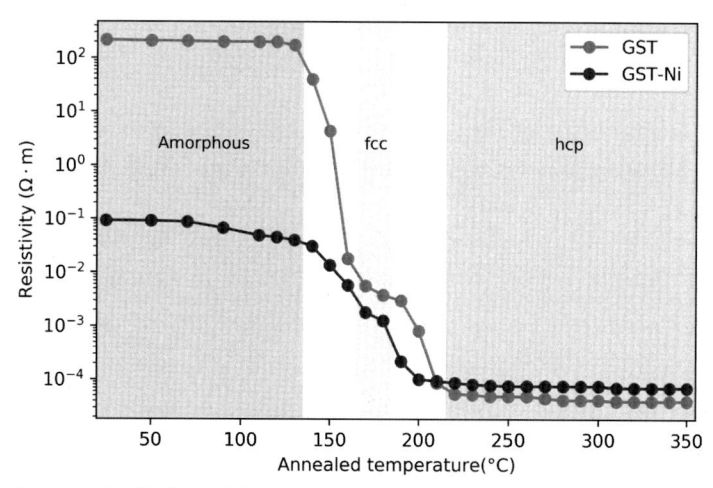

Figure 7.7 Resistivity of GST-225 with anneal temperature. The blue curve is for GST-225 doped with 2% Ni of the as-deposited pure GST. *GST, Germanium—antimony—telluride. Reproduced with permission from P. Guo, J.A. Burrow, G.A. Sevison, A. Sood, M. Asheghi, J.R. Hendrickson, et al., Improving the performance of Ge₂Sb₂Te₅ materials via nickel doping: towards RF-compatible phase-change devices, Appl. Phys. Lett. 113 (17) (2018) 171903.*

7.2 Design of thin film structures using phase change materials

In this section, we will first describe the effective reflectance index (ERI) method for designing multilayer thin film structures. Then we will examine several switching optical devices using the PCMs described previously, viz., VO_2 and GST–225.

7.2.1 Principles of thin film design using the effective reflectance index method

We will briefly derive the mathematics behind the ERI method, which is a useful technique for analyzing the impedances of multilayer optical structures. The method as described in this section is applicable only to plane waves in a linear system, although it can be modified to include nonlinear systems as well as diffracting beams.

Consider a planar dielectric interface between two materials with refractive indices of n_a (air, or input side) and n_s (substrate, or transmitted side). Representing the input field amplitude as 1, the reflected field amplitude as r and the transmitted field amplitude as t, we can express the electric field continuity across the interface as

$$e^{-jk_0 n_a z_1} + re^{+jk_0 n_a z_1} = te^{-jk_0 n_s z_1}. \tag{7.2}$$

The continuity of the magnetic fields (which is equivalent to taking the continuity of the electric field derivatives in a nonmagnetic medium) will result in

$$n_a e^{-jk_0 n_a z_1} - m_a e^{+jk_0 n_a z_1} = n_s te^{-jk_0 n_s z_1}. \tag{7.3}$$

By combining Eqs. (7.2) and (7.3), we can get

$$r = \frac{n_a - n_s}{n_a + n_s},\tag{7.4}$$

which is the well-known Fresnel reflection for plane waves across a dielectric interface.

Next, considering two parallel dielectric interfaces, due to the presence of a thin film between n_a and n_s with a refractive index of n_f, we can write four continuity equations:

$$e^{-jk_0 n_a z_1} + re^{+jk_0 n_a z_1} = Ae^{-jk_0 n_f z_1} + Be^{+jk_0 n_f z_1},\tag{7.5}$$

$$n_a e^{-jk_0 n_a z_1} - m_a e^{+jk_0 n_a z_1} = n_f Ae^{-jk_0 n_f z_1} - n_f Be^{+jk_0 n_f z_1},\tag{7.6}$$

$$Ae^{-jk_0 n_f z_2} + Be^{+jk_0 n_f z_2} = te^{-jk_0 n_s z_2},\tag{7.7}$$

$$n_f Ae^{-jk_0 n_f z_2} - n_f Be^{+jk_0 n_f z_2} = n_s te^{-jk_0 n_s z_2},\tag{7.8}$$

where z_1 and z_2 are the positions of the two dielectric interfaces as illustrated in Fig. 7.8. The thickness of this film is $h = z_2 - z_1$. These four equations can be solved for the four unknown A, B, t, and r. The reflection coefficient r is the one of most interest to us. After some lengthy algebra, this can be manipulated into an expression of the form

$$r = \frac{(n_a - n_r)}{(n_a + n_r)},\tag{7.9}$$

where the ERI n_r is defined as

$$n_r = n_f \frac{(n_s + n_f) + (n_s - n_f)e^{-j2k_0 n_f h}}{(n_s + n_f) - (n_s - n_f)e^{-j2k_0 n_f h}}.\tag{7.10}$$

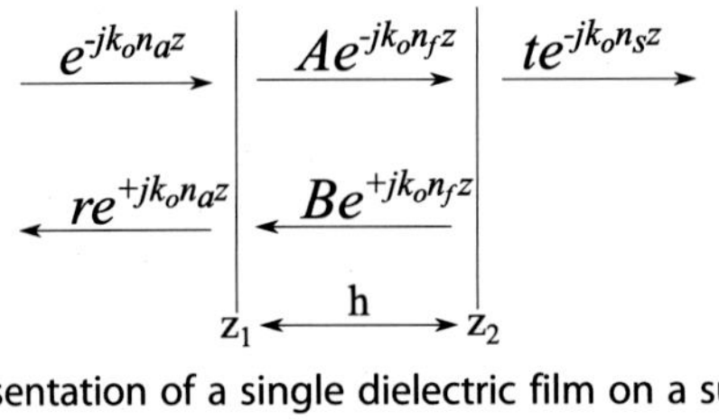

Figure 7.8 Plane wave representation of a single dielectric film on a substrate.

Recognizing that Eq. (7.9) is nearly identical to Eq. (7.4), we can interpret n_r as the equivalent refractive index of the substrate due to the presence of other films above it. This is the effective reflectance index. In general, n_r will be a complex value. However, it does not mean it exhibits a field attenuation like a complex-index material. The complex value simply represents the phase of the reflected field.

This quantity n_r is very useful in thin film design because it allows us to mathematically construct a layer structure to allow index-matching to some other medium. For example, antireflection (AR) condition is achieved by solving for $n_r = n_a$, that is, allowing the layer structure to have an equivalent refractive index that of air. We can mathematically show that this results in

$$n_f = \sqrt{n_a n_s},\qquad(7.11)$$

and

$$h = N\frac{\lambda}{4n_f},\qquad(7.12)$$

which are the well-known single-layer AR condition. Additionally, we can also plot the complex contour of n_r [using Eq. 7.10] to gain insights into the interference of the waves between the layer structure. The interested reader is referred to Ref. [6] for further details on this subject.

7.2.2 Switchable absorber using VO_2

For the device configuration, we will assume the VO_2 film to be on a sapphire substrate because it is the most common substrate lattice matched to VO_2 at the growth temperatures. In the hot state the thicker the film, the greater the absorption in the hot state. However, in the cold state, due to the lower absorption in the film, the transmission will depend on the absorption in the film as well as due to interference. We will assume a VO_2 film thickness of 200 nm, which is a fairly typical value using standard deposition methods (which are described later in this chapter).

This structure without the AR films is very straightforward to model using the plane wave transfer matrix method (TMM) [6]. Using the n and κ values shown in Fig. 7.2 and the standard refractive index values of sapphire, we can obtain the transmission and reflection from the 200 nm VO_2 film in the cold and hot states. This is shown in Fig. 7.9. We can see that the transmission switching ratio is about 12 dB and grows at longer wavelengths.

The performance of this absorption modulator can be increased further by adding AR films above the VO_2 as well as below the substrate. Fig. 7.10 illustrates the basic structure. The 200 nm VO_2 film is on the sapphire substrate, with a front-side AR film and a backside AR film. The backside AR is straightforward so we will not

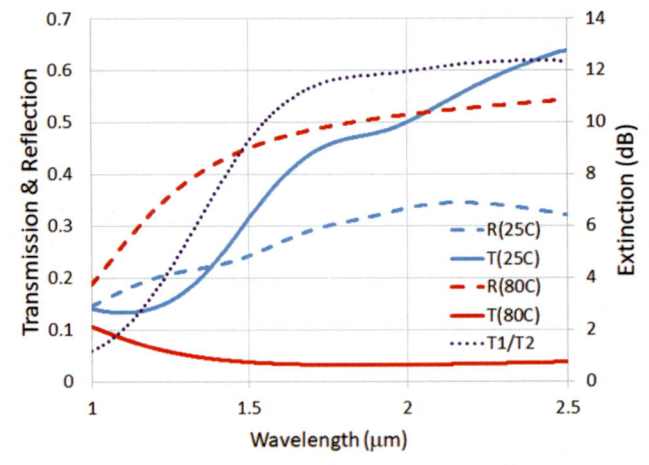

Figure 7.9 Calculated reflection, transmission, and extinction ratio for a 200 nm VO$_2$ film on a sapphire substrate. The solid lines are transmissions, and the dashed lines are reflections. The dotted line is the extinction ratio on the right-hand axis expressed in decibels (dB).

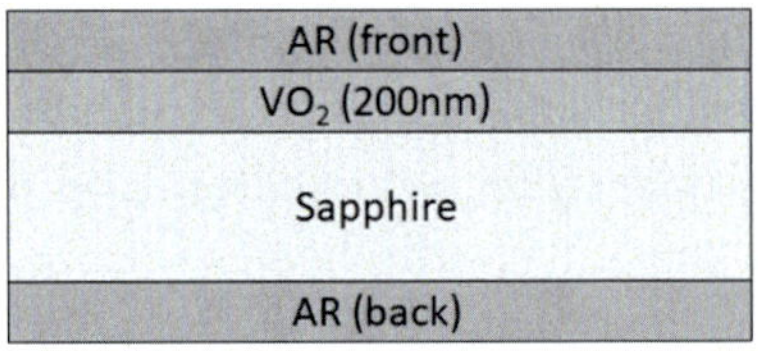

AR (front)
VO$_2$ (200nm)
Sapphire
AR (back)

Figure 7.10 Thin film configuration used for the switching absorber.

discuss it here. The front-side AR can be designed such that it minimizes the cold state reflection or the hot state reflection.

Fig. 7.11 shows two scenarios. The first one (the *blue line*) is an AR film designed to eliminate reflection in the cold state. The second one (*red line*) is for eliminating reflection in the hot state. Both of these were designed for 2.0 μm wavelength, which is a somewhat arbitrary choice in our example. The plot is a complex effective reflectance index contour n_r (Eq. 7.10). We can numerically solve for the condition where a second film above the 200 nm VO$_2$ film will produce a contour that will terminate at $n_r = n_a$. This is the required condition for AR. For the cold state, the solution gives a film index of 1.94 and a thickness of 231 nm. For the hot state, the required film index is 3.01 with a thickness of 92 nm. Details of this calculation are not described here, and the interested reader should refer to Ref. [6]. These values can be used in TMM to compute the spectral reflection and transmission. The results are shown in Fig. 7.12. We can see that the reflection of the cold state falls to zero at the design wavelength of 2.0 μm, but the hot–state reflection moves higher. The cold state transmission reaches a value of 75% at 2.0 μm. The resulting extinction ratio at 2.0 μm is 11.4 dB.

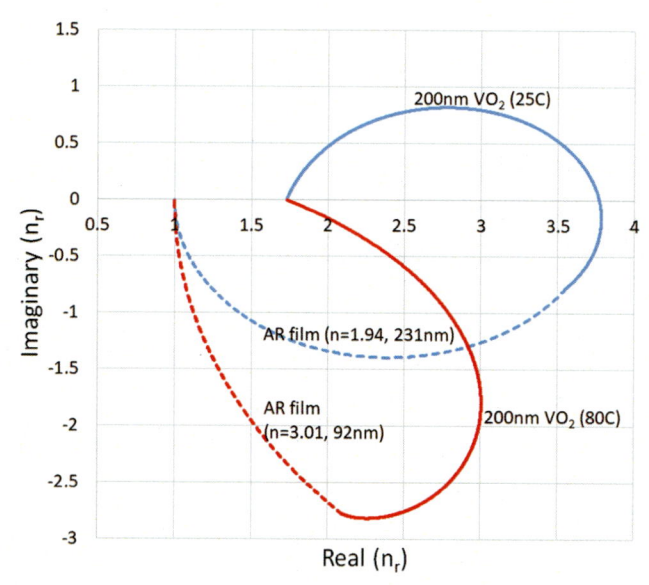

Figure 7.11 Effective reflectance index contours for the antireflection condition for the cold state (25°C) and the hot state (80°C) of a 200 nm VO_2 film.

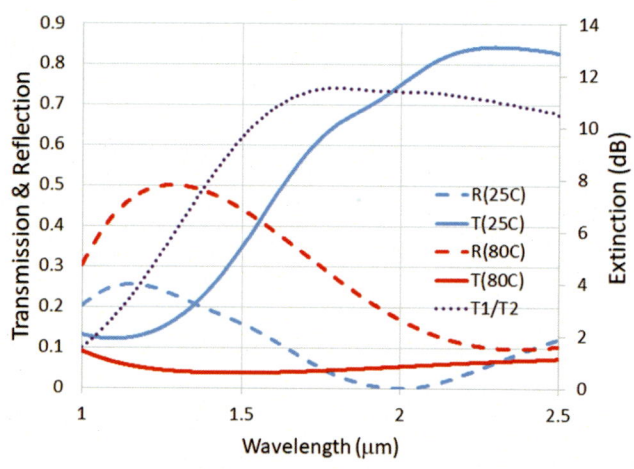

Figure 7.12 Calculated reflection, transmission and extinction ratio for a 200 nm VO_2 film on a sapphire substrate with an AR film ($n = 1.94$, $t = 231$ nm) designed for the cold state. The solid lines are transmissions, and the dashed lines are reflections. The dotted line is the extinction ratio on the right-hand axis. AR, Antireflection.

Fig. 7.13 shows the calculated spectra when the top film is designed to eliminate reflection when the VO_2 is in the hot state. In this case, the reflection of the hot state falls to zero at the design wavelength of 2.0 μm, and the transmission if the cold state is 62%. The resulting extinction coefficient at 2.0 μm is 9.8 dB, which is slightly lower than the previous case.

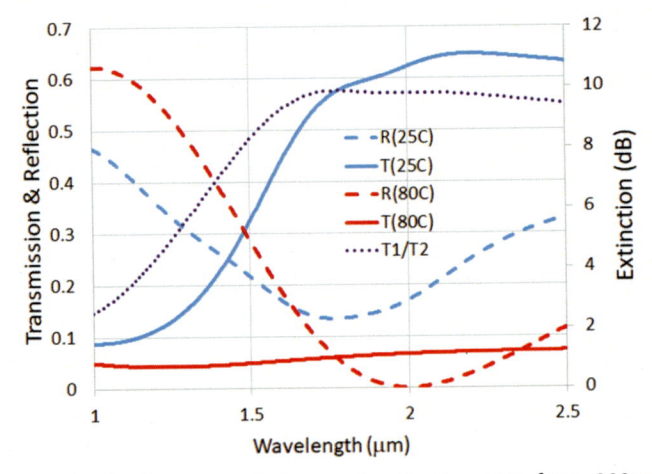

Figure 7.13 Calculated reflection, transmission, and extinction ratio for a 200 nm VO_2 film on a sapphire substrate with an AR film ($n = 3.01$, $t = 92$ nm) designed for the hot state. The solid lines are transmissions, and the dashed lines are reflections. The dotted line is the extinction ratio on the right-hand axis. *AR, Antireflection.*

A number of optical limiting absorbers have been demonstrated in the literature using this principle of semiconductor–to-metal transition of VO_2 [12–14]. In Ref. [13], the authors used a coupled array of silicon nanopillars coated with a thin layer of VO_2 to enhance the extinction ratio and to preserve the spectral bandwidth of the absorber. A resistive heater was used to enable the switching action, although in principle a laser beam can also be used. In Ref. [12], the authors used a laser beam to switch the VO_2 film, demonstrating a more realistic optical limiting effect.

7.2.3 Multilayer germanium–antimony–telluride structure for optical switching

Compared to VO_2, GST has a few favorable properties in optical switching applications. Most importantly, it is significantly easier to fabricate compared to VO_2. It can be deposited at room temperature on virtually any substrate and can be easily patterned. Second, its loss tangent values are smaller, even in the crystalline fcc state. Third, its refractive index contrast between the amorphous and fcc states (the real part) is larger than that of VO_2. This allows GST to be easily utilized in Bragg-like multilayer optical thin film structures to create tunable effects.

One possible design approach is to use a distributed Bragg reflector (DBR) using GST and another layer. Since the refractive index of GST-225 in the amorphous state is about 4, it is convenient to treat it as the high-index layer and choose a low-index material such as SiO_2 or MgF_2. Representing a quarter-wave thick SiO_2 as L (for low-index layer) and a quarter-wave thick GST-225 as H (for high-index layer), a

Bragg stack can be represented as $\left(\frac{H}{2}L\frac{H}{2}\right)^N$ where N is the number of repetitions of the symmetric unit cell $\frac{H}{2}L\frac{H}{2}$. Readers not familiar with this thin film notation should refer to Ref. [6].

For illustration purposes, we will first treat the GST-225 film as a lossless dispersion-less film with an index value of 4.0. The SiO_2 will also be treated with a constant index value of 1.5. Using a 2 μm reference wavelength, the film thickness for L and H will be $2000/4 \times 1.5 = 333$ nm and $2000/4 \times 4.0 = 125$ nm, respectively. Using $N = 4$, we can calculate a transmission and reflection spectra as shown in Fig. 7.14. We can clearly see the transmission band edge of the DBR structure shift as the refractive index contrast between the high- and low-index layers increases. This is a well-known effect of Bragg layers. The total width of the Bragg reflection spectrum is directly related to the refractive index contrast between the high- and the low-index layers. This width increases on both sides of the Bragg reflection spectrum, but in this case, we are only exploiting the long-wavelength edge of the band. Furthermore, another known effect is the oscillation strength (peak-to-valley ratio) on either side of the band. It can be shown that $\left(\frac{H}{2}L\frac{H}{2}\right)$ unit cell has a flatter spectrum on the long-wavelength edge of the band and $\left(\frac{L}{2}H\frac{L}{2}\right)$ unit cell has a flatter spectrum on the short wavelength edge of the band. This effect is also clearly evident from Fig. 7.14. The dotted line shows the transmission ratio between the two curves. We can see that the transmission ratio is as high as 25 dB near the band edge. This will be the operating window of the optical switch. This windows is shown by the dashed vertical lines in Fig. 7.14.

While Fig. 7.14 was an idealized example (because it used lossless, dispersion-less materials, and the layer thicknesses of the two structures were different to account for

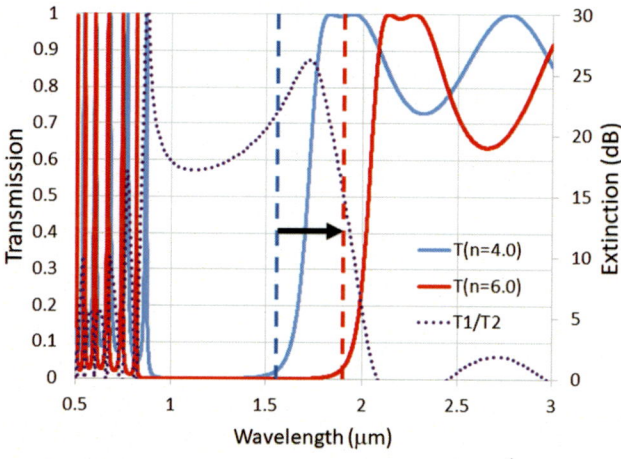

Figure 7.14 Calculated transmission and extinction ratio for a $\left(\frac{H}{2}L\frac{H}{2}\right)^4$ where L has an index of 1.5 and H has an index of 4.0 (*blue line*) and 6.0 (*red line*).

the change in refractive index), we can consider a more practical case by using the complex refractive indices as shown in Fig. 7.5. This structure was designed for a reference wavelength of 1155 nm, resulting in a layer thickness for GST-225 of 65 nm, and a layer thickness for SiO$_2$ of 200 nm. The full layer structure is shown in Fig. 7.15. In Ref. [10], the operation of this device was experimentally demonstrated and was found to exhibit an extinction ratio as high as 30 dB within a 300 nm wide operating band around 1500 nm. The measured extinction ratios compared to the calculated values are shown in Fig. 7.16.

| GST (32nm) |
| SiO$_2$ (200nm) |
| GST (64nm) |
| SiO$_2$ (200nm) |
| GST (64nm) |
| SiO$_2$ (200nm) |
| GST (64nm) |
| SiO$_2$ (200nm) |
| GST (32nm) |

Substrate

Figure 7.15 A 9-layer structure $\left(\frac{H}{2}L\frac{H}{2}\right)^N$ utilizing GST-225 as the high-index layer and SiO$_2$ as the low-index layer designed for a reference wavelength of 1150 nm. GST, Germanium—antimony—telluride.

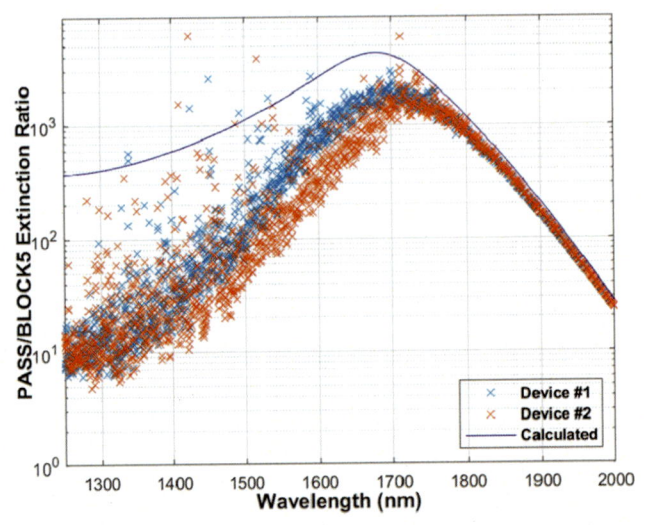

Figure 7.16 Measured and calculated extinction ratios between the amorphous state and the fcc state for the structure shown in Fig. 7.15. *Reproduced with permission from A. Sarangan, J. Duran, V. Vasilyev, N. Limberopoulos, I. Vitebskiy, I. Anisimov, Broadband reflective optical limiter using GST phase change material, IEEE Photonics J. 10 (2) (2018) 1—9.*

The forward transmission when the structure is in the amorphous state was measured to be more than 80% with an acceptance angle of $\pm 45°$. The measured transmission and reflection of the device are shown in Fig. 7.17. The switching action in this device was actuated by an external heating source.

Despite the experimental demonstration, there are some drawbacks to using GST as an optical limiter. While it is relatively easy to switch the amorphous layer to the

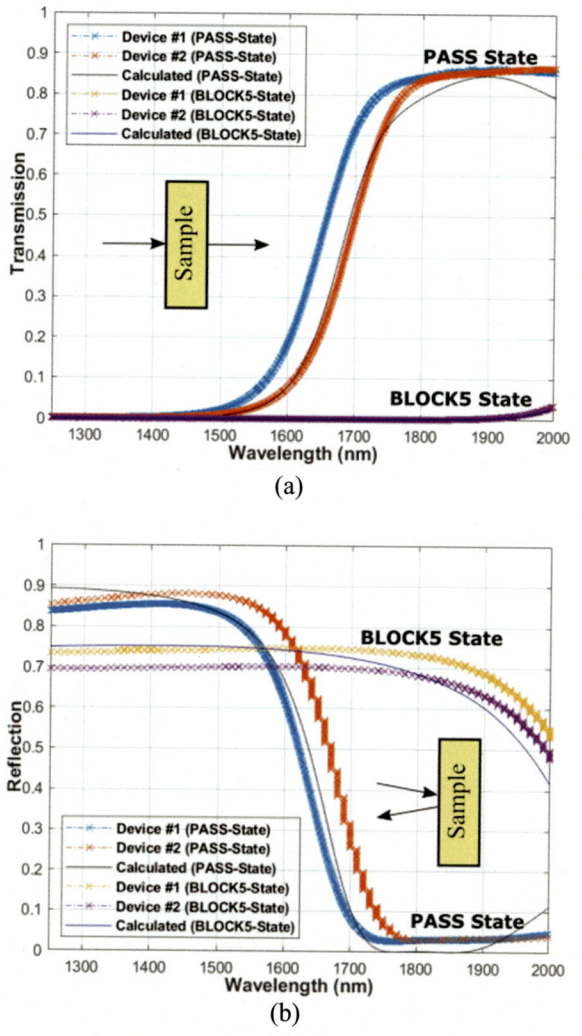

Figure 7.17 Measured and calculated transmission (A) and reflection (B) through the multilayer structure shown in Fig. 7.15. *Reproduced with permission from A. Sarangan, J. Duran, V. Vasilyev, N. Limberopoulos, I. Vitebskiy, I. Anisimov, Broadband reflective optical limiter using GST phase change material, IEEE Photonics J. 10 (2) (2018) 1—9.*

fcc crystalline layer, it is far more difficult to switch a crystalline layer back to amorphous. This would require a high-speed heat pulse and considerations for the thermal conductivities of the surrounding media. In a large-area optical aperture, these would be difficult to implement. Hence this device is limited to an one-way pass-to-block state switching only. Nevertheless in some optical limiting applications, this may be adequate where each switched region can be treated as a sacrificial pixel on a focal plane.

7.2.4 Switching transmission filters using germanium—antimony—telluride

Another type of switching mechanism can be realized by utilizing a resonant cavity configuration such as

$$\left(\frac{H}{2}\right)\left(\frac{H}{2}L\frac{H}{2}\right)^N H\left(\frac{H}{2}L\frac{H}{2}\right)^N\left(\frac{H}{2}\right).$$

We will choose germanium as the high-index H film because it is closely matched to the refractive index of GST-225 in the amorphous state. Furthermore, we can choose $N = 1$ and replace one of the reflectors with GST, such as

$$\left(\frac{H}{2}\right)\left(\frac{H}{2}L\frac{H}{2}\right)H\left(\frac{G}{2}L\frac{G}{2}\right)\left(\frac{G}{2}\right),$$

where the G is the quarter-wave layer of GST-225 in the amorphous state. When the GST layers are in the amorphous state, the cavity reflectors will be matched, and resonant transmission will occur at the reference wavelength. In this case, to allow transmission through germanium, we will need to choose a reference wavelength greater than the equivalent bandgap wavelength of 1.85 μm. For convenience, we can choose 2 μm. The resulting layer structure is shown in Fig. 7.18. As the GST layers switch to the fcc state, the rear reflector will become detuned from the front reflector, causing a small shift in the resonant wavelength. Using the experimentally measured refractive

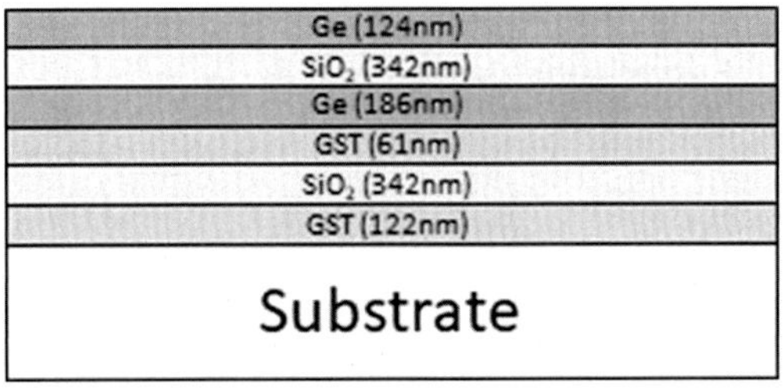

| Ge (124nm) |
| SiO₂ (342nm) |
| Ge (186nm) |
| GST (61nm) |
| SiO₂ (342nm) |
| GST (122nm) |

Substrate

Figure 7.18 Layer structure of the single cavity with a GST multilayer reflector. *GST*, Germanium—antimony—telluride.

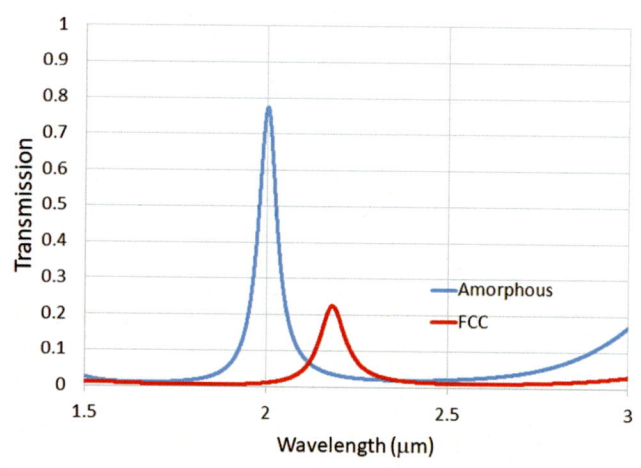

Figure 7.19 Simulated spectral transmission from the $\left(\frac{H}{2}\right)\left(\frac{H}{2}L\frac{H}{2}\right)H\left(\frac{G}{2}L\frac{G}{2}\right)\left(\frac{G}{2}\right)$ cavity filter for the amorphous state and the fcc crystalline state.

index values for GST-225, we can use the plane wave TMM to get the results shown in Fig. 7.19.

The simulated results for this filter are shown in Fig. 7.19. In the amorphous state, the cavity is very close to being in resonance (because amorphous GST and Ge have similar refractive indices), and the resonance occurs at the reference wavelength of 2.0 μm. As the GST crystallizes to the fcc state, the reflection band of the rear reflector shifts toward longer wavelengths, pulling the resonance of the cavity toward a longer wavelengths. The spectral shift in the transmission peak is about 190 nm. Unfortunately, however, the tuning also results in a decline in peak transmission, from 80% down to 22%. This arises due to the increased absorption in the fcc-GST compared to the amorphous GST. There are techniques to mitigate this effect, but they involve more complex solutions, such as a two-dimensional metasurface in addition to the multilayer configuration in the vertical direction [15].

7.3 Deposition methods

7.3.1 Vanadium dioxide (VO$_2$)

Compared to most inorganic oxides, vanadium dioxide is a nontrivial oxide film to produce. This is primarily due to the multiple stable oxidation states of VO_2. The dominant oxides of vanadium are VO, V_2O_3, VO_2, and V_2O_5. As a result, the process window for selectively producing VO_2 is fairly narrow. Most commonly, VO_2 is deposited using physical vapor deposition using reactive sputtering from a vanadium target [16], ion beam sputtering [17], or reactive evaporation [18,19]. Substrate temperature and oxygen partial pressure are critical values that must be tightly controlled.

Excessive oxygen pressure or excessive temperature will lead to higher oxidation states, and vice versa. Typical values reported in the literature are $500°C$ and $1-10$ mT pressure. VO_2 can be grown on nearly any substrate, but sapphire is the most commonly used substrate because the film is nearly lattice matched to the rutile phase of VO_2 (which is the phase that will be present at the high growth temperatures used in these processes). In most of these processes, a postdeposition anneal at a high temperature (typically $500°C$) is used to improve the performance of the VO_2 films [20]. Generally, an increase in crystallite size is observed with an increase in annealing time, as well as a small increase in the transition temperature. These anneals are performed at $2-5$ mT of oxygen. Annealing VO_2 at higher oxygen pressures results in the formation of higher oxidation states. Even exposure to $300°C$ in air seems to oxidize VO_2 into V_2O_5 [21]. More recently, atomic layer deposition (ALD) has also been successfully used to produce VO_2 [22,23] using the Tetrakis[ethylmethylamino] vanadium and water vapor as the precursor gases.

In the author's laboratory, a different method has been developed based on a thermal oxidation method of metallic vanadium [3]. In this method, pure vanadium films are deposited on the substrate using any of the commonly used methods, such as RF (Radio Frequency) or DC (Direct Current) argon sputtering or thermal evaporation. No substrate heating or reactive process is used during this deposition. After deposition, the substrates are quickly loaded into a quartz tube furnace. Exposure to air is minimized to prevent oxidation of the vanadium film. Thin films of vanadium will begin to oxidize in air to V_2O_5 in a matter of hours, consuming the entire film in a matter of days depending on the density of the vanadium film. The temperature of the tube furnace is maintained at $485°C$ using an oxygen pressure of 2 mT. The duration of the thermal oxidation process is determined based on the film thickness. For a 100 nm vanadium film, a duration of 6 h was experimentally found to be optimal. Longer durations will result in the formation of higher oxides such as V_2O_5, and shorter durations will result in lower oxides such as V_2O_3.

There are several distinct advantages to this method compared to reactive physical vapor deposition. Heating a substrate to $500°C$ inside a line-of-sight deposition chamber introduces numerous complications, such as poor thermal contact in vacuum and nonuniform substrate temperature distribution. Second, the higher substrate temperature makes lift-off lithography problematic. Because photoresists generally cannot withstand temperatures above $150°C$, the only option for patterning VO_2 is by postdeposition etching. However, plasma etching of VO_2 is nontrivial and still under development [24]. With thermal oxidation of vanadium, it is possible to prepattern the metallic vanadium using a number of different lithographic techniques, including lift-off, and then thermally oxidize the patterned films into VO_2. Third, thermal oxidation in a tube furnace is a highly scalable batch process. Many substrates could be processed simultaneously, making it a more manufacturing-friendly process.

Table 7.1 Computed values of the volume expansion of vanadium due to oxidation.

	Density g/cm³	Molecular weight g/mol	Volume change
V	5.8	50.94	1
V_2O_3	4.87	149.9	1.75
VO_2	4.57	82.94	2.05
V_2O_5	3.36	181.9	3.08

During oxidation, vanadium is known to grow in size. This can be computed precisely using the density of vanadium and its common oxides, and their molecular weights. This is shown in Table 7.1. We can see that $V \rightarrow VO_2$ results in a volume expansion factor of 2.05. The other two neighboring oxides have expansions of 1.75 and 3.08. As a result, it is relatively straightforward to measure the film thicknesses before and after the oxidation to determine the stoichiometry of the oxidized film.

7.3.2 Germanium–antimony–telluride

A wide range of physical and chemical vapor deposition methods have been explored for producing GST thin films [25]. However, the most common method used in research laboratories is by using RF or DC sputtering from a GST target in an argon ion plasma [26]. The stoichiometry of the film can be preserved quite well after a brief preconditioning stage. Initially, the removal rate of the target will be highly influenced by the different sputter yields of Ge, Sb, and Te. Using 100 eV argon ions, the sputter yield of Te can be calculated to be 0.63, followed by 0.5 for Sb and then 0.4 for Ge. As a result, the initial films on the substrate will be rich in Te and deficient in Ge. As the target conditioning proceeds, the target surface will naturally become richer in Ge and deficient in Te in the exact proportion needed to remove all three elements at the target's bulk stoichiometric ratio. Any type of surface cleaning will invalidate this target conditioning, and it will have to be re-conditioned. Pulsed laser deposition is another commonly used technique for GST [27]. GST is also deposited using thermal evaporation. However, because of its high volatility and high vapor pressures, very low levels of power have to be applied to produce a controlled evaporation rate. Plasma enhanced chemical vapor deposition is another approach [28] as well as ALD [29]. ALD techniques can yield very conformal depositions in densely packed structures using in phase change memories.

7.4 Conclusion

In this chapter, we reviewed the basic optical properties of two important PCMs (VO_2 and GST-225). Examples of tunable thin film device structures employing these PCMs and their design techniques were discussed. It was shown that absorption switching, optical limiting, and spectral tuning can be accomplished by utilizing these PCMs in the multilayer structure. Finally, the commonly used deposition methods for producing VO_2 and GST were reviewed.

References

[1] J. Lappalainen, S. Heinilehto, S. Saukko, V. Lantto, H. Jantunen, Microstructure dependent switching properties of VO_2 thin films, Sens. Actuators A Phys. 142 (1) (2008) 250−255.

[2] S. Chen, J. Liu, L. Wang, H. Luo, Y. Gao, Unraveling mechanism on reducing thermal hysteresis width of VO_2 by Ti doping: a joint experimental and theoretical study, J. Phys. Chem. C 118 (33) (2014) 18938−18944.

[3] P. Guo, Z. Biegler, T. Back, A. Sarangan, Vanadium dioxide phase change thin films produced by thermal oxidation of metallic vanadium, Thin Solid Films 707 (2020) 138117.

[4] M. Currie, M.A. Mastro, V.D. Wheeler, Characterizing the tunable refractive index of vanadium dioxide, Opt. Mater. Express 7 (5) (2017).

[5] K. Dai, J. Lian, M.J. Miller, J. Wang, Y. Shi, Y. Liu, et al., Optical properties of VO_2 thin films deposited on different glass substrates, Opt. Mater. Express 9 (2) (2019) 663.

[6] A. Sarangan, Optical Thin Film Design, CRC Press, 2020.

[7] M. Nishikawa, T. Nakajima, T. Kumagai, T. Okutani, T. Tsuchiya, Adjustment of thermal hysteresis in epitaxial VO_2 films by doping metal ions, J. Ceram. Soc. Jpn. 119 (1391) (2011) 577−580.

[8] Z. Shao, X. Cao, H. Luo, P. Jin, Recent progress in the phase-transition mechanism and modulation of vanadium dioxide materials, NPG Asia Mater. 10 (7) (2018) 581−605.

[9] S. Guerin, B. Hayden, D.W. Hewak, C. Vian, Synthesis and screening of phase change chalcogenide thin film materials for data storage, ACS Comb. Sci. 19 (7) (2017) 478−491.

[10] A. Sarangan, J. Duran, V. Vasilyev, N. Limberopoulos, I. Vitebskiy, I. Anisimov, Broadband reflective optical limiter using GST phase change material, IEEE Photonics J. 10 (2) (2018) 1−9.

[11] P. Guo, J.A. Burrow, G.A. Sevison, A. Sood, M. Asheghi, J.R. Hendrickson, et al., Improving the performance of $Ge_2Sb_2Te_5$ materials via nickel doping: towards RF-compatible phase-change devices, Appl. Phys. Lett. 113 (17) (2018) 171903.

[12] W. Wang, Y. Luo, D. Zhang, F. Luo, Dynamic optical limiting experiments on vanadium dioxide and vanadium pentoxide thin films irradiated by a laser beam, Appl. Opt. 45 (14) (2006) 3378.

[13] A. Howes, Z. Zhu, D. Curie, J.R. Avila, V.D. Wheeler, R.F. Haglund, et al., Optical limiting based on Huygens' metasurfaces, Nano Lett. 20 (6) (2020) 4638−4644.

[14] M. Maaza, D. Hamidi, A. Simo, T. Kerdja, A. Chaudhary, J. Kana Kana, Optical limiting in pulsed laser deposited VO_2 nanostructures, Opt. Commun. 285 (6) (2012) 1190−1193.

[15] M.N. Julian, C. Williams, S. Borg, S. Bartram, H.J. Kim, Reversible optical tuning of GeSbTe phase-change metasurface spectral filters for mid-wave infrared imaging, Optica 7 (7) (2020) 746.

[16] E.N. Fuls, D.H. Hensler, A.R. Ross, Reactively sputtered vanadium dioxide thin films, Appl. Phys. Lett. 10 (7) (1967) 199−201.

[17] X. Yi, C. Chen, L. Liu, Y. Wang, B. Xiong, H. Wang, et al., A new fabrication method for vanadium dioxide thin films deposited by ion beam sputtering, Infrared Phys. Technol. 44 (2) (2003) 137−141.

[18] J. Leroy, A. Bessaudou, F. Cosset, A. Crunteanu, Structural, electrical and optical properties of thermochromic VO_2 thin films obtained by reactive electron beam evaporation, Thin Solid Films 520 (14) (2012) 4823−4825.

[19] M. Zou, C. Ni, A. Sarangan, Ion-assisted evaporation of vanadium dioxide thin films, in: E.M. Campo, E.A. Dobisz, L.A. Eldada (Eds.), Nanoengineering: Fabrication, Properties, Optics, and Devices XIII, SPIE, 2016.

[20] M. Kumar, J.P. Singh, K.H. Chae, J. Park, H.H. Lee, Annealing effect on phase transition and thermochromic properties of VO_2 thin films, Superlattices Microstruct. 137 (2020) 106335.

[21] G. Fu, A. Polity, N. Volbers, B.K. Meyer, Annealing effects on VO_2 thin films deposited by reactive sputtering, Thin Solid Films 515 (4) (2006) 2519−2522.

[22] K. Zhang, M. Tangirala, D. Nminibapiel, W. Cao, V. Pallem, C. Dussarrat, et al., Synthesis of VO_2 thin films by atomic layer deposition with TEMAV as precursor, ECS Trans. 50 (13) (2013) 175−182.

[23] V. Prasadam, B. Dey, S. Bulou, T. Schenk, N. Bahlawane, Study of VO_2 thin film synthesis by atomic layer deposition, Mater. Today Chem. 12 (2019) 332−342.

[24] Y.-H. Ham, A. Efremov, N.-K. Min, H.W. Lee, S.J. Yun, K.-H. Kwon, Etching characteristics of VO_2 thin films using inductively coupled Cl_2/Ar plasma, Jpn. J. Appl. Phys. 48 (8) (2009) 08HD04.

[25] J.E. Boschker, R. Calarco, Growth of crystalline phase change materials by physical deposition methods, Adv. Phys. X 2 (3) (2017) 675−694.

[26] P. Guo, A. Sarangan, I. Agha, A review of germanium-antimony-telluride phase change materials for non-volatile memories and optical modulators, Appl. Sci. 9 (3) (2019) 530.

[27] M. Bouška, S. Pechev, Q. Simon, R. Boidin, V. Nazabal, J. Gutwirth, et al., Pulsed laser deposited GeTe-rich $GeTe-Sb_2Te_3$ thin films, Sci. Rep. 6 (1) (2016).

[28] B.J. Choi, S. Choi, Y.C. Shin, C.S. Hwang, J.W. Lee, J. Jeong, et al., Cyclic PECVD of $Ge_2Sb_2Te_5$ films using metallorganic sources, J. Electrochem. Soc. 154 (4) (2007) H318.

[29] M. Ritala, V. Pore, T. Hatanpaa, M. Heikkila, M. Leskela, K. Mizohata, et al., Atomic layer deposition of $Ge_2Sb_2Te_5$ thin films, Microelectron. Eng. 86 (7−9) (2009) 1946−1949.

CHAPTER 8

Chromogenically tunable thin film photonic crystals

Pandurang Ashrit and Tran-Vinh Son
Thin Films and Photonic Research Group (GCMP), Department of Physics and Astronomy, University of Moncton, Moncton, NB, Canada

8.1 Introduction

8.1.1 Chromogenics

The word *Chromogenics* comes from the Greek word *Chromo* for color and has been recently used to refer to the field of study of materials that show a reversible change in their optical properties (color) as a function of the ambient conditions they are subjected to. These external stimuli include light, heat, electric field, exposure to gas and many others. The chromogenic materials form an important class of *meta materials* that provide an inordinate ability to control the propagation of electromagnetic (EM) radiation through these materials. Thus these materials have become the object of an intensive research and development effort in the area of photonics in recent years. Depending on the type of external stimuli applied to bring about the reversible color change, the chromogenic materials are further subclassified as *photochromic* (light activated), *electrochromic* (electric field activated), *thermochromic* (heat activated), *gasochromic* (gas exposure activated), *piezochromic* (pressure activated), *magnetochromic* (magnetic field activated) and more [1]. Because of their interactive nature, the chromogenic materials form an important class of smart materials showing a reversible change of various physical and chemical properties as a function of the externally applied stimulus. Their dynamic and interactive nature via the external forces have made them highly sought after for various devices and applications. Their efficient response to various external forces makes it easier to build a wide range of automated smart systems where light control is needed. Their switching between different metastable states with wide-ranging change in physical and chemical properties provides for a rich platform for fundamental studies. Following are some of the well-known chromogenic types.

Liquid crystal displays that have found a wide variety of display applications are one of the most known electrochromic materials in which a reversible color change can be triggered through the application of an electric field [2]. The change from their normal transparent state to an opaque state under the application of an electric field

Thin Film Nanophotonics
DOI: https://doi.org/10.1016/B978-0-12-822085-6.00006-6

comes from a change in the orientation of the liquid crystal molecules. High power consumption to maintain the active opaque state and switching only between the "on" and "off" states can be a disadvantage for some of the applications such as energy efficiency windows. Electro or electrophoretic deposition in which tiny particles suspended in a solution present in a transparent electrochemical cell get deposited on the cell wall under the application of a small electric field is another type of electrochromic device [3]. Normally the cells are transparent and when applied with the electric field, the suspended particles deposit on the electrode changing the appearance of the cell, thus inducing a color change. The presence of a liquid component can limit this approach for many applications. Many metal-halide-based compounds change from a transparent to a mirror-like reflectance state under exposure to gases such as hydrogen and a simultaneous application of a small electric field, thus exhibiting very efficient gaso-electrochromic behavior [4]. Similarly, photochromic glasses based on the coloration change upon the impingement of light or electromagnetic radiation have been in use for a long time [5,6]. The coloration occurs through a charge transfer between two electronic sites when activated by light (invariably ultraviolet or visible) exposure. The metastable colored state is not directly reversible and other external forces such as electric field or heat may be necessary to revert to the normal state. Quite a few organic and inorganic materials are found to exhibit efficient photochromic coloration [7]. However, the inorganic materials with a more metastable colored or excited state are preferable from an application point of view compared to the less stable organic materials. Many organic and inorganic materials are also known to exhibit very efficient thermochromic behavior, both a gradual as well as a first-order (drastic) reversible change in optical properties under the influence of temperature [8,9]. These have been very successfully applied in many areas such as security, imagery, sensors, copiers and more. The piezochromic or mechanochromic materials which show a reversible change in optical properties under the influence of pressure have also been known for a long time. Various mechanisms can lead to their color change stemming from electronic excitation to crystal structure change to changes occurring at the molecular level. These materials also range from the organic mixture of polymers in cholesteryl derivatives and dimethylglyoxime [10] to the inorganic samarium monosulphide [11]. Many inorganic magnetomaterials have found many applications. The change in optical properties or optical constants arising from the application of the magnetic field is rooted in a change in the spin induced electric polarization [12] or spin induced ferroelectricity [13] or even in the simplest magnetochromic microspheres containing iron oxides (Fe_3O_4) in which the applied magnetic field brings about the orientation of the microspheres and the reversible optical change [14]. Thus a wide range of approaches can be taken to build chromogenic devices. However, an important class of materials that are being studied in recent decades for their chromogenic properties are the transition metal oxides (TMO) and their thin films as described below.

8.1.2 Transition metal oxides

Of all the known inorganic compounds, the TMO have been known for a long time for their unusual and fascinating characteristics arising from the special electronic structure of the base transition metal and its bonding with oxygen. The partially filled d orbitals in the transition metals not only lead to the formation of a wide range of oxides but also to the multiple metastable oxidation states of the TMOs. It is the charge exchange of electrons between the highly electronegative oxygen atoms and the less electronegative transition metal atoms that leads to the formation of this wide range of TMOs. In addition, it also provides for the formation of various structures and phases of the TMOs as well as for the inclusion of other compatible atoms. The TMOs exhibit a wide range of behavior from ionic to metallic. Hence, to understand the complex behavior of these oxides, that is, to correlate their physical and electronic structure with their wide-ranging properties, with a single theoretical model becomes an overwhelming task [15]. However, these wide-ranging properties and their metastable multiple oxidation states are the characteristics that make them very attractive from fundamental study point of view as well as for applications. The TMO property of relevance to chromogenics is the possibility to easily and reversibly switch these oxides from one metastable state to the other through a small external force. In this aspect of facile switching, the TMOs are comparable to semiconductors. Many TMOs are known to exhibit large reversible changes in their optical properties under the influence of the external forces such as heat, light, electric field, pressure, gas and more. Hence, the TMOs are very attractive candidates for chromogenic property study and applications. The optical changes occurring in the TMOs are accompanied also by large changes in their electrical, structural, and magnetic properties making them also interesting candidates for optoelectronic and optomagnetic studies. Many TMOs present wavelength-selective optical switching as discussed below in the specific examples. Prepared in thin film form, they offer even greater versatility in switching and control of various optical parameters such as reflectance, transmittance, and absorptance as well as of the fundamental optical constants. Hence, going from a bulk 3-D state of the TMOs to the thin film (2-D) state presents enormous possibility in addition to the reversible switching of the optical properties. The optical behavior is strongly influenced by the film thickness as well as the film nanostructure depending on the interaction light undergoes at the various film interfaces and the grain boundaries (light scattering). Further, the TMO film nanostructure can be periodic or random. A periodic TMO nanostructure in the light wavelength range leads to the formation of a photonic bandgap (PBG) as discussed below in detail. A random nanostructure on light wavelength scale, on the other hand, leads to scattering and/or more efficient interaction at the grain interfaces leading to light trapping and light-harvesting effects similar to that seen in many solar cell configurations [16–18]. A wide range of physical and chemical techniques are now available for the preparation of TMO thin films and their nanostructuring [19−21]. Some of the TMOs that have attracted a lot of research and development attention in recent decades for their efficient chromogenic switching are tungsten trioxide (WO_3), molybdenum trioxide (MoO_3), vanadium pentoxide (V_2O_5), vanadium dioxide (VO_2), and chromium oxide (Cr_2O_3). The chromogenic behavior of each of these oxides, especially from the optical switching point of view, is discussed briefly below.

8.1.3 Tungsten trioxide (WO_3)

Tungsten trioxide (WO_3) exhibiting highly efficient electrochromic and fairly good photochromic properties is the most studied TMO in the literature related to chromogenic properties. WO_3 is formed of corner-sharing WO_6 octahedral units as shown in Fig. 8.1. Each tungsten ion (W^{6+}) is surrounded by six oxygen ions (O^{2-}). The WO_6 octahedral units form a network of linear chains, through corner sharing at the oxygen atom sites as shown in Fig. 8.1. In such a network the oxygen and tungsten ions are found at alternative sites in the linear chain. Such an arrangement not only gives rise to various arrangements of the octahedra but also to a large open space in the network that facilitates the introduction of other large atoms. In WO_3, this is an important aspect for electrochromism, where a deep blue coloration from the initially transparent state occurs via intercalation of other species (H^+, Li^+, Ag^+, etc.) due to the reversible formation of tungsten bronze. The actual electrochromic coloration/decoloration happens under the double insertion/extraction of equal number of external ions (M^+) and electrons (e^-) leading to the following reversible reaction:

$$WO_3 + xe^- + xM^+ \Leftrightarrow M_xWO_3 \quad \text{(Tungsten bronze)}$$
$$\text{(Transparent)} \qquad\qquad \text{(Dark blue)}. \tag{8.1}$$

The insertion of equal number of ions and electrons into the normally transparent tungsten trioxide transforms it into dark blue-colored tungsten bronze. The extraction

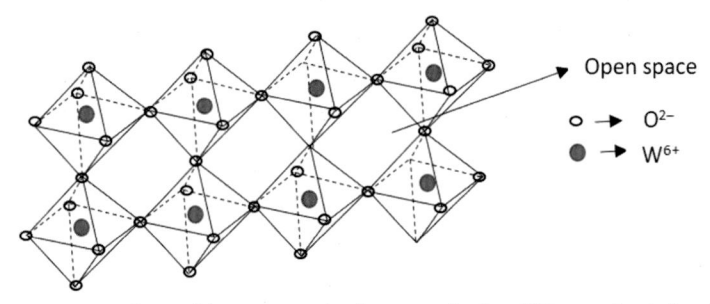

Figure 8.1 WO_3 structure formed by a network of corner-sharing WO_6 octahedral units incorporating large open space.

of these species returns it to the original state. The open structure of WO_3 as shown in Fig. 8.1 renders this intercalation and extraction of the ions facile, thus enabling high-efficiency optical switching from a highly transparent to intense colored states. High level of reversible intercalation can be induced with x values reaching up to 0.7 for lithium ions, for example, before the onset of secondary irreversible reactions [22].

Due to the wide range of structural arrangements possible with the WO_6 octahedral units, a wide range of theoretical models are proposed to account for the ensuing coloration, each fitting for a particular case [23–27]. The two most prevailing models of electrochromic coloration of this material correspond to purely crystalline and amorphous WO_3 films. In the crystalline WO_3 where the long-range periodic arrangement of the WO_6 octahedral units exists, there is an overlapping of wavefunctions associated with the neighboring tungsten atoms creating bands. Hence, the inserted electrons, shown in Eq. (8.1), become quasifree electrons and the electrochromic (coloration) behavior can be explained according to the Drude theory of materials containing large density of electrons [28]. A normally transparent dielectric polycrystalline WO_3 shows increasingly semiconductor to nearly metallic behavior under increasing free electron and ion insertion where it becomes reflective and opaque. With increasing free electron density (n) in the film, the plasma frequency (ω_p) increases in accordance with the relation [29],

$$\omega_p = 4\pi n e^2 / m, \tag{8.2}$$

with e and m being the electron charge and mass, respectively. The optical behavior of the WO_3 film being described in terms of the changing dielectric constant ($\varepsilon(\omega)$) in the optical frequency (ω) regime by,

$$\varepsilon(\omega) = 1 - \frac{\omega_p^2}{\omega^2}. \tag{8.3}$$

The reflectance of such a WO_3 film increases gradually at higher frequencies and moves toward lower wavelengths with increasing electron insertion, as shown in Fig. 8.2 for a simulated WO_3 film of 0.2 μm thickness [30].

Hence a polycrystalline WO_3 film shows a wavelength-selective reflectance modulation with increasing amount of insertion of electrons and ions. The accompanying ions occupy the open space in the octahedral periodic network as shown in Fig. 8.1. Shown in Fig. 8.3 are the experimental results for a polycrystalline WO_3 film of 0.2 μm thickness. The as-deposited film exhibits a high transmittance in the solar spectral range and under increasing lithium insertion exhibits a strong wavelength-selective coloration as the transmittance curve moves to lower wavelengths [31].

However, the coloration behavior of an amorphous WO_3 is drastically different from the polycrystalline one due to the lack of long-range ordering of the WO_6

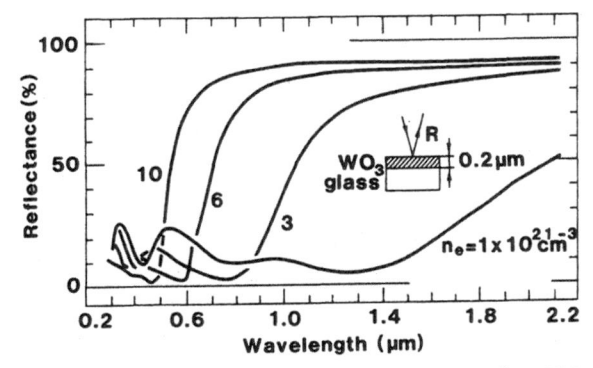

Figure 8.2 Simulated reflectance spectra of a 0.2-μm-thick polycrystalline WO_3 film on a glass substrate with increasing density of electrons (ions) inserted [30].

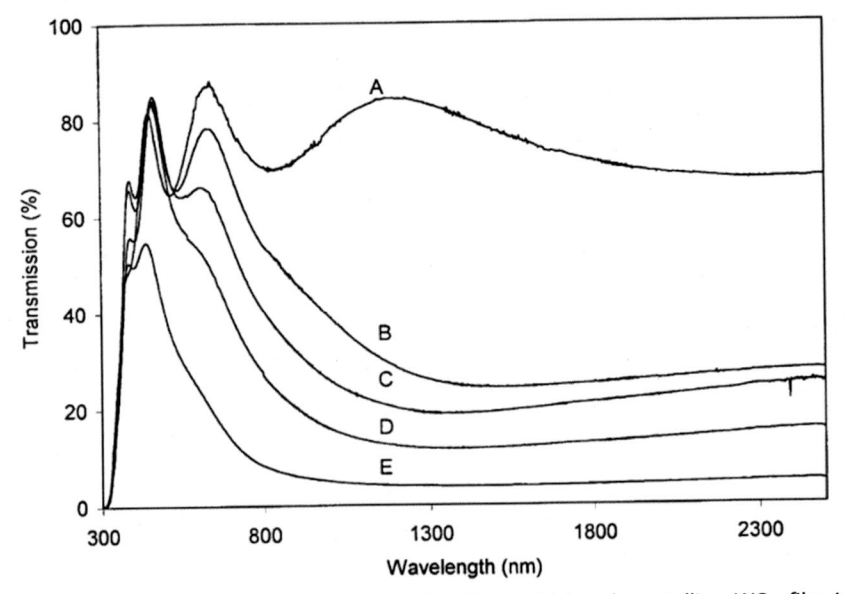

Figure 8.3 Experimental transmittance spectra of a 0.2-μm-thick polycrystalline WO_3 film inserted with various amounts of mass thickness of lithium. The curves A, B, C, D, E, and F correspond respectively to 0, 5, 7.5, 10, 15, and 20 nm of lithium [31].

octahedral units. In this case, the inserted electrons get localized on the tungsten site changing the valence from W^{6+} to W^{5+}. The electrochromic coloration is deemed to occur from the intervalence transfer of these localized electrons through the absorption of light of certain wavelengths [24]. The schematic of such an electron transfer between W sites through the absorption of light is shown in Fig. 8.4.

Hence, the electrochromic coloration in the case of an amorphous WO_3 film is through absorption modulation. With increasing number of electrons inserted in the

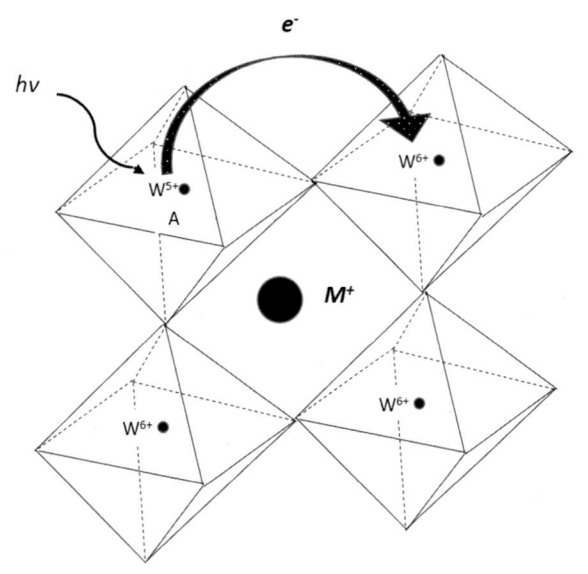

Figure 8.4 Schematic diagram of the localization of an electron (e^-) on a W^{5+} at site A and its transfer to W^{6+} by absorption of photon. The associated ion is located in the nearby open space.

film, the absorption, generally occurring in the visible and near-infrared (NIR) region, deepens but does not show a wavelength-selective behavior as in the case of crystalline WO_3. An example of the absorption modulation-based electrochromic coloration of amorphous WO_3 film is shown in Fig. 8.5 [31]. The change in transmittance in the solar spectral range is more or less uniform in this case, showing no wavelength selectivity.

This aspect of WO_3 phase–dependent reflectance or absorption modulation is very important from the electrochromic device design point of view. However, in both the types of films, the accompanying ions are situated in the large empty spaces in the structure as shown in Figs. 8.1 and 8.3 maintaining the charge neutrality of the films. Both the amorphous and polycrystalline WO_3 films undergo large reversible changes in their optical constants (n, k) with the insertion of lithium as shown in Fig. 8.6 [32].

8.1.4 Vanadium dioxide (VO_2)

In recent years VO_2, exhibiting an inordinately efficient thermochromic coloration near room temperature has become the focus of enormous applied and basic research. This material undergoes a first-order semiconductor to metal transition at the temperature of 68°C with drastic reversible changes in its optical, electrical, and magnetic properties. At this transition temperature, it also undergoes a structural change at the lattice level. As mentioned earlier, the multiple oxidation states of transition metals

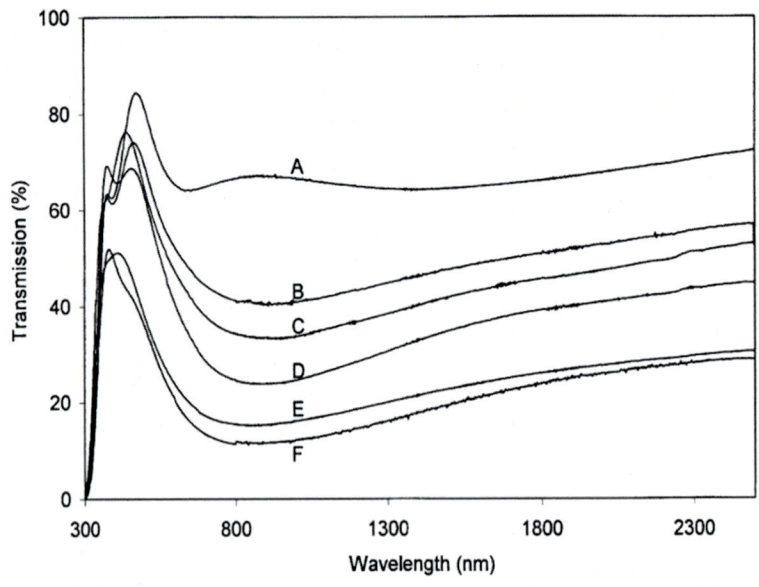

Figure 8.5 Experimental transmittance spectra of a 0.2-μm-thick amorphous WO_3 film inserted with various amounts of mass thickness of lithium. The curves A, B, C, D, E, and F correspond respectively to 0, 5, 7.5, 10, 15, and 20 nm of lithium [31].

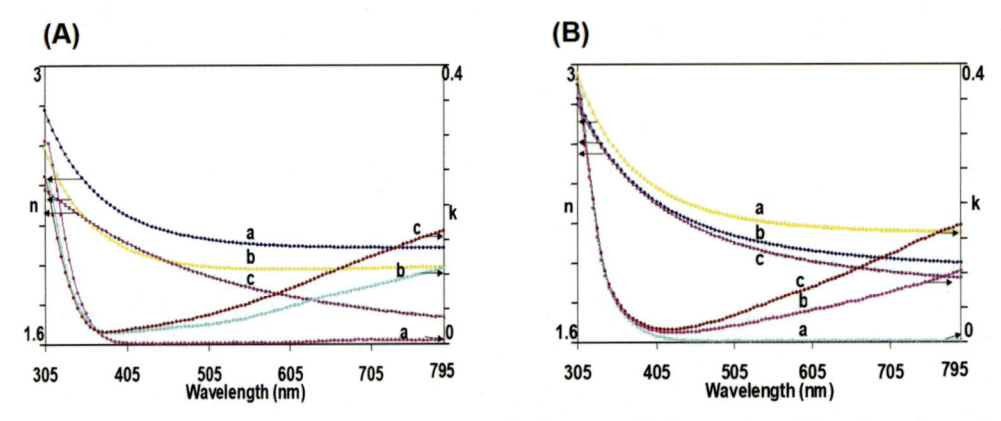

Figure 8.6 Spectral variation of optical constants (n, k) of a 0.2-μm-thick amorphous (A) and a polycrystalline (B) WO_3 film with the insertion of (a) 0 nm, (b) 2.5 nm, and (c) 5 nm mass thickness of lithium [32].

render the possibility of the formation of a wide range of TMO. Vanadium, being one such transition metal, exhibits oxidation states from 2+ to 5+ and forms a wide range of oxides from V_2O to V_2O_5 [33]. VO_2 with an oxidation state of +4 is found to form in very narrow range of conditions and presents itself in a monoclinic or triclinic or tetragonal phase depending on the temperature.

Above the transition temperature ($>68°C$), VO_2 has a tetragonal structure referred to as the rutile structure (R) as shown in Fig. 8.7B with regular distancing of the V atoms along a-axis (4.5546 Å) and c-axis (2.8514 Å) (also shown in Fig. 8.7D). The VO_2 structure is made up of corner-sharing and edge sharing VO_6 octahedra [34]. This structure leads to the liberation of electrons (shown in green in Fig. 8.7B) due to the large charge screening that occurs between the equidistant V atoms separated by oxygen atoms. Hence, in this high-temperature phase, VO_2 presents a near metallic behavior with high conductivity and reflectance. Shown in Fig. 8.7B are also the lattice vibrations (phonons) at high temperatures.

Below the transition temperature ($<68°C$), however, VO_2 has a monoclinic structure referred to as $M1$ phase, as shown in Fig. 8.7A. As can be seen in this figure, a periodic distortion between the V atoms occurs along the c-axis and the V−V distance becomes irregular due to the alternate short (2.65 Å) and long (3.12 Å) chains. This arrangement leads to an increase in the size of the unit cell, pairing up of the V−V atoms along the short arm and to the tilting of the bonds along with the c, as also seen in Fig. 8.7C. The pairing of the nearby V−V atoms also causes the localization of the d orbital electrons (shown as *green dots* in Fig. 8.7A) driving VO_2 to a low conductivity semiconductor/dielectric state. In addition to this $M1$ phase, VO_2 also exhibits another low-temperature crystalline phase, $M2$ which is a close monoclinic phase to $M1$ but with different lattice constants and slight variation in the location of the V and O atoms [35]. Both these low-temperature monoclinic phases, $M1$ and $M2$, transform to the rutile tetragonal phase R at high temperature ($>68°C$).

The simultaneous occurrence of the optical, electrical, and structural changes along with the existence of two low-temperature phases ($M1$ and $M2$) tending to the same high temperature R phase, make it a daunting task to develop a comprehensive theoretical model to explain all the reversible changes undergone by VO_2. Although, various approaches have been taken to this end, a clear single model that takes into

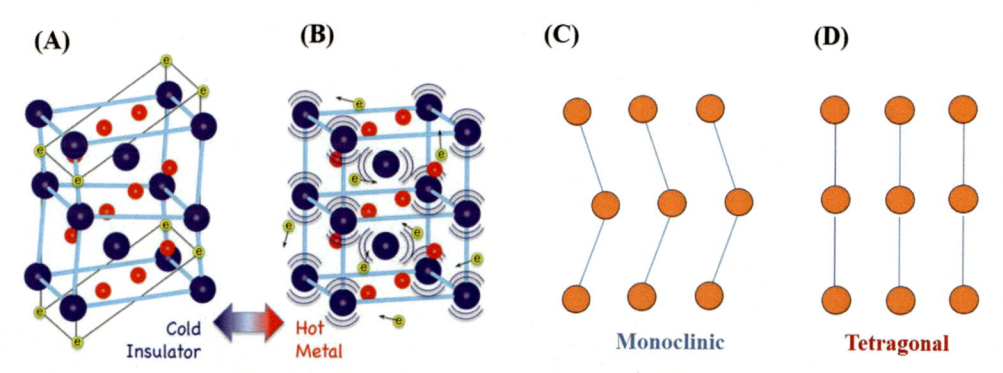

Figure 8.7 Comparison of the crystal structure (A, B) and schematic representation (C, D) of VO_2 monoclinic (below transition temperature) and tetragonal (above transition temperature) phases.

account all the changes are still eluding [36–38]. However, all the drastic physical reversible changes occurring in VO_2 with temperature and having their origin at the microscopic level have made VO_2 an object of intense fundamental study in recent years. The reversible changes, especially in the optical and electrical properties, have led to an enormous amount of applied research on this material with a number of applications in photonics, optoelectronics and other areas [39–41].

An example of the change in optical properties of 240-nm-thick VO_2 film is shown in Fig. 8.8 [42]. At room temperature, this film exhibits a fairly high reflectance over the near-infrared and infrared (IR) regions. At high temperature ($>68°C$), the film transmittance falls to near-zero and the film reflectance becomes mirror like through most of the IR spectral region.

The drastic optical changes in VO_2 are accompanied by the first-order change in the electrical resistance. With the switching of VO_2 from a room temperature ($<68°C$) semiconducting state to a high temperature ($>68°C$) metallic state, a drop in resistance of the order of 10^4-10^5 is commonly seen in bulk VO_2 [43]. However, the nature of both the optical and electrical switching in VO_2 is found to depend profoundly on the preparation parameters as well as the micro and nanostructure when prepared in thin film form. In thin film form, VO_2 is found to exhibit a hysteresis behavior with respect to temperature. An example of this is shown in Fig. 8.9 for a 300-nm VO_2 film [44].

As seen in this figure, the path of change in film transmittance and resistance during the heating cycle and the cooling cycle is not the same. In addition to hysteresis, the transition temperature is also found to depend very strongly on the VO_2 film nanostructure. As seen in Fig. 8.9, the transition temperature, normally calculated

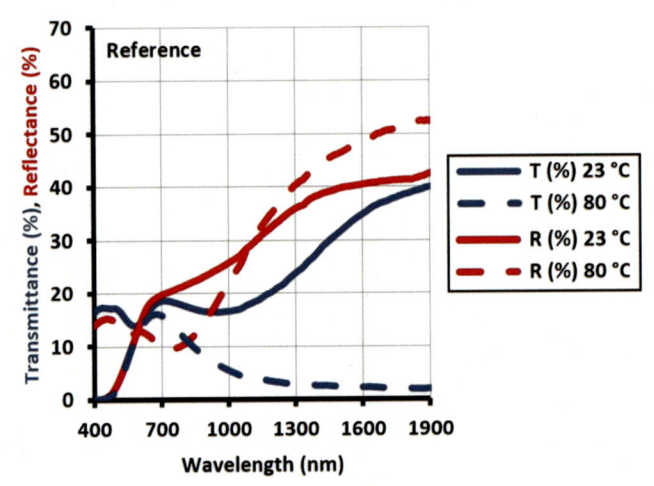

Figure 8.8 Transmittance (*T*) and reflectance (*R*) spectra of a 240-nm-thick VO_2 film on glass above and below the transition temperature [42].

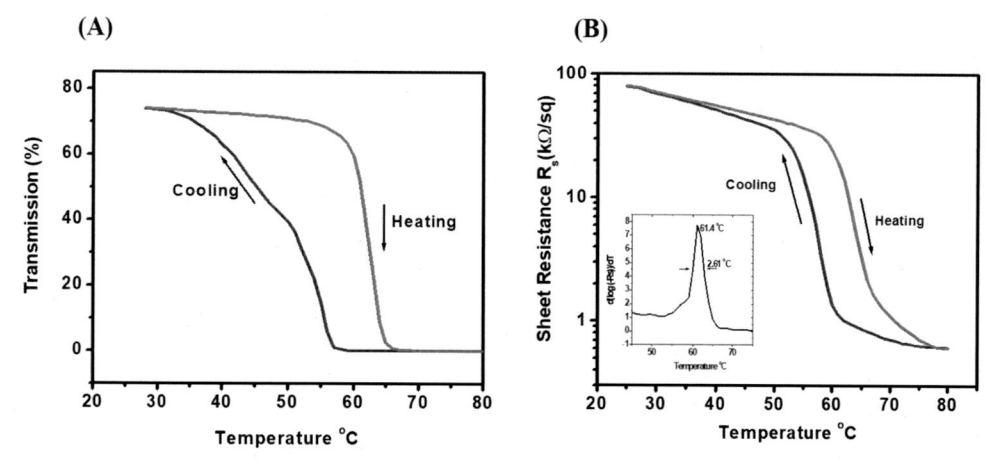

Figure 8.9 Thermal hysteresis of transmittance (A) and sheet resistance (B) of a 0.3-μm-thick VO_2 film (inset B: temperature derivative of the sheet resistance vs temperature) [44].

from the steepest change in resistance is found to be 61.4°C. A tremendous research effort is currently dedicated to diminishing the transition temperature in VO_2 thin films to bring it closer to room temperature to facilitate many energy-related applications of VO_2. The approach that is found to be the most effective and commonly used is to dope VO_2 with compatible atoms such as tungsten (W), molybdenum (Mo), and niobium (Nb) [45].

The typical and important changes occurring in the optical constants in the visible region of the spectrum, below and above the transition temperature of a 300-nm-thick VO_2 sample, are shown in Fig. 8.10. The reflectivity change ($\Delta R/R$) associated with this thermochromic switching is also shown in the bottom panel of Fig. 8.10 [46].

8.1.5 Molybdenum trioxide (MoO_3)

Molybdenum trioxide is another very interesting TMO exhibiting very efficient electrochromic and photochromic switching. It is very surprising that this material, exhibiting equally efficient chromogenic properties as tungsten trioxide (WO_3) and having equally splendid application potential, has not attracted an equal amount of research effort. Hence, tremendous possibilities exist with this material, both for fundamental as well as applied research. Similar to WO_3, MoO_3 structure is formed of octahedral units in which the molybdenum (Mo) atom is surrounded by six oxygen atoms, as shown in Fig. 8.11A. These basic octahedral units bond together in various ways to form the α (orthorhombic—double chain), β (monoclinic), and h (hexagonal) polymorphs of MoO_3, shown, respectively, in Fig. 8.11B(a−c) [47,48]. These various structures and the open space they render make MoO_3 an excellent candidate for intercalation and deintercalation studies and applications. For example, the open

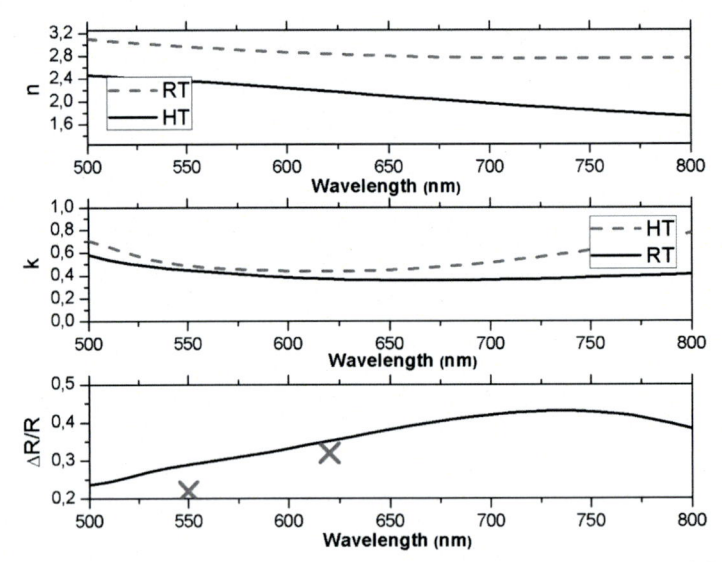

Figure 8.10 Refractive index, *n* (top) and extinction coefficient, *k* (middle) of a VO$_2$ film above and below the transition temperature. Film reflectivity change (bottom); black line-average and cross-data points relative to the sample used in the experiment [46].

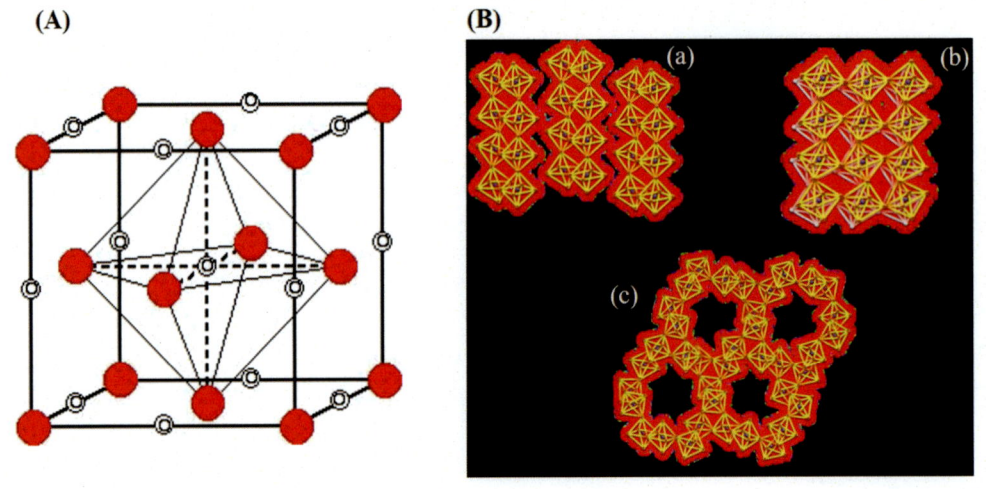

Figure 8.11 (A) MoO$_3$ octahedral unit formed by the hexacoordinated molybdenum atom by six oxygen atoms. (B) Crystal structure of the three polymorphs of MoO$_3$: α-MoO$_3$ (a), β-MoO$_3$ (b) and *h*-MoO$_3$ (c).

structure in *h*-MoO$_3$ makes it the perfect material for lithium storage batteries with high charge capacity.

Under ion/electron double insertion (electrochromic effect) or light irradiation (photochromic effect), the MoO$_3$ films undergo a significant optical change in the

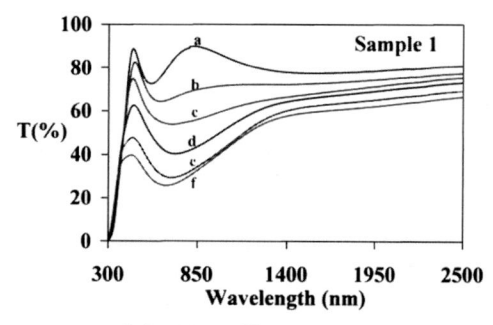

Figure 8.12 Transmittance spectra of the MoO_3 film with the insertion of various amounts of mass thickness of lithium of (a) 0 nm, (b) 2.5 nm, (c) 5 nm, (d) 10 nm, (e) 15 nm, and (f) 20 nm [49].

visible and near-infrared regions. Similar to WO_3 films, the electrochromic coloration under double insertion/extraction can be represented as follows, where M^+ can be any ion (Li^+, H^+, etc.), to form molybdenum bronze

$$MoO_3 + xe^- + xM^+ \Leftrightarrow MxMoO_3 (\text{Molybdenum bronze})$$

(Transparent)   (Deep dark blue).   (8.4)

An example of the electrochromic coloration of MoO_3 films is shown in Fig. 8.12 [49]. As can be seen from this figure, compared to the tungsten bronze (Fig. 8.5), the molybdenum bronze formed through the double insertion shows a more narrowband coloration change concentrated in the visible and lower near-infrared region. Hence, MoO_3 films—based electrochromic devices have a greater potential for display devices due to their higher coloration efficiency in the visible region [50].

8.1.6 Vanadium pentoxide (V_2O_5)

Vanadium pentoxide is another TMO that is vastly studied due to its anodic electrochromic coloration under ion/electron double insertion. The thin films V_2O_5 present a deep yellowish color in the as-deposited state due to the optical absorption band situated in the middle of the visible region (~ 500 nm). Similar to other TMOs V_2O_5 exhibits open and layered structure due to the corner-sharing oxygen units. However, unlike the other TMOs discussed earlier, the transition metal V in this case is penta-coordinated by oxygen atoms, with four oxygen atoms forming the base of the pyramid, one oxygen atom at the apex and a long-distance V—O bond forming along the c-axis, as shown in Fig. 8.13A. Hence, unlike in the VO_2 formed by the VO_6 octahedral, in V_2O_5 the building blocks are the VO_5 pyramidal units. The layered structure (Fig. 8.13B) emanates from the long and weaker V—O bonds leading to corner-sharing oxygen atoms and edge sharing VO_5 pyramids. The various bonding possibilities also lead to many polymorphs of the V_2O_5 [51].

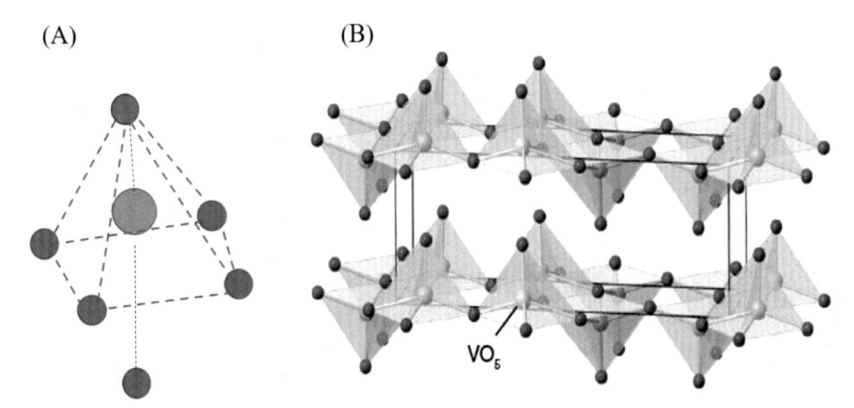

Figure 8.13 (A) VO_5 square pyramid formed by the penta-coordinated vanadium atom by four oxygen atoms forming the base of the pyramid and a vanadyl bond forming the apex. (B) Cubic structure formed by the upright and inverted VO_5 square pyramids.

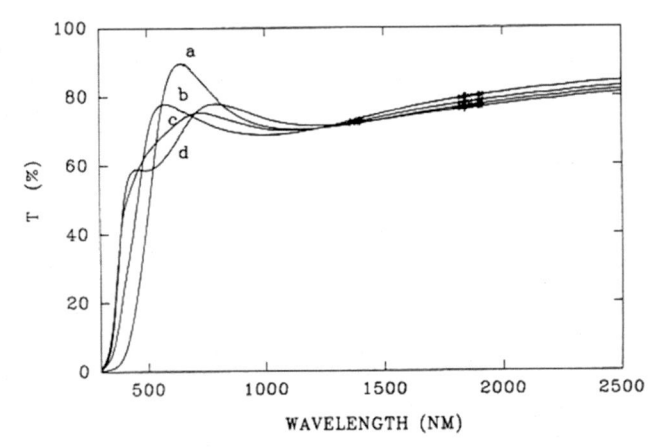

Figure 8.14 Transmission spectra of a 0.2-μm-thick V_2O_5 film with the insertion of different mass thickness of lithium: (a) 0 nm, (b) 10 nm, (c) 20 nm, and (d) 50 nm [52].

The large orthorhombic structure ($a = 11.5$ Å, $b = 3.6$ Å, $c = 4.4$ Å) along with the layered formation and open space makes V_2O_5 an excellent candidate for high energy charge storage. V_2O_5 also undergoes a large optical change (yellow to transparent) under the double ion/electron insertion:

$$V_2O_5 + xe^- + xM^+ \Leftrightarrow M_xV_2O_5 \ (\text{Vanadyl bronze}). \tag{8.5}$$

(Deep yellow)   (Transparent)

The fact that it undergoes an optical change exactly opposite to that of the cathodically coloring WO_3 makes it also a suitable candidate as a charge storage layer (counter electrode) in electrochromic devices [52]. An example of the discoloration undergone by a 200–nm-thick V_2O_5 film is shown in Fig. 8.14. Bulk of the change in transmittance occurs at lower wavelengths. The optical band of V_2O_5 centered around 500 nm wavelength moves toward lower wavelengths with increasing charge insertion changing it from a yellowish to colorless state. Hence, all the large and reversible changes exhibited by the TMOs discussed earlier make them very suitable candidates for various interactive photonic devices.

8.1.7 Photonic crystals

Photonic crystals are periodic structures in 1, 2, and 3 dimensions made of thin layers or rods or particles of nanometric size of the order of light wavelengths (few hundred nanometers), as shown in Fig. 8.15 [53]. They are analogous to electronic crystals which are periodic structures on electron wavelength scale (subnanometer). Similar to the propagation of electrons with certain energy in a semiconductor crystal amidst the ionic lattice (periodic variation of electric potential) and the development of an electronic bandgap, the propagation of light of certain wavelengths in a photonic crystal is influenced by the periodic variation of the dielectric constant. The light impinging upon a photonic crystal is subjected to a periodic variation of the refractive index in the direction of its propagation in the medium, determining forbidden and allowed wavelengths. Depending on the contrast between the high and the low refractive indices of the medium, the spacing between the two media and the direction of propagation, a complete or partial PBG, similar to the electronic bandgap in semiconductor crystals, can develop for certain light wavelengths.

If all the conditions are met, a perfect PBG can develop and these wavelengths are forbidden from propagating in any direction, getting trapped in the photonic crystals. The simplest form of photonic crystals is the well-known one-dimensional periodic structure, multilayer stacks of high and low refractive index, also known as Bragg mirror that has a stop band for certain directions and wavelengths. The spectral position of the stop band is determined by the match between the dielectric index contrast and the thickness of the layers. A schematic representation of a 1-D photonic crystal (Bragg mirror) is shown in Fig. 8.16A [54]. The resulting spectrum from the constructive and destructive interference in the case of a 60-layer multistack of high (titanium oxide) and low (silicon dioxide) refractive index films deposited on glass is shown in

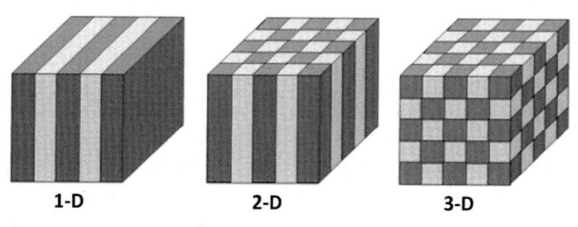

1-D   **2-D**   **3-D**

Figure 8.15 Schematic representation of 1-D, 2-D, and 3-D photonic crystals [15].

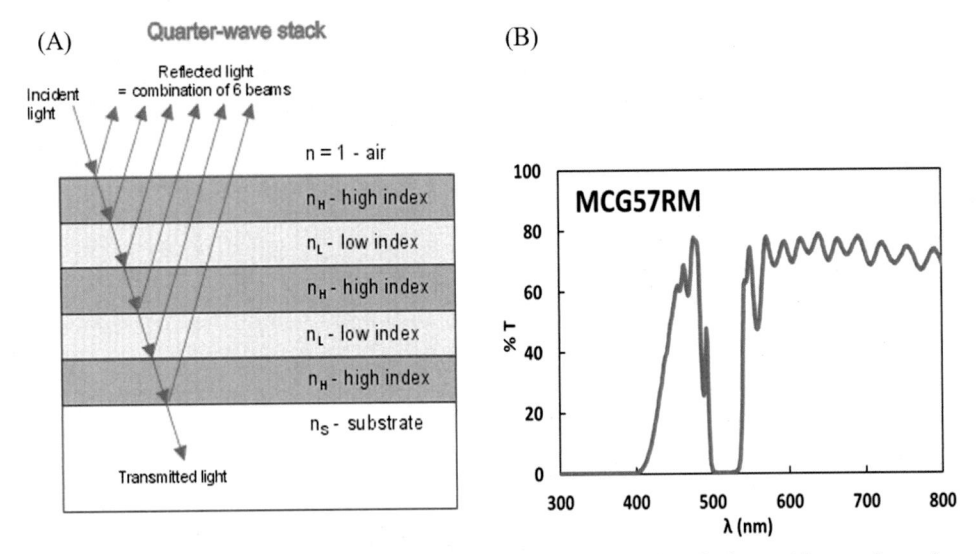

Figure 8.16 (A) Schematic representation of a 1-D photonic crystal with the incident, reflected and transmitted light [54]. (B) Transmittance spectrum of 57-layer high and low refractive index pairs with a bandgap centered around 520 nm wavelength [55].

Fig. 8.16B. The coating was designed and fabricated to block a narrow spectral band centered around 532 nm (*green light*) [55].

The simplest example of a three-dimensional photonic crystal is the synthetic opal created in the laboratory by mimicking to a large extent the natural opals. These structures are created by the periodic arrangement of high index SiO_2 particles 150–300 nm stacked along the three axes, as shown in Fig. 8.17 [56]. The periodic airgap present in between these particles creates the low refractive index layer thus rendering different hue and small degree of PBG due to the small contrast in refractive index between the two media.

Although Bragg mirrors and interference coatings have been known and developed for over a number of decades [57] and used in innumerable number of applications, it is only recently that such devices with their ability to manipulate light are being referred to as photonic crystals. This change in terminology is following the theoretical study of 3-D periodic structures by various researchers around 1987 [58,59]. The study of photonic crystals has become increasingly important in recent years due to the possibility they offer to control the flow of electromagnetic radiation through such periodic mediums. For a given wavelength of light, this flow is principally a function of the scale of the periodicity of the structure and the refractive index contrast. Thus by various materials combination and the scale of their arrangement in space, a wide range of photonic crystals can be fabricated for light propagation manipulation over a

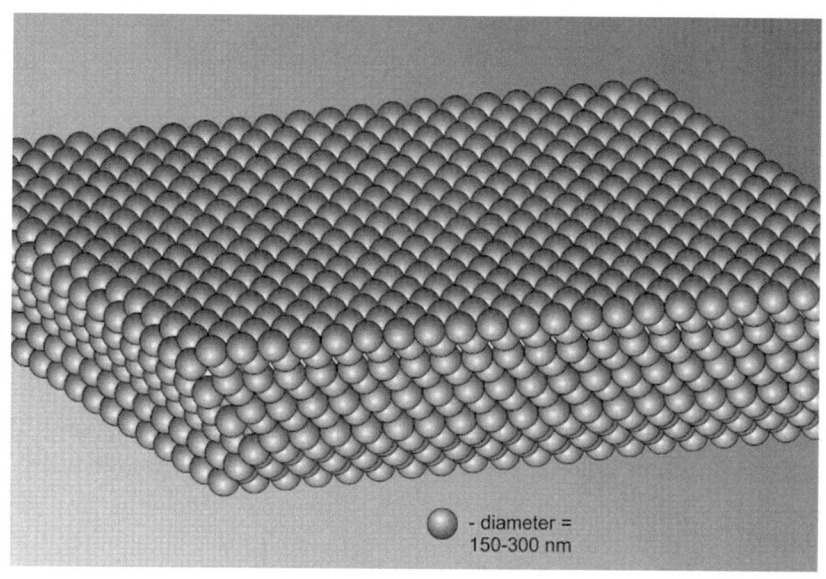

- diameter = 150-300 nm

Figure 8.17 Idealized diagram of a 3-D opal [56].

wide range of electromagnetic spectrum. These virtues of photonic crystals have given an enormous impetus for both fundamental and applied research in this area [60–63].

On the experimental side, the biggest challenge is the material realization of the photonic crystals with a long range and perfect periodicity although many techniques are being developed in recent years toward achieving this. These approaches can be broadly divided into three categories: (1) self-assembly techniques, (2) micromachining, and (3) holographic exposure [64]. In the early years of work on photonic crystals the challenges related to experimental fabrication of these structures prompted a lot of theoretical work laying the basis for the work that is being carried out in recent years. As far as the theoretical characterization of the light-photonic interaction is concerned, Maxwell's equations are used to examine the propagation of light in the photonic crystal medium defined by the periodic variation of the dielectric constants. The three main approaches in the theoretical characterization are: (1) band structure calculation of light propagation in an infinite periodic medium using various algorithms [65–69], (2) general electromagnetic approach [70], and (3) transfer matrix approach [71–74]. The wide range of possibilities that exist to create photonic crystals that are effective in various wavelength ranges of the electromagnetic spectrum for light propagation manipulation has made these structures a fertile field of study. However, most of the study in this field is concentrated on static behavior of photonic crystals, that is, once these crystals are created for a certain range of EM radiation, their PBG is fixed and cannot be dynamically altered. However, if these photonic crystals can be built out optical materials whose optical constants (n, k) can be altered, the PBG can be tuned over a certain wavelength range. Such an approach is highly desirable, as it would make the photonic crystals even more versatile for their fundamental study as well as for their applications. This would open up the possibility to build various dynamic devices such as tunable optical filters, optical switches, sensors, integrated circuits and more. As underlined earlier, the two most fundamental parameters associated with the photonic crystals are the (1) contrast in optical or dielectric constants and (2) lattice constants (scale of periodicity). Many attempts in this direction have been done to achieve the tuning of the PBG varying these parameters through mechanical force [75], electric and/or magnetic fields [76−78].

8.1.8 Transition metal oxide−based photonic crystals

The preceding discussion on the different approaches [73−76] to PBG tuning demonstrates that the PBG shift in these cases is not adequate and/or requires large external forces to induce these reversible changes and also that a fine control of the PBG is difficult. As also seen earlier, the TMOs such as WO_3, VO_2, MoO_3, and V_2O_5 showing very efficient chromogenic (electrochromic, thermochromic, and photochromic) properties undergo large reversible optical (dielectric) constants under the application of small external forces such as an electric field, temperature, and light irradiation. Hence, the TMOs open a distinct and facile way to PBG tuning. If the 1-D, 2-D, and 3-D photonic crystals can be fabricated out of the TMOs, the PBG in these structures can be tuned over a wide wavelength range in a very efficient manner through the application of small external forces. As shown previously, the TMOs undergo very large reversible optical (dielectric) constants change with these forces.

8.1.9 Electrochromic photonic crystals

These devices based on materials and thin films showing very efficient electrochromic effect such as WO_3, MoO_3, and V_2O_5 can be very interesting for photonic crystal fabrication, study, and applications. The thin films of these TMOs show very efficient coloration in the UV, visible, and near-infrared region of the EM spectrum under the double electron/ion insertion. They undergo large reversible changes in the optical (dielectric) constants. As discussed earlier, the tungsten trioxide (WO_3) thin films showing electrochromic coloration in both amorphous and polycrystalline form have been studied for a long time for their electrochromic properties. Hence, they are excellent candidates for electrochromic photonic crystal studies. Their optical switching behavior and their preparation by a wide range of methods is very well known. Many studies of WO_3-based photonic crystals have been undertaken. Kuai et al. [79] have fabricated WO_3 inverse opals. In their work, polystyrene (PS) spheres of 300 and 760 nm were first used to build the periodic structure as the template using the self-assembly method on glass substrates. The substrates were then dipped into WO_3

solution [80] to infiltrate WO_3 into the open space in the PS template. This assembly was then sintered at high temperature to burn off the PS leaving behind a WO_3 periodic frame structure (inverse opal). Hence, this structure was largely made of the empty air-filled (low refractive index) pockets surrounded by high refractive index WO_3. In Fig. 8.18 are shown the scanning electron microscope (SEM) images of the initial PS template and the resulting WO_3 inverse opal, as well as the reflectance spectra of these templates showing the location of the PBGs.

By inserting lithium atoms into the WO_3 framework and the ensuing formation of lithium tungsten bronze (Li_xWO_3—see Eq. 8.1), the refractive index contrast could be varied over a wide range. Shown in Fig. 8.19 is the shift in two PBGs with increasing lithium insertion. The spectral shift of the two PBGs as a function of lithium insertion is also shown in this figure.

The dry lithiation method [81] used in this work was, however, not conducive to the removal of the lithium to demonstrate the reversibility. Nevertheless, this work has established the feasibility of PBG tuning through electrochromic effect. The theoretical characterization of this behavior was carried out using the band structure calculation by the Finite-difference time-domain (FTDT) method [67] at various stages of the photonic crystal formation and eventually the PBG tuning from the n, k values of the crystal and the inverse opal, in the as-deposited and lithiated states. The same group later fabricated a four-layer electrochromic sandwich device with the WO_3 inverse opal as the base layer. The schematic and the operation of the device are shown in Fig. 8.20 [82].

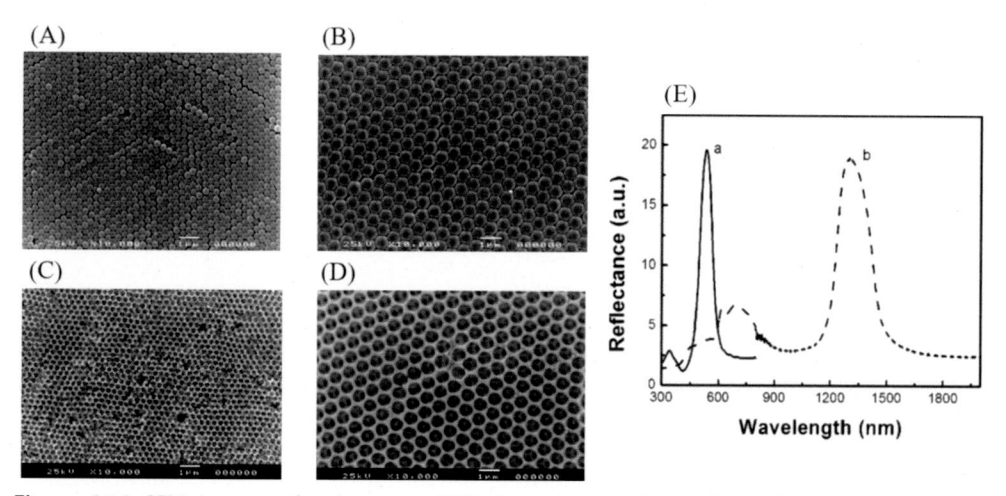

Figure 8.18 SEM images of polystyrene (PS) photonic crystal template of (A) 300 nm and (B) 760 nm spheres. SEM images of the WO_3 inverse opals derived from these templates are shown, respectively in (C) and (D). The reflectance spectra at near-normal incidence of the WO_3 inverse opals are shown in (E): (a) 300 nm inverse opal and (b) 760 nm inverse opal [79]. *SEM*, Scanning electron microscope.

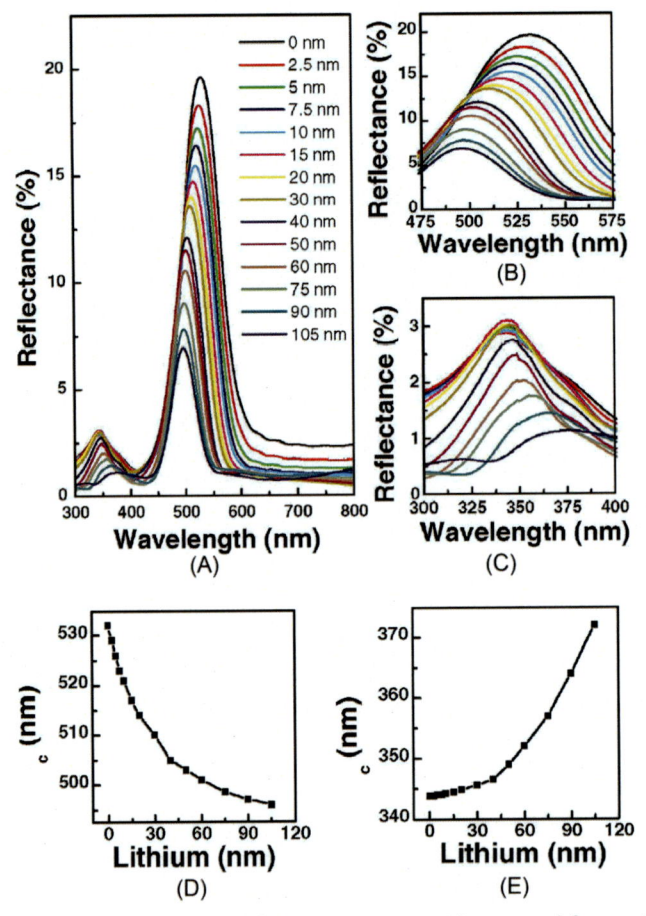

Figure 8.19 (A) Shift of reflection peaks of a WO_3 inverse opal prepared from a 300-nm template with the insertion of various amounts of lithium, (B) blue shift of the reflection peak located at 532 nm with lithium insertion, (C) red shift of the reflection peak located at 344 nm with lithium insertion, (C, E) reflection peak position shift with lithium insertion for peak located at 532 nm (D) and 344 nm (E) [79].

The WO_3 inverse opal was grown on transparent conductor (ITO)-coated glass substrate. With another ITO-coated glass substrate, a sandwich structure (electrochemical cell) as shown in Fig. 8.20A was fabricated with lithium perchlorate ($LiClO_4$) dissolved in propylene carbonate as the source of lithium. The double injection and extraction of lithium ions and electrons were carried out by applying relevant potential between the two ITO electrodes. As can be seen in Fig. 8.20B, the dip in transmittance spectrum can be tuned reversibly. Although, much work needs to be done to increase the PBG tunability and to incorporate inverse opal of polycrystalline WO_3 to induce reflectance modulation, this work initial work demonstrates the possibilities

(A) (B)

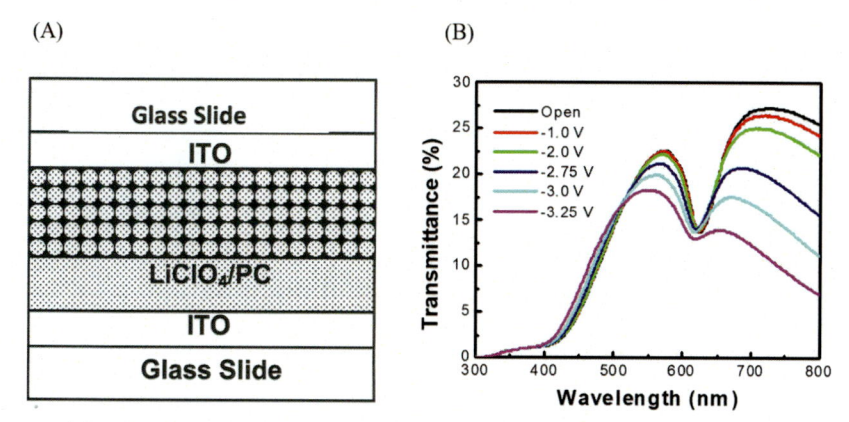

Figure 8.20 (A) Schematic representation of WO_3 with $LiClO_4$ between two Indium tin oxide (ITO)-coated glasses for photonic devices. (B) Visible-light transmittance through the device with increasing applied voltage between the ITO electrodes [82].

exist for electrochromic tunability of PBG. Similar works have been carried out on WO_3 and TiO_2 inverse opals under lithium insertion [83,84] and the potential for their application in dye-sensitized solar cells was examined. Following these earlier works demonstrating the potential of chromogenic (phase change) materials, various TMO-based photonic crystals have been fabricated and studied. The n-type WO_3 is also very well known for its chemical activity and change in its optical and/or electrical properties when exposed to many gases. Hence, it is used extensively as a gas sensor for NO_x, O_3, H_2S, CO, H_2, etc. [85]. WO_3-based photonic crystals with their sharp Bragg reflectance peak and its shift have the potential to be even more sensitive to such gases. Xie et al. [86] have fabricated WO_3 inverse opals for H_2 gas sensing. PS periodic templates were used to make the WO_3 inverse opals (Fig. 8.21A) that were then overcoated with platinum (Pt) (Fig. 8.21B). When exposed to H_2 gas, through the catalytic activity of Pt, hydrogen is transferred to WO_3 leading to the reversible formation of tungsten bronze (H_xWO_3) as shown in Eq. (8.1). With increasing H_2 exposure and denser formation of the bronze, the reflectance Bragg peak located around 592 nm rapidly shifts to lower wavelengths as shown in Fig. 8.21C. The reversibility of this shift under exposure to atmosphere and removal of hydrogen is quite evident (Fig. 8.21D). The rapid and reversible response to H_2 has prompted the authors to propose these photonic crystals for H_2 as well as other gases using appropriate catalysts.

An improved visible-light photocatalytic activity has been demonstrated by Cui et al. by working with multicomponent WO_3-based photonic crystals [87]. Heterostructures of $CdS-Au-WO_3$ were prepared using the PMMA microspheres of different sizes in order to examine the photocatalytic activity under visible-light illumination. Au and CdS were deposited on a WO_3 inverse opal. Such a structure was

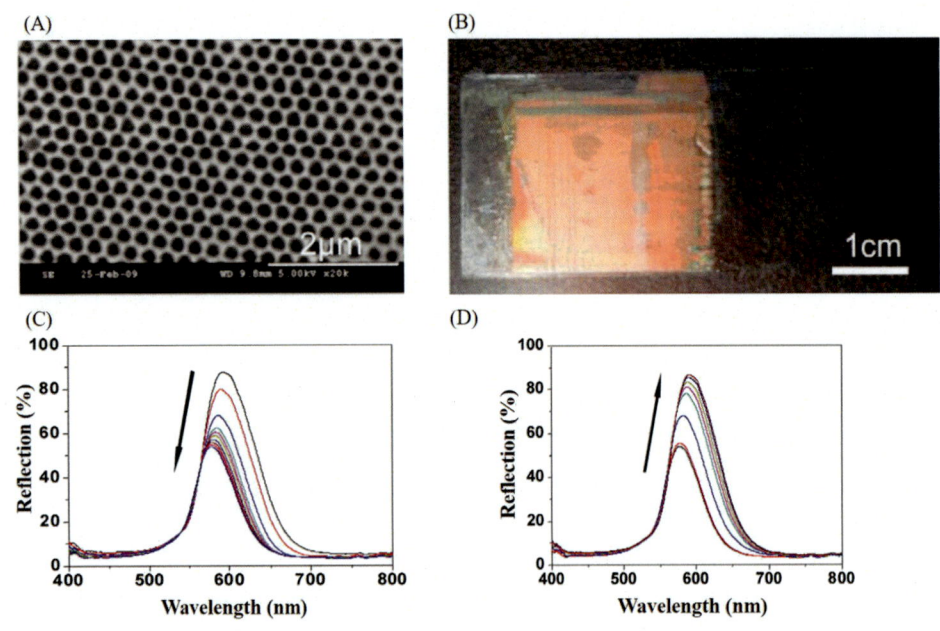

Figure 8.21 (A) SEM image of WO$_3$ inverse opal, (B) WO$_3$ inverse opal with platinum coated on the top sample, (C) blue shift of the reflectance peak of the WO$_3$ inverse opal with increasing H$_2$ exposure, (D) red shift of the reflectance peak of the WO$_3$ inverse opal with increasing removal of the H$_2$ [86]. *SEM*, Scanning electron microscope.

found to show a superior visible-light photocatalytic activity of water splitting for both oxygen and hydrogen evolution. The higher efficiency of such heterostructured photonic crystals is attributed to the improved light harvesting in the periodic structure as well as the efficient electron transfer. Large increase in visible-light photocurrent generation has also been demonstrated in WO$_3$-based inverse opals by overlapping the PBG and the electronic absorption band of WO$_3$ [88]. WO$_3$ inverse opals were prepared using different size PS microspheres giving photonic stopbands at different wavelengths. SEM images and the visual appearance of these inverse opals under white light illumination are shown in Fig. 8.22A. The results of the light absorption efficiency of the WO$_3$-based inverse opals are compared in this work with those of disordered porous WO$_3$ and unpatterned nonporous WO$_3$. Photochemical analysis of these samples was carried out under UV−visible (>300 nm) and visible-light illumination. As shown in Fig. 8.22B, under both the illuminations, the WO$_3$ inverse opals show an enhanced photon to electron conversion efficiency, that is, photocurrent generation as compared to the disordered porous WO$_3$ and the unpatterned nonporous WO$_3$. These virtues of the inverse opals are attributed to the better light harvesting and connectivity that facilitates charge transport in these structures. Amongst the WO$_3$ inverse opals, the sample with the 260-nm pore size exhibits the best electron conversion efficiency under both UV−visible and visible illuminations. This result is

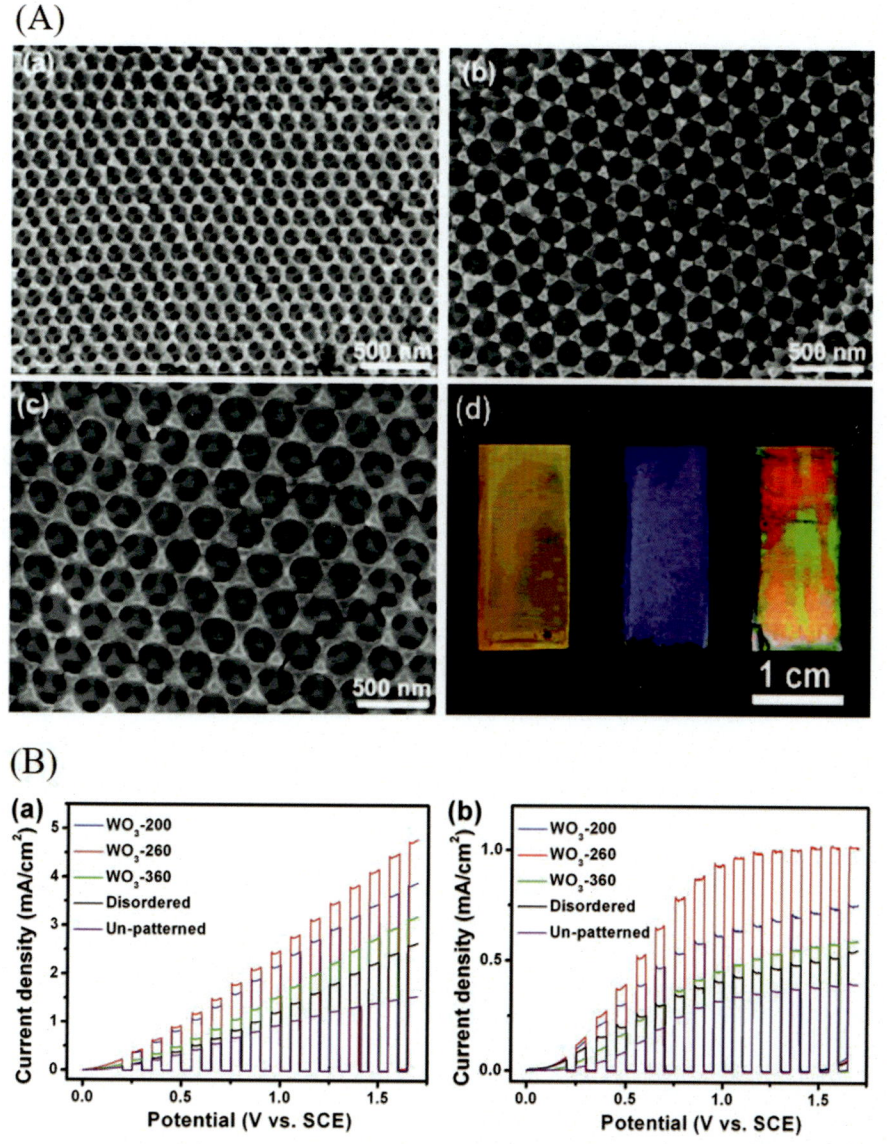

Figure 8.22 (A) SEM images of WO$_3$ inverse opals prepared using (a) 200 nm, (b) 260 nm, and (c) 360 nm pore sizes. (A(d)) photoluminescence of the samples under white light illumination. (B) photocurrent of different inverse opals, disordered porous and unpatterned nonporous WO$_3$ samples under (a) UV−visible and (b) visible illuminations [88]. *SEM,* Scanning electron microscope.

attributed to the better overlapping of the stop band and the electronic absorption edge of WO$_3$ in this structure.

The enhanced and reversible change in optical properties due to the formation of tungsten bronze (H$_x$WO$_3$), especially the sensitive shift of the Bragg peak in WO$_3$

inverse opals, has also been studied by Amrehn et al. [89]. Platinum (Pt) coated WO_3 inverse opals were studied as optical transducers (Fig. 8.23A) for gas sensing elucidating their advantages oversemiconducting ones such as simplicity of design, remote operation, and sensitivity. It was found that with increasing hydrogen content (x) in the bronze (H_xWO_3) and the ensuing coloration, the optical constants n (decrease) and k (increase) changed sensitively. The corresponding reflectance spectra calculated from these optical constants is as shown in Fig. 8.23B.

8.1.10 Thermochromic photonic crystals

The study of components and devices based on TMOs showing very efficient thermochromic change, such as vanadium dioxide (VO_2) have become very popular in recent years owing to the rich application possibilities in both optical and optoelectronic sectors. As discussed earlier, VO_2 undergoes a very efficient first-order reversible change in its optical and electrical properties around the transition temperature of 68°C in bulk form. While a change of the order of 70%−80% in near-infrared and above transmittance is seen in VO_2 thins films, a drop in film resistance of 3−4 orders occurs in the electrical properties. Further various techniques can be used to bring down the transition temperature. The film nanostructure is found to influence nearly all the parameters related to the thermochromic behavior of VO_2. All these possibilities render the work on VO_2-based photonic crystals even more interesting from the point of view of reversibly tuning the PBG as a function of temperature. For the fabrication and study of VO_2-based thermochromic photonic crystals, it is of particular interest that the optical constants (n, k) also undergo a large change [46]. Hence, the PBG tuning over a wide spectral region can be interactively controlled via Joule heating and/ or via optothermal methods (light irradiation using high power lasers). Ibstate et al.

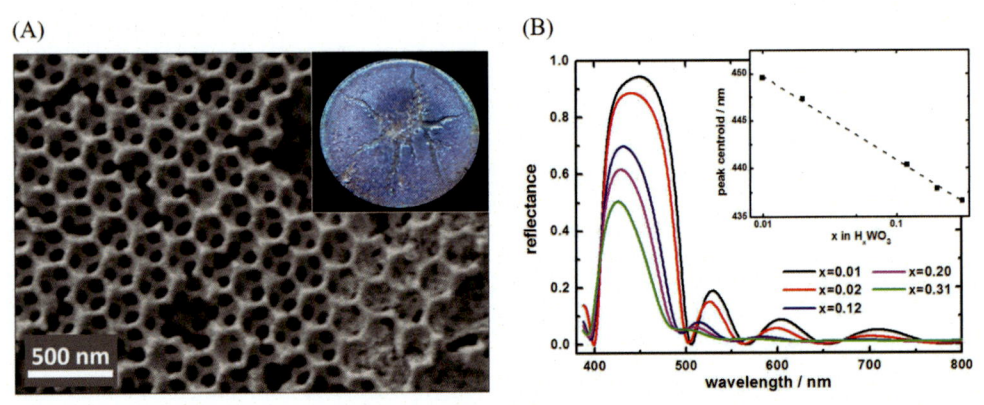

Figure 8.23 (A) SEM image and appearance (inset) of Pt coated WO_3 inverse opal for gas sensing. (B) Calculated reflectance spectra of tungsten bronze for different hydrogen content (x) in H_xWO_3 from 0.01 to 0.31 [89]. *SEM*, Scanning electron microscope.

[90] have fabricated VO$_2$-based inverse opals using SiO$_2$ templates. The SiO$_2$ photonic crystal templates made up of 610 nm SiO$_2$ particles were initially infiltrated with vanadium pentoxide (V$_2$O$_5$) by chemical vapor deposition over a number of cycles to form SiO$_2$/V$_2$O$_5$ composite crystals and later the VO$_2$ inverse opals were obtained by thermal reduction of V$_2$O$_5$ and by etching away the SiO$_2$. Shown in Fig. 8.24 are the SEM images and the optical tuning of the system at different temperatures. The VO$_2$/SiO$_2$ composite photonic crystal and the VO$_2$ inverse opal (inset) are shown in Fig. 8.24A. Although only a part of the inverse opal seems to have been formed from these images, the structure seems to be fairly effective in terms of the reversible optical changes induced as a function of temperature. In Fig. 8.24B are shown these optical changes in the form of the differential reflectance spectra at various temperatures. Significant variations in the differential reflectance can be seen near the transition temperature indicating the distinct possibility of the reversible optical tuning of the optical properties via temperature control.

Similarly, Golubev et al. [91] have fabricated and studied VO$_2$-based inverse opals obtained through infiltrating 250 nm particle size SiO$_2$ photonic crystals via chemical bath infiltration of V$_2$O$_5$. The VO$_2$ inverse opals were obtained by high–temperature sintering possessing the Bragg reflectance peak around 625 nm, as shown in Fig. 8.25A. When the samples were heated above the transition temperature of VO$_2$, the clear 25 nm blue shift of the Bragg peak is evident. At temperature higher than the transition temperature, the Bragg profile becomes narrower with decreasing intensity. The optical changes occurring can also be seen through the thermal hysteresis curve where the peak Bragg wavelength (PBG) was recorded as a function of temperature around the transition temperature.

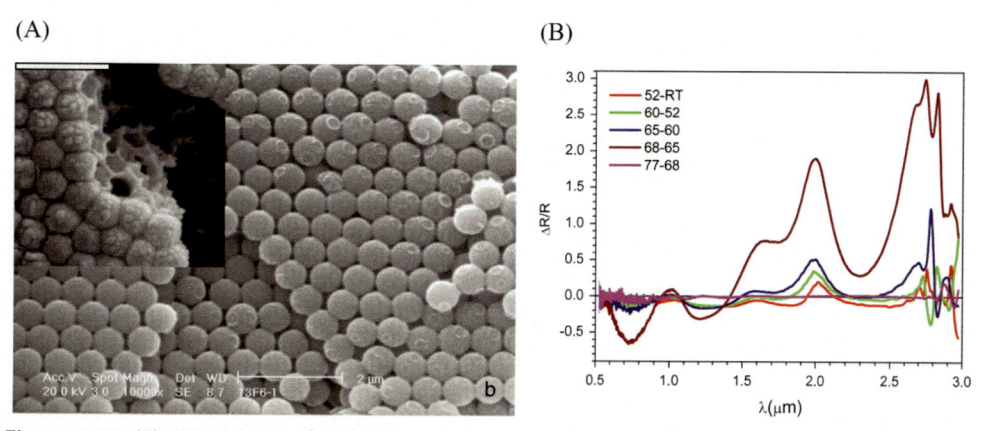

Figure 8.24 (A) SEM image of VO$_2$/SiO$_2$ composite photonic crystal with inserted image of VO$_2$ inverse opal at the top left corner. (B) Reflectance spectra of the VO$_2$ inverse opal at different temperatures [90]. *SEM*, Scanning electron microscope.

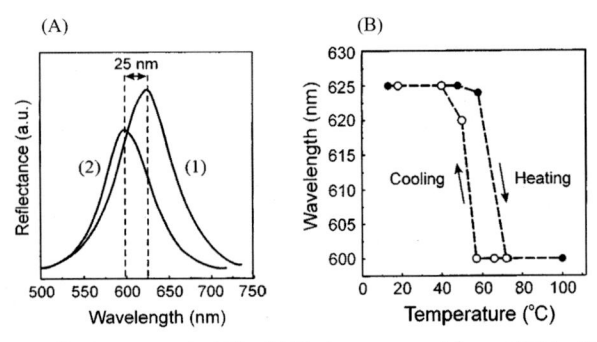

Figure 8.25 (A) Bragg reflectance peak shift of VO_2 inverse opal from 625 to 600 nm with increase to above transition temperature. (B) Thermal hysteresis curve of the VO_2 inverse opal showing the shift of the Bragg peak during the heating and cooling process [91].

Ke et al. [92] have provided an interesting prospective of using 2-D tunable photonic crystals for smart window applications. They have carried out simulation and experimental work on SiO_2/VO_2 2-D photonic crystals. In the FTDT simulation, the samples were made up of SiO_2/VO_2 core shell, that is, a monolayer of SiO_2 particles of different sizes (400−700 nm) coated with an outer layer of VO_2. By varying the SiO_2/VO_2 unit size in these crystals, a way to achieve static (as–deposited) tunability of the visible transmittance rendering different colors to the sample has been proposed. The possibility to tune the PBG across the visible spectrum is demonstrated. In addition to this visible range static tunability, with the presence of thermochromic VO_2 and its switching, the system is shown to afford an efficient optical modulation in the near-infrared region as a function of temperature. These simulation aspects are illustrated in Fig. 8.26A. In their experimental work, these authors have prepared the initial SiO_2 photonic crystal templates [monolayer colloidal crystals (MCC)] and surface coated with VO_2 by dipping the SiO_2 templates in vanadium precursors of various concentration and through thermal treatment. The SEM image of a photonic crystal prepared with a 400-nm MCC in different concentrations vanadium precursors is shown in Fig. 8.26B. Although, the experimental results, both in terms of the visible region static tunability and the NIR modulation, fall short of the predicted behavior of these periodic structures, this approach carries a rich merit and needs to be explored further. This departure of experimental results from the simulated ones is mainly due to the limitation in achieving exacting SiO_2/VO_2 core-shell structure shown in the inset of Fig. 8.26A due to the incomplete infiltration of vanadium.

Liang et al. [93] taken this idea of VO_2/SiO_2 core shell−based composite 2-D photonic crystal further by the inclusion of dynamic tuning of the PBG and optical properties both in the visible and near-infrared region. The photonic crystals, in this case, were prepared by sputter depositing vanadium metal on SiO_2 MCC to continuously cover the top surface of the MCC and form a VO_2 hemispherical shell array by

(A)

(B)

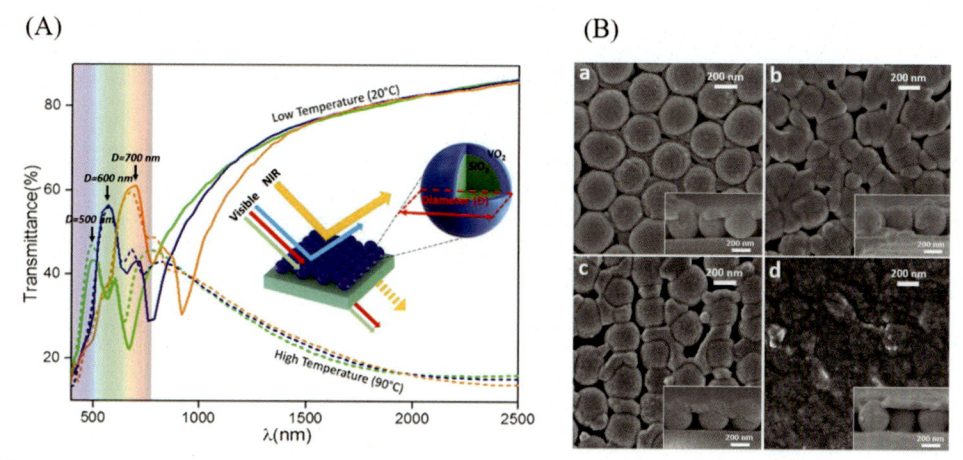

Figure 8.26 (A) Simulated transmittance of SiO_2/VO_2 layers for different size of SiO_2 particles at low (20°C) and high (90°C) temperatures. (B) SEM images of SiO_2/VO_2 samples prepared with different concentrations of vanadium precursor [92]. *SEM*, Scanning electron microscope.

rapid thermal annealing (RTA) as shown in the schematic of Fig. 8.27A(b). The SEM of the actual composite sample prepared on 600 nm SiO_2 MCC is shown in Fig. 8.27A(a). These authors have varied the refractive index (n) of the VO_2 layer by controlling the oxygen flow rate during the RTA process. The optical performance of the three different samples obtained at three different oxygen flow rates is shown in Fig. 8.27B along with the simulated transmittance values at high and low temperatures. These results clearly highlight the merit of this approach in achieving a dynamic tuning of optical properties via temperature throughout the solar energy spectrum. The experimental results conform fairly well with the simulation results showing high transmittance in the near-infrared region (1100−2500 nm) at room temperature. However, some departures are seen between the experimental results and simulation as far as the location and number of the interference peaks, degree of PBG tuning in the visible region (blue shift) as well as the thermochromic switch in the infrared region at high temperature (VO_2 switched state). Despite these departures, which are to be expected in the face of constraints related to the experimental control of such complex systems as 2-D photonic crystals, the work carried a high potential for application in smart windows based on thermochromic activation.

The PBG tuning of VO_2-based 2-D photonic crystals is studied by Ye et al. [94] by simulating VO_2 base material drilled with cylindrical holes as shown in Fig. 8.28. The effect on various geometrical parameters (cylinder radium, depth, and period) associated with the 2-D photonic crystal has been studied in depth for emissivity considerations of VO_2 in the metallic phase. The potential to control and tune the emissivity and cut-off wavelength is demonstrated. The performance of such simulated 2-D photonic crystals is compared to that of traditional emitters

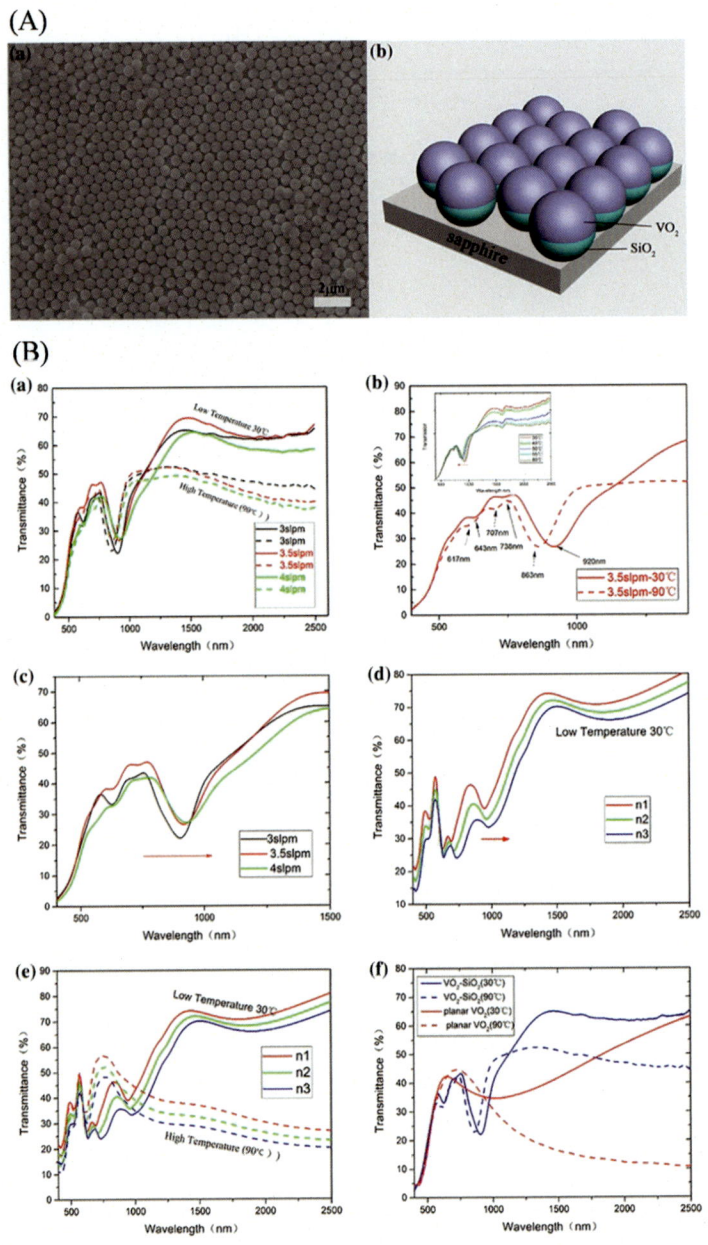

Figure 8.27 (A(a)) SEM image and (A(b)) idealized image of VO$_2$ deposited on 2-D SiO$_2$ monolayer colloidal crystal. (B(a),(b)) Transmittance of VO$_2$/SiO$_2$ samples made at different oxygen flow rates. (B(c),(d)) Red shift of the transmittance interference peaks with increasing oxygen flow rate during thermal annealing process. (B(e)) Simulated transmittance of VO$_2$/SiO$_2$ samples at high and low temperatures. (B(f)) Experimental transmittance for a planar VO$_2$ and a VO$_2$/SiO$_2$ samples at 30°C and 90°C [93]. *SEM*, Scanning electron microscope.

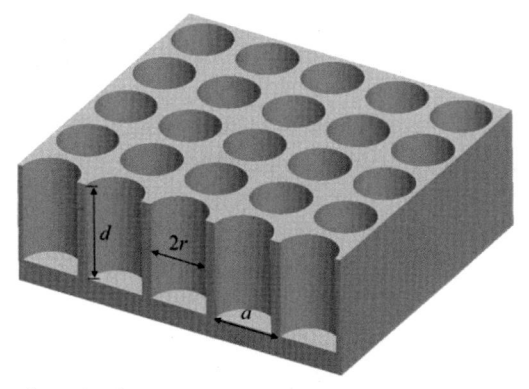

Figure 8.28 Simulation of a 2-D photonic crystal with cylindrical holes in a VO_2 film as an emitter [94].

such as SiC and InGaAs and the superior performance of the photonic crystal—based approach is demonstrated. Although this approach is very promising, the challenges related to experimental fabrication of large area 2-D photonic crystal emitters needs to be yet overcome.

Similarly, various other types of 1-D, 2-D, and 3-D photonic crystal approaches have been taken to exploit the thermochromic switching properties of VO_2 [95—97]. In these simulated and/or experimental works the photooptically induced ultrafast VO_2 thermochromic switch has been studied in various photonic crystal configurations such as VO_2 as a top layer in a 1-D photonic crystal based on multilayer SiO_2/TiO_2 for optical switching response control [96], a 3-D opal VO_2—SiO_2 composite to study the high-speed kinetics control by varying the VO_2 filling factor [97] and a chip integrated 1-D photonic crystal based on VO_2/SiO_2 heterostructure as switchable topological insulator [95].

8.1.11 Photochromic photonic crystals

Photochromic switching is another optical switching phenomenon that gives the possibility to control optical properties, especially the PBG by light sources such as laser by building these periodic structures out of photochromic materials. Many TMO exhibit an efficient and convenient form of photochromism. Amongst the TMOs, tungsten trioxide (WO_3), and molybdenum oxide (MoO_3) exhibit a very efficient photochromic effect under UV and/or visible-light irradiation [98,99]. The photochromic performance of these two oxides is indirectly related to and very similar to their efficient electrochromic coloration [100]. The UV and/or visible-light irradiation on the WO_3 or MoO_3 films results in the formation of electron—hole pairs. The holes thus created react with the structural water molecules in the film to form protons (H^+). The protons, thus created, and the electrons lead to the coloration of the films through the double insertion, similar to the electrochromic phenomenon shown in Eqs. (8.1) and (8.4). Hence, the photochromic coloration has its roots in chemical reactions occurring under light irradiation. The photochromic coloration, similar to the electrochromic coloration in WO_3 and MoO_3 discussed earlier, can lead to large changes in the optical constants (n, k). The photochromic coloration, per se, is not directly reversible like in the electrochromic devices, which occurs through the reversal of the polarity of the applied electric field. Hence, the photochromic coloration which carries the films to a metastable state can persist for a long time, in excess of a few years under right conditions. Such films and their devices showing photochromic coloration are proposed for optical memory storage applications. However, the direct irreversibility of the coloration can be an impediment in some optically interactive applications. However, a distinct possibility exists to color such devices by light irradiation (photochromic coloration) and to discolor them by the application of a small electric field (electrochromic discoloration). Irrespective of the mode of coloration and discoloration, the photochromic phenomenon in the TMOs, especially the WO_3 and MoO_3, offer an excellent potential to build photonic crystals activated by UV and or visible-light irradiation and manipulation of the PBG.

Tahar et al. [101] have fabricated and studied a 1-D Bragg mirror (photonic crystal) based on high index $(n = 2.1)$ photochromic MoO_3 and low index $(n = 1.5)$ SiO_2 multilayers. In a pump-probe experiment, these Bragg mirrors of multiple pairs (m) were irradiated by a pump beam centered around 380 nm. The optical changes occurring in these stacks after the irradiation are compared to the transparent samples of 5, 10, and 20 SiO_2/MoO_3 pairs in Fig. 8.29A. The appearance of the broad optical stop band is clearly evident with the 10 and 20 pair structures, although in the later the increasing absorption in the device suppresses and obscures the stop band. Important optical changes occur under the UV irradiation throughout the spectrum studied. Although, photochromic tunability of the stop band is not clearly evident, the results point to the potential of achieving this under more optimized conditions. An example of a sample colored by UV light exposure through a mask is shown in Fig. 8.29B. The dark area in the middle of the sample depicting the logo of Université de Moncton is clearly seen amidst the interference coating elsewhere. This photochromic coloration has sustained for over 8 years of laboratory condition exposure underlining the strong optical memory of such devices under photochromic coloration.

Li et al. [102] have fabricated hierarchically porous titanium dioxide (TiO_2) coated WO_3 2-D bilayer films and have studied the photochromic response of the device for smart window applications. Such device was found to have a high absorption in the near-infrared region and a mild absorption in the near-visible region.

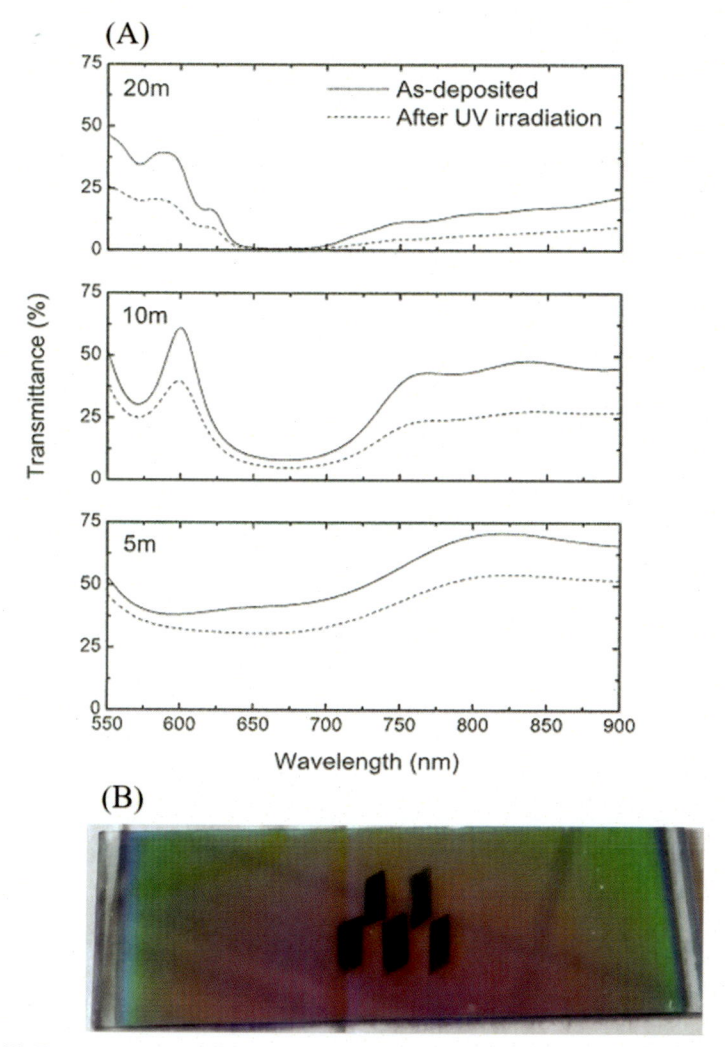

Figure 8.29 (A) Transmittance of the samples of 5, 10, and 20 SiO$_2$/MoO$_3$ pairs with and without UV irradiation [101]. (B) A photochromic sample was exposed to UV irradiation using a mask of Université de Moncton logo (dark area in the middle).

8.2 Conclusion

Photonic crystals are made up of periodic arrangement of particles on nanometric scale of a few nanometers to a few hundred nanometers. When this scale or periodicity corresponds to light wavelengths of interest, the photonic crystals represent a very interesting electromagnetic medium for propagation due to the periodic variation of the dielectric constant confronted by the incident EM wave, which gives rise to PBG at certain wavelengths. These wavelengths are forbidden from propagating through the media, like forbidden electronic bandgaps in semiconductor crystals. Identical to the semiconductors discovered a few decades back for electron flow control, from light propagation control point of view, the study of photonic crystals has become very important in recent years and holds the potential to open new vistas. The principal factors determining the PBG formation for a given wavelength band are the degree of periodic variation of the dielectric constant (contrast) and the spatial periodicity (scale) itself. By controlling these two factors, various 1-D (Bragg mirrors), 2-D, and 3-D photonic crystals with static or passive (fixed) PBG in various electromagnetic spectral regions have been proposed and studied both theoretically and experimentally over the last couple of decades ever since the inception of the idea of photonic crystals. Bragg mirrors or the interference coatings have been designed, fabricated, and used in various applications for a much longer period. More recent work in this field is focused on photonic crystals with tunable PBG. Such a dynamic and interactive tunability of PBG affords important applications in the area of photonics, optics and optoelectronics such as optical switches, tunable filters, reconfigurable optical networks and more. These tunable photonic crystals are based on optical materials whose physical properties can be interactively and reversibly controlled. Some approaches to interactively vary the spatial periodicity have been proposed with limited success due to the inherent strain the structure is subjected to. An interactive variation of the dielectric constant or optical constants (n, k) of the medium seems to be a more judicious path forward. In this context, some TMO in which reversible optical changes can be induced by various external forces are becoming increasingly important as base materials for photonic crystals. The TMOs are at the forefront in recent years in the field of chromogenics, dedicated to the study of materials showing reversible optical property change through various effects. The facile fabrication of the TMOs, especially in thin film form, and the activation of the optical change by small external forces such as heat (thermochromic), light (photochromic), electric field (electrochromic), and gas exposure (gasochromic) have made these materials extremely attractive for photonic crystal fabrication. PBG and other optical properties in such photonic crystals can be easily tuned in the spectral region of interest by building the TMO crystals accordingly by controlling the dielectric contrast and periodicity.

TMOs such as tungsten trioxide (WO_3), molybdenum oxide (MoO_3), vanadium pentoxide (V_2O_5) and some others are very well known for their electrochromic properties, undergoing drastic reversible changes in optical properties and constants (n, k) with the application of a small electric field. A wealth of knowledge exists on the electrochromic behavior of these materials and on their fabrication in device form for ease of operation. Many industrial applications such as smart windows and dimmable rear-view mirrors in vehicles have been realized. Further, depending on the phase and nanostructure of these TMOs, different forms of optical modulations can be induced. The electrochromic materials give the possibility to continuously tune the optical properties via the control of the inserted species (electrons and ions).

Vanadium dioxide (VO_2) has been studied extensively for its thermochromic properties, that is, undergoing a first-order change in its optical properties, especially in the near-infrared region, around a temperature as low as $68°C$ with the possibility to lower this transition temperature even closer to room temperature. The thermochromic effect in VO_2 can be induced by Joule heating or optothermally using lasers for study of high-speed optical effects. The thermochromic switch being a first-order one, the material only switches between two states, that is, above and below the transition temperature.

WO_3 and MoO_3 also exhibit a very efficient photochromic coloration under UV and some cases, visible-light irradiation. The degree of optical change induced can be continuously varied through the intensity of light. The photochromic change induced by the light irradiation is restrictive in terms of reversibility, but possibility exists to configure it similar to electrochromic system to revert to the original state.

Suffice it to say, working with the thin film form of these already well-known chromogenic TMOs and fabricating them in 1-D. 2-D and 3-D photonic crystal form, a wealth of fundamental as well as applied study can be carried out. The field of TMO-based chromogenically tunable photonic crystals is in its infancy and open to a vast landscape of study and applications. On the fundamental study side the challenge is to accurately represent the dynamic changes occurring in the band structure which is rooted in establishing a good correlation between the optical changes (dielectric constants) and the applied external force (heat, light, electric field, etc.). These calculations can be quite complicated. The challenge on the experimental front is related to the fabrication of the periodic samples, especially of large physical dimensions and of substantial thickness. It is a wise choice to begin both the theoretical and experimental work with 1-D TMO-based chromogenic photonic crystals and to gain their understanding through the existing theories and comportment of interference coatings (Bragg mirrors) and to extend their behavior to dynamic (colored) states and to higher dimensions (2-D and 3-D). The possibility of installing the PBG over a wide range of the EM spectrum and its tunability over a certain range by judicious TMO choice and the scale (periodicity) variation will make the field of photonic crystals a very fertile ground for growth in the near future. Further, the response time of the various TMOs in their optical switch ranging from femto- and picoseconds in thermochromic VO_2 to much slower response in electrochromic WO_3 and MoO_3 will also be a factor to consider in the design of the photonic crystals.

References

[1] C.M. Lampert, Chromogenic smart materials, Mater. Today (2004) 28−35.
[2] T.J. Sluckin, D.A. Dunmur, H. Stegemeyer, Crystals That Flow: Classic Papers From the History of Liquid Crystals, Taylor and Francis, 2004.
[3] G. Oskam, P.M. Vereecken, X. Shao, J. Fransaer (Eds.), Semiconductors, metal oxides and composites: metallization and electrodeposition of thin films and nanostructures, in: Electrochemical

Society (ECS) Transactions 25 (27) (2010). https://iopscience.iop.org/issue/1938-5862/25/27?pageTitle = IOPscience.

[4] W. Lohstroh, R.J. Westerwaal, B. Noheda, S. Enache, I.A.M.E. Giebels, B. Dam, et al., Self-organized layered hydrogenation in black Mg_2NiH_χ switchable mirrors, Phys. Rev. Lett. 93 (2004) 197404−197408.

[5] P.M.S. Monk, R.J. Mortinmer, D.R. Rosseinsky, Electrochromism: Fundamentals and Applications, VCH Inc, Weinheim, 1995.

[6] R.C. Duncan Jr., D.L. Staebler, Chapter 5—Inorganic photochromic materials. Springer Book Series, Topics in Applied Physics in: Howard M. Smith (Ed.), Holographic Recording Materials, 20, Springer, 1977, pp. 133−160.

[7] G.H. Brown, Photochromism, John Wiley and Sons Inc, 1971.

[8] C.M. Lampert, C.G. Granqvist (Eds.), Large Area Chromogenics: Materials and Devices for Transmission Control, IS4, SPIE, SPIE Institute Series, Bellingham, 1990.

[9] Pragna Kiri, Geoff Hyettb, Russell Binionsa, Solid state thermochromic materials, Adv. Mater. Lett. 1 (2) (2010) 85−106.

[10] A. Seeboth, D. Lotzsch, R. Ruhmann, Piezochromic polymer materials displaying pressure changes in bar-ranges, Am J. Mater. Sci. 1 (2) (2008) 139−142.

[11] P. Jin, S. Tanemura, Manufacturing methods of samarium sulfide thin films, United States Patent 6132568 (2000).

[12] N. Kida, S. Kurnakura, S. Ishiwata, Y. Taguchi, Y. Tokura, Gigantic terahertz magnetochromism via electromagnons in hexaferrite magnet, $Ba_2Mg_2F_{12}O_{22}$, Phys. Rev. B 83 (2011) 064422-1−064422-8.

[13] S. Toyoda, N. Abe, T. Arima, S. Kimura, Large magnetochromism in multiferroic $MnWO_4$, Phys. Rev. B 91 (2015) 054417.

[14] M. Bichurin, V. Petrov, A. Leontyev, et al., Two-range magneto-electric sensor, AIP Adv. 7 (2017) 056317.

[15] C.N.R. Rao, B. Raveau, Transition Metal Oxides, second ed., Wiley-VCH, 1998.

[16] Y.-S. Hsiao, S. Charan, F.-Y. Wu, F.-C. Chien, C.-W. Chu, P. Chen, et al., Improving the light trapping efficiency of plasmonic polymer solar cells through photon management, J. Phys. Chem. C 116 (39) (2012) 20731−20737.

[17] D. O'Carroll, C.E. Hofmann, H.A. Atwater, Conjugated polymer/metal nanowire heterostructure plasmonic antennas, Adv. Mater. (2010).

[18] N. Kalfagiannis, P.G. Karagiannidis, C. Pitsalidis, N.T. Panagiotopoulos, C. Gravalidis, S. Kassavetis, et al., Plasmonic silver nanoparticles for improved organic solar cells, Sol. Energy Mater. Sol. Cells 104 (2012) 165−174.

[19] T. Guo, M.-S. Yao, Y.-H. Lin, C.-W. Nan, A comprehensive review on synthesis methods for transition-metal oxide nanostructures, CrystEngComm 17 (2015) 3551−3585.

[20] K. Gesheva, T. Ivanova, G. Bodurov, I.M. Szilágyi, N. Justh, O. Kéri, et al., Technologies for deposition of transition metal oxide thin films: application as functional layers in "Smart windows" and photocatalytic systems, J. Phys. Conf. Ser. 682 (2015).

[21] C.T. Chen, J. Pedrini, E.A. Gaulding, et al., Very high refractive index transition metal dichalcogenide photonic conformal coatings by conversion of ALD metal oxides, Sci. Rep. 9 (2019) 2768.

[22] L. Berggren, Optical Absorption and Electrical Conductivity in Lithium Intercalated Amorphous Tungsten Oxide Films (Ph.D. thesis), Uppsala University, 2004.

[23] S.K. Deb, Optical and photoelectric properties and color centers in thin films of tungsten oxide, Philos. Mag. 27 (1973) 801−822.

[24] B.W. Faughnan, R.S. Crandall, P.M. Heyman, Electrochromism in WO3 amorphous films, RCA Rev. 36 (1975) 177−197.

[25] L.G. Austin, N.F. Mott, Polarons in crystalline and non-crystalline materials, Adv. Phys. 18 (1969) 41−102.

[26] I.F. Chang, B.L. Gilbert, T.I. Sun, Electrochromic systems for display applications, J. Electrochem. Soc. 122 (7) (1975) 955−962.

[27] C. Sunseri, F.D. Quarto, A.D. Paola, Kinetics of coloration of anodic electrochromic films of WO_3-H_2O, J. Appl. Electrochem. 10 (1980) 669−675.

[28] F. Wooten, Optical Properties of Solids, Academic Press, 1972.

[29] N.W. Ashcroft, N.D. Mermin, Solid State Physics, Saunders College, 1976.

[30] C.G. Granqvist, Handbook of Inorganic Electrochromic Materials, second ed., Elsevier, 2002.

[31] P.V. Ashrit, Dry lithiation study of nanocrystalline, polycrystalline and amorphous tungsten trioxide thin-films, Thin Solid Films 385 (2001) 81−88.

[32] Thin Films & Photonics Research Group, Université de Moncton, Unpublished work.

[33] C.H. Griffith, H.K. Eastwood, Influence of stoichiometry on the semiconductor to metal transition of vanadium dioxide, J. Appl. Phys. 45 (1974) 2201.

[34] V. Eyert, The metal-insulator transitions of VO_2−a band theoretical approach, Ann. Phys. 11 (2002) 9−61.

[35] J.P. Pouget, et al., Electron localization induced by uniaxial stress in pure VO_2, Phys. Rev. Lett. 35 (13) (1975) 873−875.

[36] S. Biermann, et al., Dynamical singlets and correlation-assisted Peierls transition in VO_2, Phys. Rev. Lett. 94 (2005) 026404.

[37] N.F. Mott, Metal-insulator transition, Rev. Mod. Phys. 40 (4) (1968) 677−683.

[38] M.M. Qazilbash, et al., Mott transition in VO_2 revealed by infrared spectroscopy and nano-imaging, Science 318 (5857) (2007) 1750−1753.

[39] R.O. Dillon, K. Le, N. Ianno, Thermochromic VO_2 sputtered by control of a vanadium-oxygen emission ratio, Thin Solid Films 398−399 (2001) 10−16.

[40] F. Guinnecton, et al., Optimized infrared switching properties in thermochromic vanadium dioxide thin films: role of deposition process and nanostructure, Thin Solid Films 446 (2004) 287−295.

[41] S. Bonora, et al., Mid-IR to near-IR image conversion by thermally induced optical switching in vanadium dioxide, Opt. Lett. 103 (2) (2010) 103−105.

[42] C.O.F. Ba, V. Fortin, S.T. Bah, R. Vallée, A. Pandurang, Formation of VO_2 by rapid thermal annealing and cooling of sputtered vanadium thin films, J. Vac. Sci. Technol. A 34 (2016) 031505.

[43] A. Pergament, et al., Vanadium dioxide: metal-insulator transition—electrical switching and oscillations—a review of the state of the art and recent progress, in: Energy Materials and Nanotechnology (EMN) Meeting on Computation and Theory, 9−12 November 2015, Istanbul, pp. 1−25.

[44] R. Balu, P.V. Ashrit, Near-zero, IR transmission in the metal-insulator transition of VO_2 thin films, Appl. Phys. Lett. 92 (2008) 021904.

[45] C.S. Blackman, et al., Atmospheric pressure chemical vapour deposition of thermochromic tungsten doped vanadium dioxide thin films for use in architectural glazing, Thin Solid Films 517 (2009) 4565−4570.

[46] S. Bonora, et al., Mid-IR laser beam quality measurement through vanadium dioxide optical switching, Opt. Lett. 38 (2013) 1554−1556.

[47] D.D. Yao, et al., Electrodeposited α and β phase MoO_3 and investigation of their gasochromic properties, Cryst. Growth Des. 12 (2012) 1865−1870.

[48] J. Zhao, et al., Synthesis of hexagonal MoO_3 nanorods and a study of their electrochemical performance as anode materials for lithium-ion batteries, J. Mater. Chem. A 3 (2015) 7463−7468.

[49] A. Taj, P.V. Ashrit, Dry lithiation studies of nanostructured sputter deposited molybdenum oxide thin films, J. Mater. Sci. 39 (2004) 3541−3544.

[50] P.M.S. Monk, R.J. Mortinmer, D.R. Rosseinsky, Electrochromism: Fundamentals and Applications, VCH Inc., Einheim, 1995.

[51] R. Shepard, M. Smeu, Ab initio investigation of α- and ζ-V_2O_5 for beyond lithium ion battery cathodes, J. Power Sources 472 (2020) 228096.

[52] P.V. Ashrit, K. Benaissa, G. Bader, F.E. Girouard, V.-V. Truong, Lithiation studies on some transition metal oxides for an all-solid thin film electrochromic system, Solid. State Ion. 59 (1993) 47−57.

[53] J.D. Joanhapoulos, S.G. Johnson, J.N. Winn, R.D. Meade, Photonic Crystals: Molding the Flow of Light, second ed., Princeton University Press, 2008.

[54] https://www.batop.de/information/r_Bragg.html.

[55] Thin Films & Photonics Research Group, Université de Moncton, Unpublished work.

[56] https://en.wikipedia.org/wiki/Opal.

[57] A. Thelen, McGraw-Hill Optical and Electro-Optical Engineering Series Design of Optical Interference Coatings, McGraw-Hill, 1989.

[58] E. Yablonovitch, Inhibited spontaneous emission in solid state physics and electronics, Phys. Rev. Lett. 58 (20) (1987) 2059−2062.

[59] S. John, Strong localization of photons in certain disordered dielectric superlattices, Phys. Rev. Lett. 58 (1987) 2486−2489.

[60] E. Yablonovitch, Photonic crystals-semiconductors of light, Sci. Am. 285 (2001) 47−55.

[61] E. Yablonovitch, T.J. Gmitter, K.M. Leung, Photonic band structure − face centered cubic case employing non-spherical atoms, Phys. Rev. Lett. 67 (17) (1991) 2295−2298.

[62] T. Sato, et al., Photonic crystals for the visible range fabricated by autocloning technique and their application, Opt. Quantum Electron. 34 (1) (2002) 63−70.

[63] https://www.cst.com/Content/Articles/article687/CST_Whitepaper_Photonic_CST.pdf.

[64] http://www.tytlabs.com/english/review/ rev394epdf/e394_033nakamura.pdf.

[65] K.M. Ho, C.T. Chan, C.M. Soukoulis, Existence of a photonic gap in periodic dielectric structures, Phys. Rev. Lett. 65 (25) (1990) 3152−3155.

[66] C.T. Chan, Q.L. Yu, K.M. Ho, Order-N spectral method for electromagnetic waves, Phys. Rev. B 51 (23) (1995) 16635−16642.

[67] J.B. Pendry, Calculating photonic band structure, J. Phys. Condens. Matter 8 (9) (1996) 1085−1108.

[68] N. Stefanou, V. Karathanos, A. Modinos, Scattering of electromagnetic waves by periodic structures, J. Phys. Condens. Matter 4 (36) (1992) 7389−7400.

[69] N. Stefanou, V. Karathanos, A. Modinos, MULTEM 2: a new version of the program for transmission and band-structure calculations of photonic crystals, Comput. Phys. Commun 132 (1−2) (2000) 189−196.

[70] K.S. Yee, Numerical solution of initial boundary value problems involving maxwell's equations in isotropic media, IEEE Trans. Antennas Propag. 14 (3) (1966).

[71] A. Taflove, S.C. Hagness, Computational Electrodynamics: The Finite-Difference Time-Domain Method, Artech House, Norwood, MA, 2000.

[72] P. Yeh, Optical Waves in Periodic Media, Wiley, New York, 1998.

[73] J.B. Pendry, A. MacKinnon, Calculation of photon dispersion relations, Phys. Rev. Lett. 69 (19) (1992) 2772−2775.

[74] P.M. Bell, J.B. Pendry, L.M. Moreno, A.J. Ward, A program for calculating photonic band structures and transmission coefficients of complex structures, Comput. Phys. Commun. 85 (2) (1995) 306−322.

[75] K. Sumioka, H. Kayashima, T. Tsutsui, Tuning the optical properties of inverse opal photonic crystals by deformation, Adv. Mater. 14 (18) (2002) 1284−1286.

[76] P. Sheng, et al., Multiply coated microspheres. A platform for realizing fields-induced structural transition and photonic bandgap, Pure Appl. Chem. 72 (2000) 309−315.

[77] K. Busch, S. John, Liquid-crystal photonic-band-gap materials: the tunable electromagnetic vacuum, Phys. Rev. Lett. 83 (5) (1999) 967−970.

[78] G. Martens, et al., Shift of the photonic bandgap in two photonic crystal/liquid crystal composites, Appl. Phys. Lett. 80 (11) (2002) 1885−1887.

[79] S.L. Kuai, G. Bader, P.V. Ashrit, Tunable electrochromic photonic crystals, Appl. Phys. Lett. 86 (22) (2005) 221110/1−221110/3.

[80] J.P. Cronin, D.J. Tarico, A. Agarwal, L. Zhang, United States Patent No. 5.277.986, 1994.

[81] P.V. Ashrit, G. Bader, F.E. Girouard, V.-V. Truong, Electrochromic materials for smart window application, Proc. SPIE 1401 (1990) 119−129.

[82] P. Ashrit, S. Kuai, Chromogenically Tunable Photonic Crystals, United States Patent 7660029 B2, 2010.

[83] Y. Lili, et al., Improved electrochromic performance of ordered macro-porous tungsten oxide films for IR electrochromic device, Sol. Energy Mater. Sol. Cell 100 (2012) 251−257.

[84] L. Kavan, M. Zukalova, M. Kalbac, M. Graetzel, Lithium insertion into anatase inverse opal, J. Electrochem. Soc. 151 (8) (2004) A1301−A1307.

[85] M. Penza, C. Martucci, G. Cassano, NO_x gas sensing characteristics of WO_3 thin films activated by noble metals (Pd, Pt, Au) layers, Sens. Actuators B 50 (1998) 5259.

[86] Z. Xie, H. Xu, F. Rong, L. Sun, S. Zhang, Z.-Z. Gu, Hydrogen activity tuning of Pt-doped WO_3 photonic crystal, Thin Solid Films 520 (2012) 4063−4067.

[87] C. Xiaofeng, et al., A photonic crystal-based CdS−Au−WO_3 heterostructure for efficient visible-light photocatalytic hydrogen and oxygen evolution, RSC Adv. 4 (2014) 15689−15694.

[88] X. Chen, et al., Enhanced incident photon-to-electron conversion efficiency of tungsten trioxide photoanodes based on 3D-photonic crystal design, ACS Nano 5 (6) (2011) 4310−4318.

[89] S. Amrehn, et al., Tungsten Oxide photonic crystals as optical transducer for gas sensing, ACS Sens. 3 (2018) 191−199.

[90] M. Ibisate, D. Golmayo, C. Lopez, Vanadium dioxide thermochromic opals grown by chemical vapour deposition, J. Opt. Pure Appl. Opt. 10 (12) (2008) 125202/1−125202/6.

[91] V.G. Golubev, et al., Phase transition-governed opal VO_2 photonic crystal, Appl. Phys. Lett. 79 (14) (2001) 2127−2129.

[92] Y. Ke, et al., Two-dimensional SiO_2/VO_2 photonic crystals with statically visible and dynamically infrared modulated for smart window deployment, ACS Appl. Mater. Interfaces 8 (2016) 33112−33120.

[93] J. Liang, P. Li, X. Song, L. Zhou, The fabrication and visible−near-infrared optical modulation of vanadium dioxide/silicon dioxide composite photonic crystal structure, Appl. Phys. A 123/794 (2017).

[94] H. Ye, H. Wang, Q. Cai, Two-dimensional VO2 photonic crystal selective emitter, J. Quant. Spectrosc. Radiat. Transf. 158 (2015) 119−126.

[95] C. Li, X. Hu, W. Gao, Y. Ao, S. Chu, H. Yang, et al., Thermo-optical tunable ultracompact chip-integrated 1D photonic topological insulator, Adv. Opt. Mater. 1701071 (2018) 1−6.

[96] Arezou Rashidi, A. Hatef, A. Namdar, Modeling photothermal induced bistability in vanadium dioxide/1D photonic crystal composite nanostructures, Appl. Phys. Lett. 113 (10) (2018) 1010103-1−1010103-5.

[97] A.B. Pevtsov, D.A. Kurdyukov, V.G. Golubev, A.V. Akimov, A.A. Meluchev, A.V. Sel'kin, et al., Ultrafast stop band kinetics in a three-dimensional opal-VO_2 photonic crystal controlled by a photoinduced semiconductor-metal phase transition, Phys. Rev. B 75 (15) (2007) 153101.

[98] B.S. Acharya, B.B. Nayak, Microstructural studies of nanocrystalline thin films of V_2O_5-MoO_3 using X-ray diffraction, optical absorption and laser micro Raman spectroscopy, Indian J. Pure Appl. Phys. 46 (2008) 866−875.

[99] T. He, J. Yao, Photochromic materials based on tungsten oxide, J. Mater. Chem. 17 (2007) 4547−4557.

[100] T. He, Y. Ma, Y. Cao, P. Jiang, X. Zhang, W. Yang, et al., Enhancement effect of gold nanoparticles on the UV-light photochromism of molybdenum trioxide thin films, Langmuir 17 (26) (2001) 8024−8027.

[101] T. Ben-Messaoud, J. Riordon, A. Melanson, P.V. Ashrit, A. Haché, Photoactive periodic media, Appl. Phys. Lett. 94 (2009) 111904.

[102] H. Li, H. Wu, J. Xiao, Y. Su, J. Robichaud, R. Brüning, et al., A hierarchically porous anatase TiO_2 coated-WO_3 2D IO bilayer film and its photochromic properties, Chem. Commun 52 (2016) 892−895.

CHAPTER 9

Thin film solar cells with graded-bandgap photon-absorbing layer

Faiz Ahmad[1], Akhlesh Lakhtakia[1] and Peter B. Monk[2]
[1]Department of Engineering Science and Mechanics, Pennsylvania State University, University Park, PA, United States
[2]Department of Mathematical Sciences, University of Delaware, Newark, DE, United States

9.1 Energy security and climate change

The global energy demand continues to rise, with the need for electrical power generation expected to reach 23.5 terawatt (TW) in the year 2040, an increase of 27% from the current requirement of 18.5 TW [1,2]. This increase will be mainly due to the population increase (with 1.7 billion people expected to be added to the developing economies) and the economic growth of developing countries in Asia (especially China and India) [2]. Today, 85% of the energy requirements are fulfilled by burning fossil fuels such as oil, coal, and natural gas. However, there are three major concerns over burning these carbonaceous fuels.

Climate emergency is arguably the most significant global concern. It is due to the greenhouse effect, which is a direct consequence of adding CO_2 and other gases to the atmosphere by burning fossil fuels. Without reducing the consumption of fossil fuels, the annually and globally averaged temperature by 2050 is likely to exceed $2°C$ above its value estimated for the year 1850. That rise would have a considerable impact on the landscape and sea levels, with its effects being already felt through changes in weather patterns, melting of polar ice, more frequent and more intense heat waves, and stronger hurricanes [3].

Energy security is the second major concern, as the supply of fossil fuels remains volatile. The fossil-fuel resources on our planet are limited and shrinking with every passing day. It is projected that these resources will get depleted within the next hundred years or so [4,5], but approximately 12% of the world's population still has no access to electricity [2]. Therefore, it will be challenging to fulfill the ever-increasing demand for energy with only fossil fuels.

The third major concern is the degradation of air quality due to the burning of fossil fuels. With the increased release of toxic gases such as sulfur dioxide and nitrogen oxides into the atmosphere, diseases such as burning lung tissue, asthma, bronchitis, pulmonary inflammation, and chronic respiratory illness are on the rise [5]. Air pollution continues to result in millions of premature deaths each year [1].

Thin Film Nanophotonics
DOI: https://doi.org/10.1016/B978-0-12-822085-6.00007-8

Therefore, scientists and engineers need to pay more attention toward eco-responsible (i.e., less polluting as well as renewable) sources of energy to continue to fulfill the continually rising energy demand as well as to tackle the problems of climate emergency and environmental degradation.

9.1.1 Eco-responsible sources of energy

The chief eco-responsible sources of energy are hydropower, wind, biomass, geothermal, and the Sun [2]. Hydropower- and wind-based technologies need favorable locations to harness the energy, but these locations are geographically restricted. Also, these technologies are grid based and transportation of energy to remote areas is costly [6]. Geothermal energy is very location specific and is not often competitive due to hefty costs associated with both geothermal power plants and geothermal heating/cooling systems [7]. Biomass technology is based on biomaterials, but the availability of these materials is also limited; furthermore, biofuels production puts undue pressure on agricultural activities necessary to feed the rising population.

Nuclear energy is another carbon-free source of energy. It can be an essential part of the energy mix necessary to meet the future demand of energy. Currently, nuclear technology provides 5% of the global energy consumption [8]. But, nuclear energy can only be a short-term solution because the fissile nature of nuclear fuels creates a gigantic security problem [8].

Yet another promising carbon-free energy source is nuclear fusion [9]. The fusion fuel cycle is sustainable and nearly inexhaustible due to the abundant availability of deuterium from seawater and tritium from breeder reactors. The major challenges for harnessing fusion power are plasma confinement and control, as well as the maintenance of a stable plasma–material interface [9]. As no fusion experiment has ever produced more power than the input power [10], nuclear fusion for energy generation remains a topic of research.

The most promising nonpolluting source of energy is the Sun. Energy falling on the Earth in 1 h from the Sun is more than the global energy consumption in one year [11]. With rapid reduction in the cost of photovoltaic solar cells [2,12], solar energy has emerged as a significant force to control global climate change and environmental degradation, and it can also help to tackle the ever-increasing energy demand. Yet, efficiently harvesting and distributing solar energy at an affordable cost is a significant challenge. Indeed, although more widely used than ever, photovoltaic solar cells contributed to <1.7% of the global electricity production in 2017 [13].

9.2 Photovoltaic solar cells

Photovoltaic solar cells, simply known as solar cells, directly convert sunlight into electricity via the photoelectric effect. The first solar cell demonstrated at Bell Labs in

1954 was made of crystalline silicon (c–Si) and had 6% power-conversion efficiency η [14]. Solar-cell technology made significant advances in the last two decades, the current record efficiency for single-junction c–Si solar-cell modules being 26.7% [15]. More advanced designs incorporating multijunction solar cells exposed to concentrated sunlight allow η to be as high as 46% [15,16]. However, the efficiencies of commercially available solar cells (15%–20%) are well below the maximum laboratory solar-cell-module efficiency due to factors such as degradation of material properties and increased electron–hole-pair recombination rate upon prolonged insolation [17].

The installed photovoltaic electricity-generation capacity increased more than a hundredfold from 2007 to 2018, but the share of solar cells in global energy production in 2018 was just 1.3% [2]. Although the energy produced using solar cells in some countries almost matches the energy produced using conventional sources in consumer price, a further cost reduction is required for solar cells to be globally competitive with conventional sources. Large-scale adoption of photovoltaic technology is hindered by factors such as low efficiency, scarcity of some materials needed to manufacture solar cells, and low industrial-scale production. However, it is projected that the global market share of photovoltaic technology will increase to 70% by 2050 [18]. In line with such projections, the price of solar-cell modules has gone down considerably and is expected to go down further as the installed capacity increases [19].

9.2.1 Important considerations for photovoltaic electricity generation

The most critical factor for the large-scale adoption of photovoltaic technology is the reduction of the unit cost. Two methods can reduce it. The first method is to increase the efficiency of the solar cell. However, there is a maximum upper limit on η called the Shockley–Queisser limit [20]. This limit is based on the maximum radiative recombination, and practically most materials can not reach the maximum radiative recombination limit. For a single-junction c–Si solar cell with a bandgap of 1.12 eV, the Shockley–Queisser limit is 33%; so, the maximum achieved efficiency (26.7%) is in the ballpark [15].

The second method is to reduce the manufacturing cost while maintaining the efficiency. The manufacturing cost can be reduced by using less expensive materials, reducing the thickness of the photon-absorbing layer, and adopting less costly production methods [21]. Thin film solar cells can provide all these features and thus have the potential to fulfill a significant fraction of the global energy demand.

9.2.2 Thin film solar cells

Thin film solar cells are a viable option compared to c–Si solar cells due to their low cost and ease of manufacturing. The c–Si solar cells are made of highly pure silicon wafers that account for 50% of the total cost of a solar-cell module. Thin film solar cells can be manufactured using physical vapor deposition and chemical vapor deposition [21]. Also, thin film solar cells are made of materials with high optical absorption coefficients compared to c-Si. As a result, they require the photon-absorbing layer to be only a few hundred nanometers in thickness. In contrast, the typical thickness of the c-Si photon-absorbing layer in a solar cell is a few hundred micrometers [22].

The energy-payback time for a thin film solar cell is almost half compared to that of a c-Si solar cell [23]. Also, the flexibility of some thin film solar cells makes them attractive for portable purposes such as camping and installation on nonplanar surfaces such as car roofs [24]. Another benefit of thin film solar cells of some types is the control over the composition of the photon-absorbing material, that is, the bandgap of the material can be adjusted by fixing the material composition appropriately.

Solar cells are named after the material used for the photon-absorbing layer, which is the major contributor to charge-carrier generation. Currently, thin film solar cells containing photon-absorbing layers made of either Cu(In, Ga)Se$_2$ (abbreviated as CIGS) or cadmium telluride (CdTe) are commercially dominant [22]. Another type of thin film solar cell in the market uses hydrogenated amorphous silicon (a-Si:H). CIGS, CdTe, and a-Si:H thin film solar cells have a 6% share of the global photovoltaic solar-cell market [25]. The maximum reported efficiency is 10.2% for a-Si:H, 22.6% for CIGS, and 21.0% for CdTe solar cells [15,16].

9.2.3 Challenges of thin film solar-cell technologies

Thin film solar cells must be made of materials that are abundant on our planet, and the materials of choice must be those that can be extracted, processed, and discarded with low environmental impact. CIGS and CdTe are commercially dominant (terrestrial) thin film solar-cell technologies; however, there are serious concerns about the planetwide availability of indium and tellurium [26]. Hence, thickness reduction of the photon-absorbing layer is necessary to tackle the scarcity of indium in CIGS solar cells and of tellurium in CdTe solar cells. Furthermore, both indium and cadmium are toxic, leading to environmental concerns about their impact following disposal after use.

The low efficiencies of thin film solar cells compared to that of c-Si solar cells are also problematic. Cu$_2$ZnSn(S,Se)$_4$ (abbreviated as CZTSSe) and a-Si:H solar cells contain Earth-abundant and nontoxic materials. But the efficiencies of CZTSSe and a-Si: H solar cells are only 12.6% and 10.2%, respectively, both of which are considerably lower than the approximately 22% efficiencies of the CdTe and CIGS solar cells [16]. The primary reason for the lower efficiency of CZTSSe solar cells is the lower lifetime of minority carriers due to higher trap/defect density, which shortens their diffusion length, thereby limiting the collection of minority carriers deep inside the CZTSSe photon-absorbing layer [27]. The reasons for low efficiency in a-Si:H solar cells are

(i) the high electron—hole recombination rate; (ii) low charge-carrier diffusion lengths [28]; and (iii) the phenomenon of light-induced degradation, also known as the Staebler—Wronski effect [17]. Therefore, the reduction of the thickness of the photon-absorbing layer is desirable for better collection of charge carriers in both CZTSSe and a-Si:H thin film solar cells.

Thin film solar cells containing gallium arsenide (GaAs) as the photon-absorbing material have excellent efficiency (28.8%) and significantly lower weight-to-output-power ratio than c-Si solar cells [29]. GaAs solar-cell technology is the current market leader for space applications, but it is prohibitively expensive for terrestrial applications [30]. Cost reduction by thinning the expensive GaAs photon-absorbing layer is necessary to make this technology viable for terrestrial applications. The other thin film solar-cell technologies, that is, perovskites, organic, and dye-sensitized solar cells, have not penetrated the market yet in large volumes [15,16].

Reduction of the thickness of the photon-absorbing layer is therefore desirable to reduce the cost, improve charge-carrier collection, and tackle the scarcity of scarce materials. Furthermore, a thinner photon-absorbing layer will improve the industrial throughput of thin film solar-cell technology. However, a thinner photon-absorbing layer will reduce the absorption of incident solar photons and, therefore, the optical short-circuit current density J_{sc}^{opt} and the open-circuit voltage V_{oc}; accordingly, η will also be reduced.

9.2.4 Photon trapping in thin film solar cells

Trapping of incident solar photons is necessary to tackle the problem of reduced absorption and to enhance the efficiency of thin film solar cells. The most straightforward technique of trapping the light is to add a backreflector which effectively doubles the optical thickness of the absorber layer and enhances the charge-carrier-generation rate inside the solar cell. If metallic, the backreflector can also serve as an electrical contact layer.

Several other techniques under investigation to offset the problem of low optical absorption include the use of light-trapping nanostructures [31—35]; textured front faces [36,37]; antireflection coatings [38—40]; plasmonic particles [41]; back-surface passivation layers [42]; and metallic periodically corrugated backreflectors for surface plasmonics [43—48], multiplasmonics [49—51], and waveguide-mode excitation [52—54]. These techniques are at best in the research stages.

9.2.5 Bandgap grading of photon-absorbing layer

In a typical solar cell, there is a thick photon-absorbing layer that is the major contributor to the electrical current developed inside the device. This layer can be made of either an n-type or a p-type semiconductor. c-Si, CIGS, and CZTSSe solar cells have a single p-type photon-absorbing layer, whereas GaAs and (Al,Ga)As (abbreviated as AlGaAs) solar cells have a single n-type photon-absorbing layer.

Typically, the bandgap is uniform throughout the photon-absorbing layer and plays a crucial role in setting the efficiency of the solar cell. The electric current density and the voltage developed across the front- and the back-contact layers of the solar cell are directly related to the bandgap of the photon-absorbing layer. The current density is high/low but the voltage is low/high when the bandgap is low/high [19]. Hence, the performance of a solar cell can be improved by the proper choice of the bandgap, especially when the photon-absorbing layer is made of a compound semiconductor exemplified by CIGS, CZTSSe, and AlGaAs [55,56].

One way to improve the efficiency of a thin film solar cell is to grade the bandgap of the photon-absorbing layer in the thickness direction [55–58]. The bandgap grading can be accomplished by dynamically controlling the composition of the compound semiconductor during fabrication [59–62]. Bandgap grading can allow photon absorption over a wider frequency range. Also, bandgap grading can increase the efficiency by creating a drift electric field that accelerates photogenerated holes toward the p–n junction inside the solar cell [61]. Linear bandgap-grading has been shown experimentally to increase the open-circuit voltage in thin film solar cells [63,64], which should assist in enhancing efficiency; however, suboptimal bandgap-grading can reduce the short-circuit current density J_{sc} to offset the increase in V_{oc}. Optimal grading of the bandgap is required to maintain J_{sc} while enhancing V_{oc}.

9.3 Coupled optoelectronic modeling and optimization of thin film solar cells

It is common in optics literature to optimize the photon-absorbing characteristics as quantified by J_{sc}^{opt} [34,65,66]; however, this is a deficient technique since the optimization of J_{sc}^{opt} can reduce V_{oc} [67]. Another figure-of-merit is the maximum power density P_{sup} supplied by the solar cell. Its optimization is better than that of J_{sc}^{opt} [68], but that too is deficient in terms of modeling V_{oc} and η.

Any reasonably proper model of a thin film solar cell, therefore, requires the coupling of (i) an optical submodel capable of simulating the absorption of photons and (ii) an electronic submodel capable of simulating the transport of charge carriers throughout the semiconducting regions of the solar cell [69,70]. Both submodels need to be computationally fast, accurate, and robust across the relevant parameter space for the optimization of solar-cell designs. Hence, a rigorous optoelectronic model is necessary for graded-bandgap thin film solar cells.

For more than 40 years, it has been thought that periodicity on the scale of a wavelength in the solar spectrum must be incorporated in the transverse plane to assist in photon absorption [36,46–49,54,66]. As part of a research project funded by the

United States National Science Foundation, an optoelectronic model was therefore developed for thin film photovoltaic solar cells containing periodic structures [71] and was used together with the differential evolution algorithm (DEA) [72–74] to find optimal thin film solar-cell designs [57,58,75,76].

The structures of the CIGS and CZTSSe solar cells are shown in Fig. 9.1A and B, respectively. The structures of these two–solar cells are identical except that a p-type photon-absorbing layer is made of either CIGS or CZTSSe. Solar cells of both types have a molybdenum (Mo) back–contact layer, an aluminum oxide (Al_2O_3) back–surface passivation layer, the photon-absorbing layer of the p-type semiconductor, a layer of n-type cadmium sulfide (CdS), an oxygen–deficient zinc-oxide (od-ZnO) front–surface passivation layer, a front contact layer made of aluminum-doped zinc oxide (AZO), and an antireflection coating of magnesium fluoride (MgF_2). The front- and the back–surface passivation layers improve J_{sc} by depressing the electron–hole recombination rate [77]. All layers in the solar cell must be considered in the optical submodel, but only the region between the two electrical contact layers has to be considered in the electronic submodel.

The thicknesses of the various layers in CIGS solar cells other than the CIGS layer were fixed as follows: $L_{MgF_2} = 110$ nm, $L_{AZO} = 100$ nm, $L_{ZnO} = 80$ nm, $L_{CdS} = 70$ nm, $L_{Al_2O_3} = 50$ nm, and $L_{Mo} = 500$ nm. The thicknesses of the various layers in CZTSSe solar cells other than the CZTSSe layer were fixed as follows: $L_{MgF_2} = 110$ nm, $L_{AZO} = 100$ nm, $L_{ZnO} = 100$ nm, $L_{CdS} = 50$ nm, $L_{Al_2O_3} = 20$ nm, and $L_{Mo} = 500$ nm. The thickness L_1 of the CIGS layer or L_2 of the CZTSSe layer, as appropriate, is left as a geometric parameter to be determined so as to maximize η.

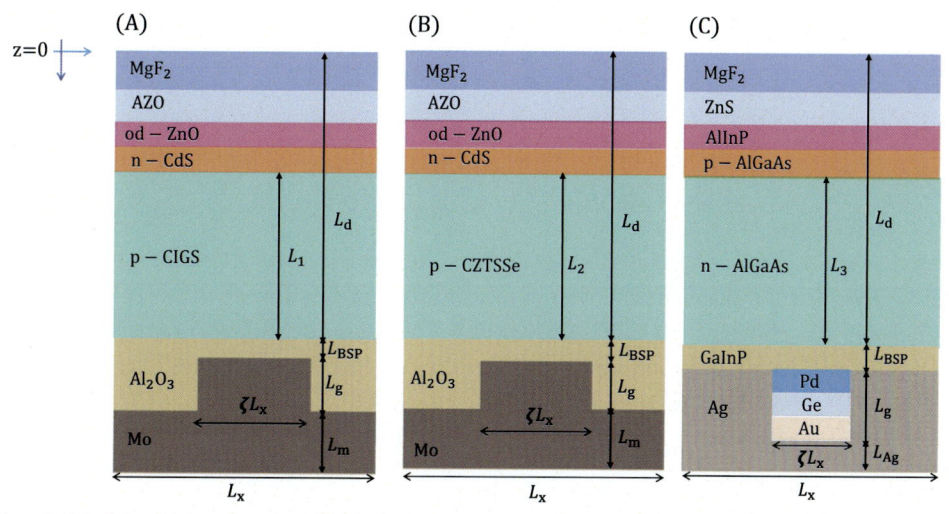

Figure 9.1 Schematic of a thin film solar cell containing a (A) CIGS, (B) CZTSSe, and (C) AlGaAs photon-absorbing layer. *CIGS*, Cu(In,Ga)Se$_2$; *CZTSSe*, Cu$_2$ZnSn(S,Se)$_4$; *AlGaAs*, (Al,Ga)As.

The back–contact layer is periodic along the x-axis with period L_x, so that the CIGS or CZTSSe solar cell has a metal/dielectric grating. The permittivity in the reference unit cell $|x| \leq L_x/2$ of the grating region is given by

$$\varepsilon_g(x, z, \lambda_0) = \begin{cases} \varepsilon_m(\lambda_0), & 0 \leq |x| < \zeta L_x/2, \\ \varepsilon_d(\lambda_0), & \zeta L_x/2 < |x| \leq L_x/2, \end{cases} \quad (9.1)$$

$$z \in (L_d + L_{BSP}, L_d + L_{BSP} + L_g),$$

where $L_d = L_{MgF_2} + L_{AZO} + L_{ZnO} + L_{CdS} + L_1$ for the CIGS solar cell and $L_d = L_{MgF_2} + L_{AZO} + L_{ZnO} + L_{CdS} + L_2$ for the CZTSSe solar cell, $L_{BSP} = L_{Al_2O_3}$ and $L_m = L_{Mo}$, λ_0 is the free-space wavelength, $\varepsilon_m(\lambda_0)$ is the permittivity of molybdenum, and $\varepsilon_d(\lambda_0)$ is the permittivity of Al_2O_3. The height L_g and the duty cycle $\zeta \in [0, 1]$ of the grating are geometric parameters to be determined so as to maximize η.

In the CIGS solar cell, the bandgap $\mathcal{E}_g$ varies with z in the CIGS layer. The linearly forward-graded bandgap is modeled as [57]

$$\mathcal{E}_g(z) = \mathcal{E}_{a1} + A_1(\mathcal{E}_{b1} - \mathcal{E}_{a1})\frac{z - (L_d - L_1)}{L_1}, \quad z \in [L_d - L_1, L_d], \quad (9.2)$$

where $\mathcal{E}_{a1}$ is the minimum bandgap, $\mathcal{E}_{b1}$ is the maximum bandgap, and $A_1 > 0$ is an amplitude parameter. The linearly backward-graded bandgap is modeled as [57]

$$\mathcal{E}_g(z) = \mathcal{E}_{b1} - A_1(\mathcal{E}_{b1} - \mathcal{E}_{a1})\frac{z - (L_d - L_1)}{L_1}, \quad z \in [L_d - L_1, L_d], \quad (9.3)$$

A more complicated grading of bandgap in the CIGS layer was modeled as [57]

$$\mathcal{E}_g(z) = \mathcal{E}_{a1} + A_1(\mathcal{E}_{b1} - \mathcal{E}_{a1}) \quad (9.4)$$

$$\times \left(\frac{1}{2}\left\{\sin\left[2\pi K_1 \frac{z - (L_d - L_1)}{L_1} - 2\pi\psi_1\right] + 1\right\}\right)^{\alpha_1},$$

$$z \in [L_d - L_1, L_d],$$

where $\psi_1 \in [0, 1]$ quantifies a relative phase shift, K_1 is the number of periods in the CIGS layer, and $\alpha_1 > 0$ is a shaping parameter. Since the right side of Eq. (9.4) involves a sinusoidal function, this grading is classified as sinusoidal. The bandgap-grading parameters $\mathcal{E}_{a1}$, $\mathcal{E}_{b1}$, A_1, ψ_1, K_1, and α_1 are to be determined during the optimization process to maximize η.

Eqs. (9.2)–(9.4) hold for bandgap grading in the CZTSSe layer in a CZTSSe solar cell after the replacement of the thickness L_1 by L_2 and the bandgap parameters $\mathcal{E}_{a1}$, $\mathcal{E}_{b1}$, A_1, ψ_1, K_1, and α_1 by $\mathcal{E}_{a2}$, $\mathcal{E}_{b2}$, A_2, ψ_2, K_2, and α_2, respectively [58]. The bandgap-grading parameters $\mathcal{E}_{a2}$, $\mathcal{E}_{b2}$, A_2, ψ_2, K_2, and α_2 are determined during the optimization process to maximize η.

The structure of the AlGaAs solar cell is shown in Fig. 9.1C. It contains a silver (Ag) backreflector along with Ag and a Pd/Ge/Au trilayer localized ohmic back-contact [75,78], which is periodic along the x-axis with period L_X and duty cycle $\zeta \in (0, 1)$. In addition, there is a back-surface passivation layer of gallium indium phosphide (GaInP), the photon-absorbing layer of n-type AlGaAs, a layer of p-type AlGaAs with homogeneous bandgap $\mathcal{E}_{g,w}$, an aluminum indium phosphide (AlInP) front-surface passivation layer, and a ZnS/MgF$_2$ double-layer antireflection coating. The thicknesses of the various layers in the AlGaAs solar cell other than the n-type AlGaAs photon-absorbing layer were fixed as follows: $L_{MgF_2} = 110$ nm, $L_{ZnS} = 40$ nm, $L_{AlInP} = 20$ nm, $L_{pAlGaAs} = 50$ nm, $L_{GaInP} = 20$ nm, $L_{Pd} = 20$ nm, $L_{Ge} = 50$ nm, $L_{Au} = 100$ nm, and $L_{Ag} = 100$ nm. The thickness L_3 of the AlGaAs photon-absorbing layer and the duty cycle ζ are left unspecified as optimization parameters. The front contact is not shown in the figure, but the Dirichlet boundary condition is applied in the electronic submodel at the ZnS/AlInP interface and the GaInP/Ag interface, the parameter ζ being small.

Eqs. (9.2)−(9.4) also hold for bandgap grading in the AlGaAs layer in an AlGaAs solar cell after the replacement of the thickness L_1 by L_3 and the bandgap-grading parameters $\mathcal{E}_{a1}$, $\mathcal{E}_{b1}$, A_1, ψ_1, K_1, and α_1 by $\mathcal{E}_{a3}$, $\mathcal{E}_{b3}$, A_3, ψ_3, K_3, and α_3, respectively [75]. Note that $L_d = L_{MgF_2} + L_{ZnS} + L_{AlInP} + L_{pAlGaAs} + L_3$ for the AlGaAs solar cell. The bandgap-grading parameters $\mathcal{E}_{a3}$, $\mathcal{E}_{b3}$, A_3, ψ_3, K_3, and α_3 are determined during the optimization process to maximize η. The bandgap $\mathcal{E}_{g,w}$ of the homogeneous p-type AlGaAs layer is also considered as a parameter for optimization to maximize η.

The coupled optoelectronic model used to simulate and optimize all three types of thin film solar cells [57,58,75] comprises a photonic submodel and an electronic sub-model [71]. In the photonic submodel, the rigorous coupled-wave approach [79,80] is used to determine the electric and magnetic fields everywhere inside the solar cell due to normally incident monochromatic radiation. The λ_0-dependent relative permittivity of every material in the solar cell is published elsewhere [57,58,75,76]. With the assumption that the absorption of every photon in each semiconductor layer excites an electron−hole pair, the 2D electron−hole-pair generation rate $G(x,z)$ is determined in those layers [57], assuming normal illumination by unpolarized polychromatic light endowed with the AM1.5G solar spectrum [81]. Since L_x is on the order of λ_0 whereas electronic processes occur at steady state, $G(x,z)$ is averaged over $x \in [-L_x/2, L_x/2]$ to obtain the 1D electron−hole pair generation rate $G(z)$.

In the electronic submodel, $G(z)$ is used as an input to the 1D drift-diffusion equations for the charge carriers and the Poisson equation for the electrostatic potential [69], with the Dirichlet boundary condition applied to the front- and back-contact layers. In a CIGS solar cell, the od-ZnO front-surface passivation layer, the n-type CdS layer, and the p-type CIGS photon-absorbing layer together constitute the electrical domain. Similarly, in a CZTSSe solar cell, the electrical domain comprises the od–ZnO front-surface passivation layer, the n-type CdS layer, and the p-type CZTSSe photon-absorbing layer. In an AlGaAs solar cell, the p^+-type AlInP front-surface passivation layer, the p-type AlGaAs layer, the n-type AlGaAs photon-absorbing layer, and the n^+-type GaInP back-surface passivation layer must be considered in the electronic submodel.

The 1D drift-diffusion equations for the gradients of the electron current density $J_n(z)$ and the hole current density $J_p(z)$ also contain the electron–hole pair recombination rate $R(z)$. The nonlinear Shockley–Read–Hall (SRH), Auger, and radiative contributions to $R(z)$ can be incorporated for every material between the two contact layers. In addition, the 1D Poisson equation for the electrostatic potential is simultaneously considered with the electron density $n(z)$, the hole density $p(z)$, and the fixed charge density (including the doping density and trap/defect density) contributing to the source term [76]. The Boltzmann approximation is adopted. Both the front- and the back-contact layers are modeled as ideally ohmic and local quasithermal equilibrium is assumed near the electrical contact layers [71]. The electrical properties of every material in the solar cell are available elsewhere [57,58,75,76].

A set of six nonlinear differential equations [71,82] is solved using a hybridizable discontinuous Galerkin scheme [71,83] to determine the z-independent current density $J_{dev} = J_n(z) + J_p(z)$ and the electrical power density $P = J_{dev} V_{ext}$ as functions of the bias voltage V_{ext} under steady-state conditions. In turn, the $J_{dev}-V_{ext}$ and the $P-V_{ext}$ curves yield J_{sc}, V_{oc}, η, and a figure-of-merit called the fill factor FF $\in$ [0, 1] which should be as high as possible [69]. The model has been validated against experimental results for all three types of solar cells with a homogeneous photon-absorbing layer [57,58,75].

Finally, the DEA [74] is used to maximize η as a function of various geometric and bandgap-grading parameters. Given an initial guess in the search space, the underlying strategy in the metaheuristic DEA is to improve the candidate solution at every iteration step and does not require explicit gradients of the cost function (i.e., η). The DEA was used with MATLAB version R2018a-R2019a for the numerical results presented in the remainder of this chapter.

9.3.1 Homogeneous photon-absorbing layer

All three types of thin film solar cells have been considered with a homogeneous photon-absorbing layer. The values of J_{sc}, V_{oc}, FF, and η for the optimal solar-cell design predicted by the coupled optoelectronic model are presented in Table 9.1. Values of $\mathcal{E}_{an}$, L_g, ζ, L_x, and L_n for each of the three optimal designs are also provided in the same table. Furthermore, the optimal value of $\mathcal{E}_{g,w}$ for the AlGaAs solar cell is provided in the same table.

Table 9.1 Predicted parameters of the optimal CIGS, CZTSSe, and AlGaAs solar cells when the photon-absorbing layer of thickness L_n, $n \in \{1, 2, 3\}$, is homogeneous (i.e., $A_n = 0$) and the backreflector is periodically corrugated.

Cell type	n	L_n (nm)	$\mathcal{E}_{g,w}$ (eV)	$\mathcal{E}_{an}$ (eV)	L_g (nm)	ζ	L_x (nm)	J_{sc} (mA cm^{-2})	V_{oc} (mV)	FF (%)	η (%)
CIGS	1	2200	—	1.24	101	0.49	502	31.11	742	82	18.93
CZTSSe	2	1200	—	1.21	101	0.51	500	30.13	558	70	11.84
AlGaAs	3	2000	2.09	1.424	—	0.05	510	30.20	1090	87	28.80

9.3.1.1 Cu(In, Ga)Se₂ solar cell

Optimization of the CIGS solar cell was carried out in the parameter space defined as follows [57]: $\mathcal{E}_{a1} \in [0.947, 1.626]$ eV, $L_g \in [0, 550]$ nm, $\zeta \in (0, 1)$, $L_x \in [100, 1000]$ nm, and $L_1 \in [100, 2200]$ nm. The highest value of η predicted for the CIGS solar cell with a homogeneous photon–absorbing layer is 18.93%; correspondingly, $J_{sc} = 31.11$ mA cm^{-2}, $V_{oc} = 742$ mV, and FF $= 82\%$. The optimal thickness of the photon–absorbing layer is 2200 nm and the corresponding bandgap parameter $\mathcal{E}_{a1} = 1.24$ eV.

9.3.1.2 Cu₂ZnSn(S,Se)₄ solar cell

For the CZTSSe solar cell, optimization was carried out in the parameter space defined as follows [58]: $\mathcal{E}_{a2} \in [0.91, 1.49]$ eV, $L_g \in [0, 550]$ nm, $\zeta \in (0, 1)$, $L_x \in [100, 1000]$ nm, and $L_2 \in [100, 2200]$ nm. The highest value of η predicted for the CZTSSe solar cell with a homogeneous photon–absorbing layer is 11.84%; correspondingly, $J_{sc} = 30.13$ mA cm^{-2}, $V_{oc} = 558$ mV, and FF $= 70\%$. The optimal thickness of the photon–absorbing layer is 1200 nm and its bandgap $\mathcal{E}_{a2} = 1.21$ eV.

9.3.1.3 (Al,Ga)As solar cell

Optimization of the AlGaAs solar cell was carried out in the parameter space defined as follows [75]: $\mathcal{E}_{a3} \in [1.424, 2.09]$ eV, $\mathcal{E}_{g,w} \in [1.424, 2.09]$ eV, $\zeta \in (0, 1)$, $L_x \in [100, 1000]$ nm, and $L_3 \in [100, 2000]$ nm. The highest efficiency predicted is 28.80%, along with $J_{sc} = 30.20$ mA cm^{-2}, $V_{oc} = 1090$ mV, and FF $= 87\%$. The optimal thickness of the photon–absorbing layer is 2000 nm and the corresponding bandgap parameters are $\mathcal{E}_{g,w} = 2.09$ eV and $\mathcal{E}_{a3} = 1.424$ eV.

The optimal thickness of the photon–absorbing layer is several times the grating height L_g for the CIGS and CZTSSe solar cells. Calculations show that setting $L_g = 0$ does not reduce η significantly. Thus the Mo backreflector need not be corrugated for these two types of thin film solar cells.

9.3.2 Linearly graded photon-absorbing layer

All three types of solar cells with linearly graded photon-absorbing layers have been optimized. The values of J_{sc}, V_{oc}, FF, and η for the optimal solar cell predicted by the coupled optoelectronic model are presented in Table 9.2. The corresponding values of $\mathcal{E}_{an}$, $\mathcal{E}_{bn}$, A_n, L_g, ζ, L_x, and L_n for each of the three optimal designs are also provided in the same table. The optimal value of $\mathcal{E}_{g,w}$ for the AlGaAs solar cell is provided in the same table.

Table 9.2 Predicted parameters of the optimal CIGS, CZTSSe, and AlGaAs solar cells when the photon-absorbing layer of thickness L_n, $n \in \{1, 2, 3\}$, is linearly graded (i.e., $A_n \neq 0$) and the backreflector is periodically corrugated.

Cell type	n	Grading	L_n (nm)	$\mathcal{E}_{g,w}$ (eV)	A_n	$\mathcal{E}_{an}$ (eV)	$\mathcal{E}_{bn}$ (eV)	L_g (nm)	ζ	L_x (nm)	J_{sc} (mA cm^{-2})	V_{oc} (mV)	FF (%)	η (%)
CIGS	1	Forward	2200	—	0	1.24	1.62	101	0.49	502	31.11	742	82	18.93
CIGS	1	Backward	2200	—	0.75	0.95	1.62	101	0.50	500	24.09	1039	77	19.27
CZTSSe	2	Forward	2200	—	0.99	0.91	1.49	100	0.51	500	36.72	628	74	17.07
CZTSSe	2	Backward	1200	—	0	1.21	1.49	101	0.51	500	30.13	558	70	11.84
AlGaAs	3	Forward	—	—	—	—	—	—	—	—	—	—	—	—
AlGaAs	3	Backward	2000	1.424	1.0	1.424	1.98	—	0.05	500	24.70	1507	88	33.10

9.3.2.1 Cu(In, Ga)Se₂ solar cell

Optimization of the CIGS solar cell was carried out in the parameter space defined as follows [57]: $\mathcal{E}_{a1} \in$ [0.947, 1.626] eV, $\mathcal{E}_{b1} \in$ [0.947, 1.626] eV with $\mathcal{E}_{b1} > \mathcal{E}_{a1}$, $A_1 \in$ [0, 1], $L_g \in$ [1, 550] nm, $\zeta \in$ (0, 1), $L_x \in$ [100, 1000] nm, and $L_1 \in$ [100, 2200] nm.

With the linearly forward-graded CIGS photon-absorbing layer [Eq. (9.2)], optoelectronic optimization yielded $A_1 = 0$. Thus the optimal results obtained with the homogeneous CIGS layer (Section 9.3.1.1) hold with the linearly forward-graded CIGS layer.

Optoelectronic optimization with the linearly backward-graded CIGS photon-absorbing layer [Eq. (9.3)] yielded $A_1 \neq 0$. The highest value of η predicted for the CIGS solar cell with the linearly (backward) graded photon-absorbing layer is 19.27%; correspondingly, $J_{sc} = 24.07$ mA cm^{-2}, $V_{oc} = 1039$ mV, and FF = 77%. The optimal thickness of the photon-absorbing layer is $L_1 = 2200$ nm. The corresponding bandgap parameters are as follows: $\mathcal{E}_{a1} = 0.95$ eV, $\mathcal{E}_{b1} = 1.62$ eV, and $A_1 = 0.75$.

9.3.2.2 Cu₂ZnSn(S,Se)₄ solar cell

For the CZTSSe solar cell, optimization was carried out in the parameter space defined as follows [58]: $\mathcal{E}_{a2} \in$ [0.91, 1.49] eV, $\mathcal{E}_{b2} \in$ [0.91, 1.49] eV with $\mathcal{E}_{b2} > \mathcal{E}_{a2}$, $A_2 \in$ [0, 1], $L_g \in$ [1, 550] nm, $\zeta \in$ (0, 1), $L_x \in$ [100, 1000] nm, and $L_2 \in$ [100, 2200] nm.

With the linearly backward-graded CZTSSe photon-absorbing layer [Eq. (9.3)], optoelectronic optimization yielded $A_2 = 0$. Thus the optimal results obtained with the homogeneous CZTSSe layer (Section 9.3.1.2) hold with linearly backward-graded CZTSSe photon-absorbing layer.

Optoelectronic optimization with the linearly forward-graded CZTSSe photon-absorbing layer [Eq. (9.2)] yielded $A_2 \neq 0$. The highest value of η predicted for the CZTSSe solar cell with the linearly (forward) graded photon-absorbing layer is 17.07%; correspondingly, $J_{sc} = 36.72$ mA cm^{-2}, $V_{oc} = 628$ mV, and FF = 74%. The optimal thickness of the photon-absorbing layer is $L_2 = 2200$ nm and its bandgap-grading parameters are as follows: $\mathcal{E}_{a2} = 0.91$ eV, $\mathcal{E}_{b2} = 1.49$ eV, and $A_2 = 0.99$.

9.3.2.3 (Al,Ga)As solar cell

Optimization of the AlGaAs solar cell was carried out in the parameter space defined as follows [75]: $\mathcal{E}_{g,w} \in$ [1.424, 2.09] eV, $\mathcal{E}_{a3} \in$ [1.424, 2.09] eV, $\mathcal{E}_{b3} \in$ [1.424, 2.09] eV with $\mathcal{E}_{b3} > \mathcal{E}_{a3}$, $A_3 \in$ [0, 1], $\zeta \in$ (0, 1), $L_x \in$ [100, 1000] nm, and $L_3 \in$ [100, 2200] nm.

Optoelectronic optimization did not perform satisfactorily with the linearly forward-graded AlGaAs photon-absorbing layer [Eq. (9.2)]. With the linearly backward-graded AlGaAs photon-absorbing layer [Eq. (9.3)], optoelectronic optimization yielded $A_3 = 1$. The highest value of η predicted with the linearly (backward) graded photon-absorbing layer is 33.10%; correspondingly, $J_{sc} = 24.70$ mA cm^{-2}, $V_{oc} = 1507$ mV, FF = 88%, $L_3 = 2000$ nm, $\mathcal{E}_{g,w} = 1.424$ eV, $\mathcal{E}_{a3} = 1.424$ eV, $\mathcal{E}_{b3} = 1.98$ eV, and $A_3 = 1$.

Calculations show that setting $L_g = 0$ does not reduce η significantly for the CIGS and CZTSSe solar cells. Hence, the Mo backreflector need not be corrugated for thin film solar cells of both types.

9.3.3 Sinusoidally graded photon-absorbing layer

Optoelectronic optimization with the sinusoidally graded photon-absorbing layer has been carried for CIGS, CZTSSe, and AlGaAs solar cells with and without periodically corrugated backreflector. The predicted values of J_{sc}, V_{oc}, FF, and η for the optimal solar cell are presented in Table 9.3. The corresponding values of $\mathcal{E}_{an}$, $\mathcal{E}_{bn}$, A_n, K_n, ψ_n, α_n, L_g, ζ, L_x, and L_n for the optimal designs are also provided in the same table. Furthermore, the optimal value of $\mathcal{E}_{g,w}$ for the AlGaAs solar cell is provided in the same table.

9.3.3.1 Cu(In, Ga)Se₂ solar cell

Optimization of the CIGS solar cell was carried out in the parameter space defined as follows [57]: $\mathcal{E}_{a1} \in [0.947, 1.626]$ eV, $\mathcal{E}_{b1} \in [0.947, 1.626]$ eV with $\mathcal{E}_{b1} > \mathcal{E}_{a1}$, $A_1 \in [0, 1]$, $K_1 \in [0, 8]$, $\psi_1 \in [0, 1]$, $\alpha_1 \in [0, 8]$, $L_g \in [1, 550]$ nm, $\zeta \in (0, 1)$, $L_x \in [100, 1000]$ nm, and $L_1 \in [100, 2200]$ nm. The highest value of η predicted is 27.70%; correspondingly, $J_{sc} = 33.16$ mA cm^{-2}, $V_{oc} = 1070$ mV, FF = 78%, $L_1 = 2200$ nm, $\mathcal{E}_{a1} = 0.95$ eV, $\mathcal{E}_{b1} = 1.626$ eV, $A_1 = 0.98$, $K_1 = 1.5$, $\psi_1 = 0.75$, and $\alpha_1 = 6$.

With a view to reduce the thickness of the sinusoidally graded photon-absorbing layer, $L_1 = 600$ nm was fixed and the maximum efficiency was determined to be 22.89%; correspondingly, $J_{sc} = 29.18$ mA cm^{-2}, $V_{oc} = 1045$ mV, FF = 75%, $\mathcal{E}_{a1} = 0.95$ eV, $\mathcal{E}_{b1} = 1.626$ eV, $A_1 = 0.99$, $K_1 = 1.5$, $\psi_1 = 0.75$, and $\alpha_1 = 6$ [57]. Thus a 600-nm-thick sinusoidally graded layer is predicted to deliver 21% higher efficiency than its 2200-nm-thick homogeneous counterpart (Table 9.1), a saving of 64% of the photon-absorbing material.

A detailed study was performed for the CIGS solar cell with the ultrathin photon-absorbing layer. The variations of $\mathcal{E}_g(z)$, $G(z)$, and $R(z)$ versus z in the semiconductor layers of the optimal solar cell are depicted in Fig. 9.2A–C. Although $\mathcal{E}_g$ is obviously independent of z in both the od-ZnO and CdS layers, it varies with z in the CIGS layer. This variation comprises constant-$\mathcal{E}_g$ regions separated by regions with large $\mathcal{E}_g$ gradients. The bandgap is low in the constant-$\mathcal{E}_g$ regions, these regions being responsible for elevating the electron–hole-pair generation rate because less energy is required to excite an electron–hole pair across a narrower bandgap [70]. Whereas G is higher in regions with lower $\mathcal{E}_g$ and vice versa, R is higher in regions with higher $\mathcal{E}_g$ due to the higher fixed charge density caused by higher gallium content in those regions.

The $J_{dev}–V_{ext}$ and $P–V_{ext}$ characteristics of the same solar cell are shown in Fig. 9.2D. The optoelectronic model predicts $J_{dev} = 24.72$ mA cm^{-2} and $V_{ext} = 926$ mV for this solar cell. Given that the thickness of the CIGS layer is just 600 nm, this value of efficiency

Table 9.3 Predicted parameters of the optimal CIGS, CZTSSe, and AlGaAs solar cells when the photon-absorbing layer of thickness L_n, $n \in \{1, 2, 3\}$, is sinusoidally graded and the backreflector is periodically corrugated.

Cell type	n	L_n (nm)	$\mathcal{E}_{g,w}$ (eV)	A_n	$\mathcal{E}_{an}$ (eV)	$\mathcal{E}_{bn}$ (eV)	K_n	ψ_n	α_n	L_g (nm)	ζ	L_x (nm)	J_{sc} (mA cm^{-2})	V_{oc} (mV)	FF (%)	η (%)
CIGS	1	2200	—	0.98	0.95	1.626	1.5	0.75	6	106	0.48	510	33.16	1070	78	27.70
CZTSSe	2	870	—	0.98	0.92	1.49	2	0.75	6	100	0.50	500	37.39	772	75	21.74
AlGaAs	3	2000	2.09	0.99	1.424	1.98	3	0.75	6	—	0.05	550	24.80	1556	89	34.50

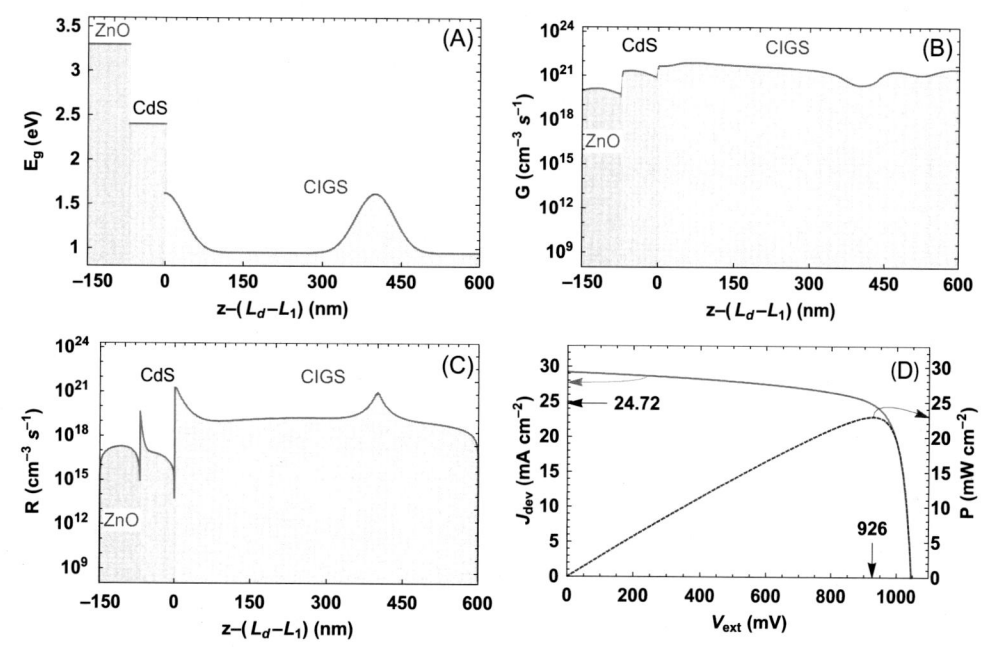

Figure 9.2 Plots of (A) $\mathcal{E}_g(z)$, (B) $G(z)$, and (C) $R(z)$ in the semiconductor layers of the optimal CIGS solar cell with a 600-nm-thick sinusoidally graded photon-absorbing layer. (D) Plots of J_{dev} and P versus V_{ext} of the optimal CIGS solar cell. The values of J_{dev} and V_{ext} for maximum P are identified. *CIGS*, Cu(In, Ga)Se$_2$.

compares favorably with the 22% efficiency reported for solar cells with a 2200 nm–thick homogeneous CIGS layer [15,84].

9.3.3.2 Cu$_2$ZnSn(S,Se)$_4$ solar cell

Optimization of the CZTSSe solar cell with a sinusoidally graded-bandgap photon-absorbing layer was carried out in the parameter space defined as follows [58]: $\mathcal{E}_{a2} \in [0.91, 1.49]$ eV, $\mathcal{E}_{b2} \in [0.91, 1.49]$ eV with $\mathcal{E}_{b2} > \mathcal{E}_{a2}$, $A_2 \in [0, 1]$, $K_2 \in [0, 8]$, $\psi_2 \in [0, 1]$, $\alpha_2 \in [0, 8]$, $L_g \in [1, 550]$ nm, $\zeta \in (0, 1)$, $L_x \in [100, 1000]$ nm, and $L_2 \in [100, 2200]$ nm. The highest value of η predicted is 21.74%; correspondingly, $J_{sc} = 37.39$ mA cm^{-2}, $V_{oc} = 772$ mV, and FF = 75%. The optimal thickness of the sinusoidally graded photon–absorbing layer is $L_2 = 870$ nm. The corresponding band-gap parameters are as follows: $\mathcal{E}_{a2} = 0.91$ eV, $\mathcal{E}_{b2} = 1.49$ eV, $A_2 = 0.98$. $K_2 = 2$, $\psi_2 = 0.75$, and $\alpha_2 = 6$.

A detailed study was performed for the optimal CZTSSe solar cell. The variations of $\mathcal{E}_g(z)$, $G(z)$, and $R(z)$ in the semiconductor layers are depicted in Fig. 9.3A−C. As for the optimal CIGS solar cell, G is higher in regions with lower $\mathcal{E}_g$ and vice versa, R

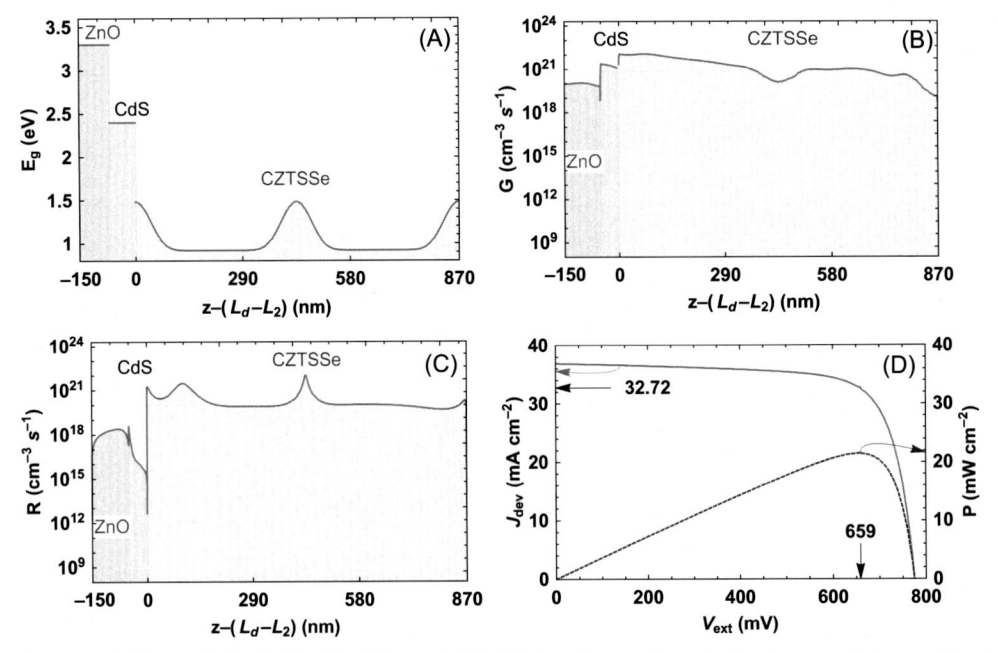

Figure 9.3 Plots of (A) $\mathcal{E}_g(z)$, (B) $G(z)$, and (C) $R(z)$ in the semiconductor layers of the optimal CZTSSe solar cell. (D) Plots of J_{dev} and P versus V_{ext} of the optimal CZTSSe solar cell. The values of J_{dev} and V_{ext} for maximum P are identified. CZTSSe, Cu$_2$ZnSn(S,Se)$_4$.

is higher in regions with higher $\mathcal{E}_g$ due to the higher fixed charge density caused by higher sulfur content in those regions. The bandgap is low in the constant-$\mathcal{E}_g$ regions, these regions being responsible for elevating the electron–hole-pair generation rate because less energy is required to excite an electron–hole pair across a narrower bandgap [70].

The $J_{dev}-V_{ext}$ and $P-V_{ext}$ characteristics of the optimal CZTSSe solar cell are shown in Fig. 9.3D. The optoelectronic model predicts $J_{dev} = 32.72$ mA cm^{-2}, $V_{ext} = 659$ mV, and FF = 75.2% for best performance.

9.3.3.3 (Al,Ga)As solar cell

Optimization of the AlGaAs solar cell with a sinusoidally graded-bandgap photon-absorbing layer was carried out in the parameter space defined as follows [75]: $\mathcal{E}_{g,w} \in$ [1.424, 2.09] eV, $\mathcal{E}_{a3} \in$ [1.424, 2.09] eV, $\mathcal{E}_{b3} \in$ [1.424, 2.09] eV with $\mathcal{E}_{b3} > \mathcal{E}_{a3}$, A_3 $\in$ [0, 1], $K_3 \in$ [0, 8], $\psi_3 \in$ [0, 1], $\alpha_3 \in$ [0, 8], $\zeta \in$ (0, 1), $L_x \in$ [100, 1000] nm, and $L_3 \in$ [100, 2000] nm. The highest value of η predicted is 34.50%; correspondingly, $J_{sc} = 24.80$ mA cm^{-2}, $V_{oc} = 1556$ mV, and $FF = 89\%$. The optimal thickness of the absorber layer is $L_3 = 2000$ nm. The corresponding bandgap-grading parameters are as

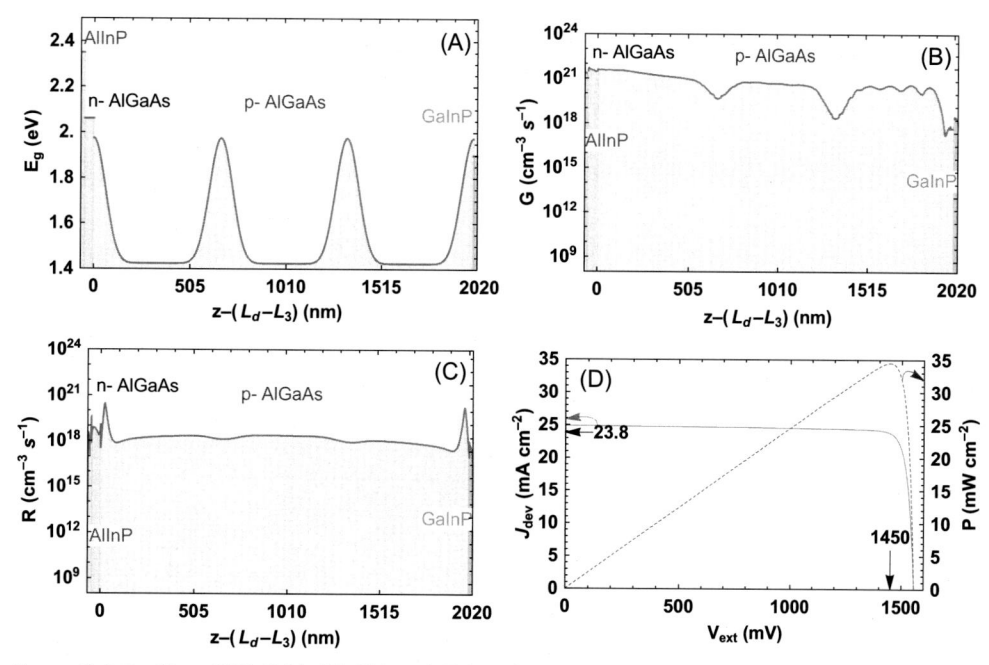

Figure 9.4 Profiles of (A) $\mathcal{E}_g(z)$, (B) $G(z)$ and $R(z)$ in the AlInP/p-AlGaAs/n-AlGaAs/GaInP region of the optimal AlGaAs solar cell; (D) Plots of J_{dev} and P versus V_{ext} of the optimal AlGaAs solar cell. The values of J_{dev} and V_{ext} for maximum P are identified. *AlGaAs, (Al,Ga)As.*

follows: $\mathcal{E}_{g,w} = 2.09$ eV, $\mathcal{E}_{a3} = 1.424$ eV, $\mathcal{E}_{b3} = 1.98$ eV, $A_3 = 0.99$, $K_3 = 3$, $\psi_3 = 0.75$, and $\alpha_3 = 6$.

A detailed study was performed for the solar cell with the thickest ($L_s = 2000$ nm) sinusoidally graded n-AlGaAs photon-absorbing layer, because it delivers the highest efficiency. The variation of $\mathcal{E}_g(z)$ with z in the semiconductor layers is provided in Fig. 9.4A. The magnitude of $\mathcal{E}_g$ is large near both faces of the n-AlGaAs photon-absorbing layer, which features elevate V_{oc} [57]. The regions in which $\mathcal{E}_g$ is small are of substantial thickness, these regions being responsible for elevating $G(z)$ [70]. The variations of $G(z)$ and $R(z)$ versus z are shown in Fig. 9.4B and C, respectively. The generation rate is higher in regions with lower bandgap and vice versa. The $J_{dev}-V_{ext}$ and $P-V_{ext}$ characteristics of the solar cell shown in Fig. 9.4D predict $J_{dev} = 23.8$ mA cm^{-2} and $V_{ext} = 1450$ mV for best performance.

9.3.4 Thin film solar cells with two photon-absorbing layers

Finally, a two-terminal solar cell with a single $p-n$ junction but containing a CIGS photon-absorbing layer above a CZTSSe photon-absorbing layer was optoelectronically optimized [85]. As shown in Fig. 9.5, the Mo backreflector was taken to be uncorrugated because the periodic corrugation does not enhance η significantly when

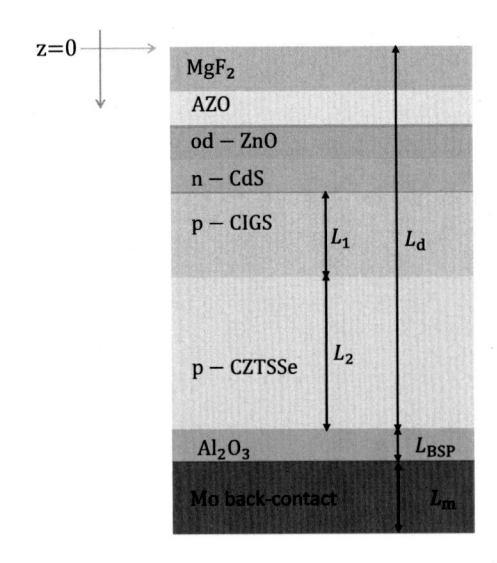

Figure 9.5 Schematic of a thin film solar cell containing a CIGS photon-absorbing layer above a CZTSSe photon-absorbing layer. The Mo backreflector is uncorrugated. *CIGS*, Cu(In,Ga)Se$_2$; *CZTSSe*, Cu$_2$ZnSn(S,Se)$_4$.

the photon–absorbing region of the solar cell is thick enough. Optimization was carried out in the parameter space defined as follows: $\mathcal{E}_{a1} \in [0.947, 1.626]$ eV, $\mathcal{E}_{b1} \in [0.947, 1.626]$ eV with $\mathcal{E}_{b1} > \mathcal{E}_{a1}$, $A_1 \in [0, 1]$, $K_1 \in [0, 8]$, $\psi_1 \in [0, 1]$, $\alpha_1 \in [0, 8]$, $\mathcal{E}_{a2} \in [0.91, 1.49]$ eV, $\mathcal{E}_{b2} \in [0.91, 1.49]$ eV with $\mathcal{E}_{b2} > \mathcal{E}_{a2}$, $A_2 \in [0, 1]$, $K_2 \in [0, 8]$, $\psi_2 \in [0, 1]$, $\alpha_2 \in (0, 8)$, $L_1 \in [0, 2200]$ nm, $L_2 \in [0, 2200]$ nm, and $0 < L_1 + L_2 \leq 2200$ nm.

The highest value of η predicted for the CIGS−CZTSSe solar cell is 34.45%; correspondingly, $J_{sc} = 38.11$ mA cm^{-2}, $V_{oc} = 1085$ mV, and FF = 83%. The optimal thicknesses of the photon-absorbing layers are $L_1 = 300$ nm and $L_2 = 870$ nm. The corresponding bandgap parameters are as follows: $\mathcal{E}_{a1} = 0.95$ eV, $\mathcal{E}_{b1} = 1.626$ eV, $A_1 = 0.91$, $K_1 = 1.88$, $\psi_1 = \psi_2 = 0.75$, $\alpha_1 = \alpha_2 = 6$, $\mathcal{E}_{a2} = 0.91$ eV, $\mathcal{E}_{b2} = 1.49$ eV, $A_2 = 0.99$, and $K_2 = 2$.

The variations of $\mathcal{E}_g(z)$, $G(z)$, and $R(z)$ with z in the semiconductor layers of the optimal solar cell are depicted in Fig. 9.6A−C. Whereas G is higher/lower in regions with lower/higher $\mathcal{E}_g$, R is higher in regions with higher $\mathcal{E}_g$ due to the higher fixed charge density caused by higher gallium or sulfur content in those regions.

The $J_{dev}−V_{ext}$ and $P−V_{ext}$ characteristics of the optimal solar cell are shown in Fig. 9.6D. The optoelectronic model predicts that the solar cell should be operated with $V_{ext} = 965$ mV to deliver $J_{dev} = 35.71$ mA cm^{-2}; then 34.45 mW cm^{-2} is predicted as the maximum extractable power density when the incident solar flux is 100 mW cm^{-2}.

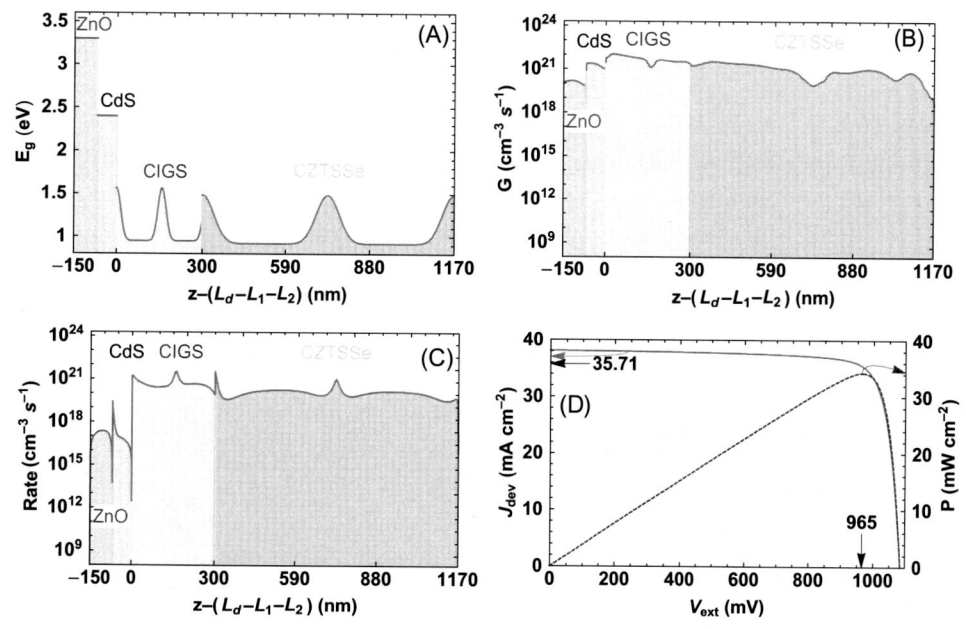

Figure 9.6 Profiles of (A) $\mathcal{E}_g(z)$, (B) $G(z)$, and (C) $R(z)$ in the od-ZnO/CdS/CIGS−CZTSSe region of the optimal CIGS−CZTSSe double photon-aborber solar cell; (D) plots of J_{dev} and P versus V_{ext} of the optimal CIGS−CZTSSe double photon-aborber solar cell. The values of J_{dev} and V_{ext} for maximum P are identified. *CIGS*, Cu(In, Ga)Se$_2$; *CZTSSe*, Cu$_2$ZnSn(S,Se)$_4$.

9.4 Concluding remarks

Optoelectronic optimization was carried out for CIGS, CZTSSe, and AlGaAs solar cells with: (i) a homogeneous photon–absorbing layer, (ii) a linearly graded photon-absorbing layer, and (iii) a sinusoidally graded photon–absorbing layer. A periodically corrugated backreflector was considered and optimized for CIGS and CZTSSe solar cells, whereas a localized ohmic back-contact was considered for AlGaAs solar cell.

An efficiency of 27.7% can be achieved with a 2200-nm-thick CIGS layer that is sinusoidally graded, whether the backreflector is periodically corrugated or planar. In comparison, the efficiency is 18.93%, when the bandgap is homogeneous and the backreflector is plane. Optoelectronic optimization also indicates that 22.89% efficiency is achievable with a solar cell with a 600-nm-thick sinusoidally graded CIGS layer. This efficiency compares favorably with the 22% efficiency demonstrated with homogeneous CIGS layers that are more than three times thicker.

Optoelectronic optimization indicates that 21.74% efficiency can be achieved for CZTSSe solar cells with an 870 nm-thick CZTSSe layer. This efficiency is significantly higher compared to 12.6% efficiency demonstrated with homogeneous CZTSSe layers that are more than two times thicker. For AlGaAs thin film solar cells, optoelectronic optimization indicates that 34.5% efficiency can be achieved with a 2000 nm-thick sinusoidally graded n-AlGaAs photon-absorbing layer. This efficiency is significantly higher compared to 27.4% efficiency demonstrated with a homogeneous n-AlGaAs absorber layer.

Finally, if photon-absorbing layers of both CIGS and CZTSSe are used in a single solar cell, then the efficiency can be boosted as high as 34.45% with a fill factor of 83%, provided that the bandgap is optimally graded, with the CIGS photon-absorbing layer being only 300-nm thick and the CZTSSe photon-absorbing layer being 870 nm thick.

Thus optimal bandgap grading is predicted to be very effective in thin film solar cells that can help tackle the scarcity of certain materials and provides a way for the ubiquitous small-scale harnessing of solar energy, for example, to power IoT (Internet of Things) devices. The model-predicted results are expected to provide impetus to devise efficacious techniques for bandgap grading.

Acknowledgments

AL thanks the Charles Godfrey Binder Endowment at Penn State for ongoing support of his research activities. This work was supported in part by the United States National Science Foundation under Grant Nos. DMS-1619901, DMS-1619904, DMS-2011996, and DMS-2011603.

References

[1] International Energy Agency, World energy outlook 2018, executive summary. https://webstore. iea.org/download/summary/190?fileName = English-WEO-2018-ES.pdf. (accessed 30.08.20).

[2] International Energy Agency, World energy outlook 2018. https://www.iea.org/weo2018/. (accessed 30.08.20).

[3] National Aeronautics and Space Administration, Global climate change, facts. https://climate.nasa. gov/evidence/. (accessed 30.08.20).

[4] Ecotricity, When will fossil fuels run out? https://www.ecotricity.co.uk/our-green-energy/energy-independence/the-end-of-fossil-fuels. (accessed 30.08.20).

[5] Union of Concerned Scientists, The hidden costs of fossil fuels. https://www.ucsusa.org/clean-energy/coal-and-other-fossil-fuels/hidden-cost-of-fossils. (accessed 30 August 2020).

[6] T.H. Anderson, Optoelectronic simulation of nonhomogeneous solar cells (Ph.D. Thesis), University of Edinburgh, Edinburgh (2016).

[7] SolarReviews, Geothermal energy pros and cons. http://energyinformative.org/geothermal-energy-pros-and-cons/. (accessed 30.08.20).

[8] International Energy Agency (IEA) and Nuclear Energy Agency (NEA), Technology Roadmap: Nuclear Energy. https://www.oecd-nea.org/pub/techroadmap/techroadmap-2015.pdf. (accessed 30.08.20).

[9] C.L. Smith, D. Ward, The path to fusion power, Philos. Trans. R. Soc. Lond. A 365 (2007) 945−956. Available from: https://doi.org/10.1098/rsta.2006.1956.

[10] World Nuclear Association, Nuclear fusion power. http://www.world-nuclear.org/info/current-and-future-generation/nuclear-fusion-power/. (accessed 30.08.20).

[11] National Renewable Energy Laboratory, Solar energy basics. https://www.nrel.gov/research/re-solar.html. (accessed 13 April 2020).

[12] M.A. Green, How did solar cells get so cheap? Joule 3 (2019) 631−633. Available from: https://doi.org/10.1016/j.joule.2019.02.010.

[13] International Energy Agency, Key world energy statistics 2019. https://webstore.iea.org/key-world-energy-statistics-2019. (accessed 31.08.20).

[14] D.M. Chapin, C.S. Fuller, G.L. Pearson, A new silicon p-n junction photocell for converting solar radiation into electrical power, J. Appl. Phys. 25 (1954) 676−677. Available from: https://doi.org/10.1063/1.1721711.

[15] M.A. Green, Y. Hishikawa, E.D. Dunlop, D.H. Levi, J. Hohl-Ebinger, A.W.Y. Ho-Baillie, Solar cell efficiency tables (version 51), Prog. Photovolt. Res. Appl. 26 (2018) 3−12. Available from: https://doi.org/10.1002/pip.2978.

[16] National Renewable Energy Laboratory, Photovoltaic research, efficiency chart. https://www.nrel.gov/pv/assets/images/efficiency-chart.png. (accessed 30.08.20).

[17] A. Kolodziej, Staebler−Wronski effect in amorphous silicon and its alloys, Optoelectron. Rev. 12 (2004) 21−32. https://www.wat.edu.pl/review/optor/12(1)21.pdf.

[18] M. Ram, D. Bogdanov, A. Aghahosseini, A. Gulagi, A.S. Oyewo, M. Child, et al., Global energy system based on 100% renewable energy—power, heat, transport and desalination sectors. Global. https://www.researchgate.net/publication/332865748. (accessed 30.08.20).

[19] M.A. Green, Photovoltaic technology and visions for the future, Prog. Energy 1 (2019) 013001. https://iopscience.iop.org/article/10.1088/2516−1083/ab0fa8.

[20] W. Shockley, H.J. Queisser, Detailed balance limit of efficiency of p-n junction solar cells, J. Appl. Phys. 32 (1961) 510−519. Available from: https://doi.org/10.1063/1.1736034.

[21] K.L. Chopra, P.D. Paulson, V. Dutta, Thin-film solar cells: an overview, Prog. Photovolt. Res. Appl. 12 (2004) 69−92. Available from: https://doi.org/10.1002/pip.541.

[22] M.A. Green, Thin-film solar cells: review of materials, technologies and commercial status, J. Mater. Sci. Mater. Electron. 18 (2007) 15−19. Available from: https://doi.org/10.1007/s10854-007-9177-9.

[23] K.P. Bhandari, J.M. Collier, R.J. Ellingson, D.S. Apul, Energy payback time (EPBT) and energy return on energy invested (EROI) of solar photovoltaic systems: a systematic review and meta-analysis, Renew. Sustain Energy Rev. 47 (2015) 133−141. Available from: https://doi.org/10.1016/j.rser.2015.02.057.

[24] PowerFilm Solar, About powerfilm. http://www.powerfilmsolar.com. (accessed 30.08.20).

[25] Fraunhofer Institute for Solar Energy Systems, Photovoltaics report, 2020. https://www.ise.fraunhofer.de/content/dam/ise/de/documents/publications/studies/Photovoltaics-Report.pdf. (accessed 30.08.20).

[26] C. Candelise, M. Winskel, R. Gross, Implications for CdTe and CIGS technologies production costs of indium and tellurium scarcity, Prog. Photovolt. Res. Appl. 20 (2012) 816−831. Available from: https://doi.org/10.1002/pip.2216.

[27] T. Gokmen, O. Gunawan, D.B. Mitzi, Minority carrier diffusion length extraction in Cu2ZnSn(Se, S)4 solar cells, J. Appl. Phys. 114 (2013) 114511. Available from: https://doi.org/10.1063/1.4821841.

[28] D.E. Carlson, C.R. Wronski, Amorphous silicon solar cell, Appl. Phys. Lett. 28 (1976) 671−673. Available from: https://doi.org/10.1063/1.88617.

[29] A. van Geelen, P.R. Hageman, G.J. Bauhuis, P.C. van Rijsingen, P. Schmidt, L.J. Giling, Epitaxial lift-off GaAs solar cell from a reusable GaAs substrate, Mater. Sci. Eng. B 45 (1997) 162−171. Available from: https://doi.org/10.1016/S0921-5107(96)02029-6.

[30] K.A.W. Horowitz, T. Remo, B. Smith, and A. Ptak, Techno-economic analysis and cost reduction roadmap for III-V solar cells (NREL/TP- 6A20−72103, 2018). https://www.nrel.gov/docs/fy19osti/72103.pdf. (accessed 08.10.19).

[31] R. Singh, G.F. Alapatt, A. Lakhtakia, Making solar cells a reality in every home: opportunities and challenges for photovoltaic device design, IEEE J. Electron. Dev. Soc. 1 (2013) 129−144. Available from: https://doi.org/10.1109/jeds.2013.2280887.

[32] R.J. Martín-Palma, A. Lakhtakia, Progress on bioinspired, biomimetic, and bioreplication routes to harvest solar energy, Appl. Phys. Rev. 4 (2017) 021103. Available from: https://doi.org/10.1063/1.4981792.

[33] M. Schmid, Review on light management by nanostructures in chalcopyrite solar cells, Semicond. Sci. Technol. 32 (2017) 043003. Available from: https://doi.org/10.1088/1361-6641/aa59ee.

[34] J. Goffard, C. Colin, F. Mollica, A. Cattoni, C. Sauvan, P. Lalanne, et al., Light trapping in ultrathin CIGS solar cells with nanostructured back mirrors, IEEE J. Photovolt. 7 (2017) 1433–1441. Available from: https://doi.org/10.1109/JPHOTOV.2017.2726566.

[35] C. van Lare, G. Yin, A. Polman, M. Schmid, Light coupling and trapping in ultrathin Cu(In,Ga)Se$_2$ solar cells using dielectric scattering patterns, ACS Nano 9 (2015) 9603–9613. Available from: https://doi.org/10.1021/acsnano.5b04091.

[36] W.H. Southwell, Pyramid-array surface-relief structures producing anti-reflection index matching on optical surfaces, J. Opt. Soc. Am. A 8 (1991) 549–553. Available from: https://doi.org/10.1364/josaa.8.000549.

[37] K.C. Sahoo, M.-K. Lin, E.-Y. Chang, T.B. Tinh, Y. Li, J.-H. Huang, Silicon nitride nanopillars and nanocones formed by nickel nanoclusters and inductively coupled plasma etching for solar cell application, Jpn. J. Appl. Phys. 48 (2009) 126508. Available from: https://doi.org/10.1143/jjap.48.126508.

[38] İ.G. Kavakli, K. Kantarli, Single and double-layer antireflection coatings on silicon, Turk. J. Phys. 26 (2002) 349–354. http://journals.tubitak.gov.tr/physics/abstract.htm?id=5759.

[39] S.K. Dhungel, J. Yoo, K. Kim, S. Jung, S. Ghosh, J. Yi, Double-layer antireflection coating of MgF$_2$/SiN$_x$ for crystalline silicon solar cells, J. Korean Phys. Soc. 49 (2006) 885–889. http://inis.iaea.org/search/search.aspx?orig_q=RN:43010932.

[40] S.A. Boden, D.M. Bagnall, Sunrise to sunset optimization of thin film antireflective coatings for encapsulated, planar silicon solar cells, Prog. Photovolt. Res. Appl. 17 (2009) 241–252. Available from: https://doi.org/10.1002/pip.884.

[41] Y. Zhang, B. Jia, Z. Ouyang, M. Gu, Influence of rear located silver nanoparticle induced light losses on the light trapping of silicon wafer-based solar cells, J. Appl. Phys. 116 (2014) 124303. Available from: https://doi.org/10.1063/1.4896486.

[42] B. Vermang, J.T. Wätjen, V. Fjällström, F. Rostvall, M. Edoff, R. Kotipalli, et al., Employing Si solar cell technology to increase efficiency of ultra-thin Cu(In,Ga)Se$_2$ solar cells, Prog. Photovolt. Res. Appl. 22 (2014) 1023–1029. Available from: https://doi.org/10.1002/pip.2527.

[43] P. Sheng, A.N. Bloch, R.S. Stepleman, Wavelength-selective absorption enhancement in thin-film solar cells, Appl. Phys. Lett. 43 (1983) 579–581. Available from: https://doi.org/10.1063/1.94432.

[44] C. Heine, R.H. Morf, Submicrometer gratings for solar energy applications, Appl. Opt. 34 (1995) 2476–2482. Available from: https://doi.org/10.1364/ao.34.002476.

[45] M. Solano, M. Faryad, A.S. Hall, T.E. Mallouk, P.B. Monk, A. Lakhtakia, Optimization of the absorption efficiency of an amorphous-silicon thin-film tandem solar cell backed by a metallic surface-relief grating, Appl. Opt. 52 (2013) 966–979. Available from: https://doi.org/10.1364/ao.52.000966. errata: 54 (2015) 398–399; https://doi.org/10.1364/ao.54.000398.

[46] L.M. Anderson, Parallel-processing with surface plasmons, a new strategy for converting the broad solar spectrum, in: Proceedings of 16th IEEE Photovoltaic Specialists Conference, Vol. 1, San Diego, CA, September 27–30, 1982, pp. 371–377. Available from: http://ntrs.nasa.gov/search.jsp?R=19840040219.

[47] L.M. Anderson, Harnessing surface plasmons for solar energy conversion, Proc. SPIE 408 (1983) 172–178. Available from: https://doi.org/10.1117/12.935723.

[48] M.G. Deceglie, V.E. Ferry, A.P. Alivisatos, H.A. Atwater, Design of nanostructured solar cells using coupled optical and electrical modeling, Nano Lett. 12 (2012) 2894–2900. Available from: https://doi.org/10.1021/nl300483y.

[49] M. Faryad, A. Lakhtakia, Enhancement of light absorption efficiency of amorphous-silicon thin-film tandem solar cell due to multiple surface-plasmon-polariton waves in the near-infrared spectral regime, Opt. Eng. 52 (2013) 087106. Available from: https://doi.org/10.1117/1.oe.52.8.087106. errata: 53 (2014) 129801. Available from: https://doi.org/10.1117/1.OE.53.12.129801.

[50] M.E. Solano, G.D. Barber, A. Lakhtakia, M. Faryad, P.B. Monk, T.E. Mallouk, Buffer layer between a planar optical concentrator and a solar cell, AIP Adv. 5 (2015) 097150. Available from: https://doi.org/10.1063/1.4931386.

[51] L. Liu, G.D. Barber, M.V. Shuba, Y. Yuwen, A. Lakhtakia, T.E. Mallouk, et al., Planar light concentration in micro-Si solar cells enabled by a metallic grating–photonic crystal architecture, ACS Photonics 3 (2016) 604–610. Available from: https://doi.org/10.1021/acsphotonics.5b00706.

[52] L. Liu, M. Faryad, A.S. Hall, G.D. Barber, S. Erten, T.E. Mallouk, et al., Experimental excitation of multiple surface-plasmon-polariton waves and waveguide modes in a one-dimensional photonic crystal atop a two-dimensional metal grating, J. Nanophotonics 9 (2015) 093593. Available from: https://doi.org/10.1117/1.jnp.9.093593.

[53] F.-J. Haug, K. Söderström, A. Naqavi, C. Ballif, Excitation of guided-mode resonances in thin film silicon solar cells, MRS Proc. 1321 (2011) 123–128. Available from: https://doi.org/10.1557/opl.2011.946.

[54] T. Khaleque, R. Magnusson, Light management through guided-mode resonances in thin-film silicon solar cells, J. Nanophotonics 8 (2014) 083995. Available from: https://doi.org/10.1117/1.jnp.8.083995.

[55] C. Frisk, C. Platzer-Björkman, J. Olsson, P. Szaniawski, J.T. Wätjen, V. Fjällström, et al., Optimizing Ga-profiles for highly efficient Cu(In, Ga)Se$_2$ thin film solar cells in simple and complex defect models, J. Phys. D: Appl. Phys. 47 (2014) 485104. Available from: https://doi.org/10.1088/0022-3727/47/48/485104.

[56] K. Woo, Y. Kim, W. Yang, K. Kim, I. Kim, Y. Oh, et al., Band-gap-graded Cu$_2$ZnSn(S$_{1-x}$,Se$_x$)$_4$ solar sells fabricated by an ethanol-based, particulate precursor ink route, Sci. Rep. 3 (2013) 03069. Available from: https://doi.org/10.1038/srep03069.

[57] F. Ahmad, T.H. Anderson, P.B. Monk, A. Lakhtakia, Efficiency enhancement of ultrathin CIGS solar cells by optimal bandgap grading, Appl. Opt. 58 (2019) 6067–6078. errata: 59 (2020) 2615. Available from: https://doi.org/10.1364/AO.58.006067. Available from: https://doi.org/10.1364/ao.389988.

[58] F. Ahmad, A. Lakhtakia, T.H. Anderson, P.B. Monk, Towards highly efficient thin-film solar cells with a graded bandgap CZTSSe layer, J. Phys. Energy 2 (2020) 025004. errata: 2 (2020) 039501. Available from: https://doi.org/10.1088/2515-7655/ab6f4a. Available from: https://doi.org/10.1088/2515-7655/ab8913.

[59] T.H. Anderson, M. Faryad, T.G. Mackay, A. Lakhtakia, R. Singh, Combined optical-electrical finite-element simulations of thin-film solar cells with homogeneous and nonhomogeneous intrinsic layers, J. Photon. Energy 6 (2016) 025502. Available from: https://doi.org/10.1117/1.jpe.6.025502.

[60] T.H. Anderson, T.G. Mackay, A. Lakhtakia, Enhanced efficiency of Schottky-barrier solar cell with periodically nonhomogeneous indium gallium nitride layer, J. Photon. Energy 7 (2017) 014502. Available from: https://doi.org/10.1117/1.jpe.7.014502.

[61] J.A. Hutchby, High-efficiency graded band-gap Al$_x$Ga$_{1-x}$As–GaAs solar cell, Appl. Phys. Lett. 26 (1975) 457–459. Available from: https://doi.org/10.1063/1.88208.

[62] I.M. Dharmadasa, Third generation multi-layer tandem solar cells for achieving high conversion efficiencies, Sol. Energy Mater. Sol. Cell 85 (2005) 293–300. Available from: https://doi.org/10.1016/j.solmat.2004.08.008.

[63] I.M. Dharmadasa, A.A. Ojo, H.I. Salim, R. Dharmadasa, Next generation solar cells based on graded bandgap device struc- tures utilising rod-type nano-materials, Energies 8 (2015) 5440–5458. Available from: https://doi.org/10.3390/en8065440.

[64] J.F. Rasheed, V.S. Babu, Performance evaluation of composition graded layer of aSi$_{1-x}$Ge$_x$: H in n$^+$aSi:H/i-aSi:H/p$^+$aSi$_{1-x}$Ge$_x$:H graded band gap single junction solar cells, Mater. Today Proc. 27 (2020) 26–31. Available from: https://doi.org/10.1016/j.matpr.2019.08.178.

[65] Y. Chen, P. Kivisaari, M.-E. Pistol, N. Anttu, Optimization of the short-circuit current in an InP nanowire array solar cell through opto-electronic modeling, Nanotechnology 27 (2016) 435404. Available from: https://doi.org/10.1088/0957-4484/27/43/435404.

[66] R. Dewan, M. Marinkovic, R. Noriega, S. Phadke, S.A. Salleo, D. Knipp, Light trapping in thin-film silicon solar cells with submicron surface texture, Opt. Express 17 (2009) 23058–23065. Available from: https://doi.org/10.1016/0927-0248(95)80004-2.

[67] P. Baruch, A.D. Vos, P.T. Landsberg, J.E. Parrott, On some thermody- namic aspects of photovoltaic solar energy conversion, Sol. Energy Mater. Sol. Cell 36 (1995) 201–222. Available from: https://doi.org/10.1364/OE.17.023058.

[68] B.J. Civiletti, T.H. Anderson, F. Ahmad, P.B. Monk, A. Lakhtakia, Optimization approach for optical absorption in three-dimensional structures including solar cells, Opt. Eng. 57 (2018) 057101. Available from: https://doi.org/10.1117/1.oe.57.5.057101.

[69] J. Nelson, The Physics of Solar Cells, Imperial College Press, London, 2003.

[70] S.J. Fonash, Solar Cell Device Physics, second ed., Academic Press, Burlington, MA, 2010.

[71] T.H. Anderson, B.J. Civiletti, P.B. Monk, A. Lakhtakia, Coupled optoelectronic simulation and optimization of thin-film photovoltaic solar cells, J. Comput. Phys. 407 (2020) 109242. errata: 418 (2020) 109561. Available from: https://doi.org/10.1016/j.jcp.2020.109242. Available from: https://doi.org/10.1016/j.jcp.2020.109561.

[72] R. Storn, K. Price, Differential evolution—a simple and efficient heuristic for global optimization over continuous spaces, J. Glob. Optim. 11 (1997) 341−359. Available from: https://doi.org/10.1023/A:1008202821328.

[73] S. Das, P.N. Suganthan, Differential evolution: a survey of the state-of-the-art, IEEE Trans. Evolut. Comput. 15 (2011) 4−31. Available from: https://doi.org/10.1109/TEVC.2010.2059031.

[74] K.V. Price, R.M. Storn, J.A. Lampinen, Differential Evolution: A Practical Approach to Global Optimization, Springer, Berlin, 2005.

[75] F. Ahmad, P.B. Monk, A. Lakhtakia, Optoelectronic optimization of graded-bandgap thin-film AlGaAs solar cells, Appl. Opt. 59 (2020) 1018−1027. Available from: https://doi.org/10.1016/S0040-6090(00)01726-0.

[76] F. Ahmad, Optoelectronic Modeling and Optimization of Graded-Bandgap Thin-Film Solar Cells (Ph.D. Dissertation), Pennsylvania State University, University Park, PA (2020).

[77] S.R. Kurtz, J.M. Olson, D.J. Friedman, J.F. Geisz, K.A. Bertness, A.E. Kibbler, Passivation of interfaces in high-efficiency photovoltaic devices, Proc. MRS 573 (1999) 95−106. Available from: https://doi.org/10.1557/proc-573-95.

[78] N. Vandamme, H.-L. Chen, A. Gaucher, B. Behaghel, A. Lemaître, A. Cattoni, et al., Ultrathin GaAs solar cells with a silver back mirror, IEEE J. Photovolt. 5 (2015) 565−570. Available from: https://doi.org/10.1109/pvsc.2015.7356352.

[79] E.N. Glytsis, T.K. Gaylord, Rigorous three-dimensional coupled-wave diffraction analysis of single and cascaded anisotropic gratings, J. Opt. Soc. Am. A 4 (1987) 2061−2080. Available from: https://doi.org/10.1364/JOSAA.4.002061.

[80] J.A. Polo Jr., T.G. Mackay, A. Lakhtakia, Electromagnetic Surface Waves: A Modern Perspective, Elsevier, Waltham, MA, 2013.

[81] National Renewable Energy Laboratory, Reference solar spectral irradiance: ASTM G-173, 2003. https://www.nrel.gov/grid/solar-resource/spectra-am1.5.html. (accessed 30.08.20).

[82] F. Brezzi, L.D. Marini, S. Micheletti, P. Pietra, R. Sacco, S. Wang, Discretization of semiconductor device problems (I), in: W.H.A. Schilders, E.J.W. ter Maten (Eds.), Handbook of Numerical Analysis: Numerical Methods for Electrodynamic Problems, Elsevier, Amsterdam, 2005, pp. 317−342.

[83] D. Brinkman, K. Fellner, P. Markowich, M.-T. Wolfram, A drift-diffusion-reaction model for excitonic photovoltaic bilayers: Asymptotic analysis and a 2−D HDG finite-element scheme, Math. Models Methods Appl. Sci. 23 (2013) 839−872. Available from: https://doi.org/10.1142/S0218202512500625.

[84] P. Jackson, R. Wuerz, D. Hariskos, E. Lotter, W. Witte, M. Powalla, Effects of heavy alkali elements in Cu(In,Ga)Se$_2$ solar cells with efficiencies up to 22.6%, Phys. Status Solidi RRL 10 (2016) 583−586. Available from: https://doi.org/10.1002/pssr.201600199.

[85] F. Ahmad, A. Lakhtakia, P.B. Monk, Double-absorber thin-film solar cell with 34% efficiency, Appl. Phys. Lett. 117 (2020) 033901. Available from: https://doi.org/10.1063/5.0017916.

Index